Advanced Sensing in Image Processing and IoT

Innovations in Multimedia, Virtual Reality and Augmentation

Series Editors:
Lalit Mohan Goyal, J. C. Bose University of Science & Technology YMCA
Rashmi Agrawal, J. C. Bose University of Science & Technology YMCA

Advanced Sensing in Image Processing and IoT
Rashmi Gupta, Korhen Cengiz, Arun Rana, Sachin Dhawan

Advanced Sensing in Image Processing and IoT

Edited by
Rashmi Gupta,
Arun Kumar Rana,
Sachin Dhawan,
and Korhan Cengiz

CRC Press is an imprint of the
Taylor & Francis Group, an **informa** business

First edition published 2022
by CRC Press
6000 Broken Sound Parkway NW, Suite 300, Boca Raton, FL 33487-2742

and by CRC Press
2 Park Square, Milton Park, Abingdon, Oxon, OX14 4RN

CRC Press is an imprint of Taylor & Francis Group, LLC

Library of Congress Cataloging-in-Publication Data
Names: Gupta, Rashmi, editor. | Cengiz, Korhan, editor. | Rana, Arun Kumar, editor. | Dhawan, Sachin, editor.
Title: Advanced sensing in image processing and IoT / edited by Rashmi Gupta, Korhen Cengiz, Arun Kumar Rana, Sachin Dhawan.
Description: First edition. | Boca Raton : CRC Press, 2022. | Includes bibliographical references and index. | Summary: "This book presents the advances and applications in the area of image processing with IoT and medical image processing and analysis techniques. It spotlights image processing and analysis techniques for IoT applications in CT images of COVID-19, remote sensing and medical engineering fields"-- Provided by publisher.
Identifiers: LCCN 2021043771 (print) | LCCN 2021043772 (ebook) | ISBN 9781032117379 (hardback) | ISBN 9781032117515 (paperback) | ISBN 9781003221333 (ebook)
Subjects: LCSH: Image processing--Digital techniques. | Internet of things. | Remote sensing. | Diagnostic imaging--Digital techniques.
Classification: LCC TA1637 .A326 2022 (print) | LCC TA1637 (ebook) | DDC 621.36/7--dc23/eng/20211104
LC record available at https://lccn.loc.gov/2021043771
LC ebook record available at https://lccn.loc.gov/2021043772

ISBN: 978-1-032-11737-9 (hbk)
ISBN: 978-1-032-11751-5 (pbk)
ISBN: 978-1-003-22133-3 (ebk)

DOI: 10.1201/9781003221333

Typeset in Times
by Deanta Global Publishing Services, Chennai, India

Contents

Editors

Rashmi Gupta earned her ME and PhD degrees in Electronics & Communication Engineering from Delhi College of Engineering, Delhi University, India. She is presently working as a Professor in the Electronics & Communication Engineering Department in Ambedkar Institute of Advanced Communication Technologies & Research under the Government of National Capital Territory of Delhi, India. She is also the Founder/Director of the Ambedkar Institute of Advanced Communication Technologies and Research–Incubation Research Foundation funded by the Delhi Government, India. Prof. Gupta has authored over 90 research papers in various renowned international journals and conferences. Her primary research interests are machine learning, computer vision, and signal and image processing.

Arun Kumar Rana earned his B.Tech degree from Kurukshetra University and M.Tech. Degree from Maharishi Markandeshwar (Deemed to be University), Mullana, India. Mr. Rana is currently pursuing his Ph.D. from Maharishi Markandeshwar (Deemed to be University), Mullana, India. His areas of interest include image processing, wireless sensor networks, Internet of Things, AI, and machine learning and embedded systems. Mr. Rana is currently working as Assistant Professor at Panipat Institute of Engineering & Technology, Samalkha with more than 13 years of experience. He has published around 70 SCI/ESCI/Scopus/others papers in national and international journals and also in conferences. He has also published six books with national and international publishers and many times members of Sci Scopus indexed international conferences/symposiums. Also, he has attended nine workshops and nine FDP. He has guided six M.Tech candidates and one candidate is in process. He serves as a reviewer for several journals and international conferences. He is also a member of the Asia Society of Research. He has also published six national and international patents. He has received many international awards from various international organizations. He is listed in the world scientist Ranking 2021.

Sachin Dhawan earned his BTech degree from Doon Valley Institute of Engineering and Technology at Kurukshetra University, Kurukshetra, India, in 2008 and his MTech degree in electronics and communication engineering from the University Institute of Engineering and Technology under Kurukshetra University, Kurukshetra, India, in 2011. Professor Dhawan is currently pursuing his PhD degree in electronics and communication engineering from the Ambedkar Institute of Advanced Communication Technologies and Research under Guru Gobind Singh Indraprastha University, Dwarka, New Delhi, India. He was an Assistant Professor at Geeta Institute of Management and Technology, Kanipla, Kurukshetra from July 2011 to July 2012. Currently, he has been working with Panipat Institute of Engineering and Technology, Samalkha since July 2012 as an Assistant Professor. He has authored over 20 research papers in various international journals and conferences. He has also published his papers in Institute of Electrical and Electronics Engineers conferences and Scopus/SCI journals. His current research interests include signal and image processing, Internet of Things, artificial intelligence, and machine learning.

Korhan Cengiz, PhD, SMIEEE was born in Edirne, Turkey in 1986. He earned BS degrees in Electronics and Communication Engineering from Kocaeli University, Turkey and Business Administration from Anadolu University, Turkey in 2008 and 2009, respectively. He also earned his MS degree in Electronics and Communication Engineering from Namik Kemal University, Turkey in 2011 and his PhD degree in Electronics Engineering from Kadir Has University, Turkey in 2016. Since 2018, he has been an Assistant Professor with the Electrical-Electronics Engineering Department, Trakya University, Turkey. He is the author of over 40 articles, including in the *IEEE Internet of Things Journal*, *IEEE Access*, *Expert Systems with Applications*, and *Knowledge Based Systems*, three book chapters, two international patents, and one book in Turkish. His research interests include wireless sensor networks, wireless communications, statistical signal processing, indoor positioning systems, power electronics, and 5G. He is also an Associate Editor of *Interdisciplinary Sciences: Computational Life Sciences*, Springer, Handling Editor of *Microprocessors and Microsystems*, Elsevier, Associate Editor of *IET Electronics Letters* and *IET Networks*, and Editor of *AEÜ – International Journal of Electronics and Communications*, Elsevier. He has held guest editorial positions in *IEEE Internet of Things Magazine* and *CMC-Computers, Materials & Continua.*

Contributors

Abhishek Agarwal
Faculty of IT and Computer Science
Parul Institute of Engineering and Technology
Vadodara, India

Nitish Agarwal
Department of Computer Science and Engineering
Panipat Institute of Engineering and Technology
Haryana, India
and
Department of Computer Engineering
Chandigarh University
Mohali, India

Alankrita Aggarwal
Assistant Professor
PIET Samalkha
Haryana, India

Anish
Department of Bio Engineering
Integral University
Lucknow, India

Ratish Agarwal
Department of Information Technology, UIT(RGPV)
Bhopal, India

Kartik Bansal
Department of Computer Engineering
Chandigarh University
Mohali, India

Erkan Bostanci
Department of Computer Engineering
Ankara University
Ankara, Turkey

Gazi Erkan Bostancı
Ankara University
Ankara, Turkey

Prodip Kumar Chakraborty
Department of Agricultural Meteorology and Physics
Bidhan Chandra Krishi Viswavidyalaya
Haringhata, India

Hina J. Chokshi
Faculty of IT and Computer Science
Parul Institute of Computer Application
Vadodara, India
and
Department of Computer Application
Parul University
Vadodara, India

Radhika G. Deshmukh
Shri Shivaji Science College
Amravati, India

Shivani Gaba
Department of Computer Science and Engineering
Panipat Institute of Engineering and Technology
Haryana, India

Rati Goel
Inderprastha Engineering College
Ghaziabad, India

Asutosh Goswami
Department of Geography and Environment Management
Vidyasagar University
Midnapore, India

Shivam Gupta
Manav Rachna International Institute of Research and Studies
Faridabad, Haryana, India

Nimisha Gutte
All India Shri Shivaji Memorial Society Institute of Information Technology
Pune, India

Mehmet Serdar Güzel
Department of Computer Engineering
Ankara University
Ankara, Turkey

Bhagyalaxmi Jena
Silicon Institute of Technology
Bhubaneswar, India

Erinç Karataş
Department of Computer Engineering
Ankara University
Ankara, Turkey

Vedita Kharabe
All India Shri Shivaji Memorial Society Institute of Information Technology
Pune, India

Ram Krishan
Department of Computer Science
Mata Sundri University Girls College
Mansa, India

Abhishek Mehta
Assistant Professor
Parul University
Vadodara, India

Maya Mehta
Faculty of IT and Computer Science
Parul Institute of Engineering and Technology
Vadodara, India
and
Faculty of IT and Computer Science
Parul University
Vadodara, India

Asmita A. Moghe
Assistant Professor
Department of Information Technology, UIT(RGPV)
Bhopal, India

Anita Mohanty
Silicon Institute of Technology
Bhubaneswar, India

Subrat Kumar Mohanty
College of Engineering Bhubaneswar
Bhubaneswar, India

Sonika Nagar
Assistant Professor
Inderprastha Engineering College
Ghaziabad, India

Shally Nagpal
Department of Computer Science and Engineering
Panipat Institute of Engineering and Technology
Haryana, India

Bhavana Nerkar
NIELIT
Aurangabad, India

Payal Parekh
Department of Computer Application
Parul University
Vadodara, India

Varsha K. Patil
All India Shri Shivaji Memorial Society Institute of Information technology
Pune, India

Vijaya Pawar
Bharati Vidyapeeth's College of Engineering
New Delhi, India

Akanksha Pinjarkar
Shri Shivaji College of Arts, Commerce and Science
Akola, India

Arun Kumar Rana
Panipat Institute of Engineering and Technology
Samalkha, India

Dharmendrasinh Rathod
Parul Institute of Computer Application
Parul University
Vadodara, India

Mustafa Sameer
National Institute of Technology
Patna, India

Anshu Saxena
Assistant Professor
Department of Information Technology, UIT(RGPV)
Bhopal, India

Suhel Sen
Department of Geography
Vivekananda College
Madhyamgram, India

Punit Kumar Singh
Department of Bio Engineering
Integral University
Lucknow, India

Sudhakar Singh
Department of Biomedical Engineering
Lovely Professional University
Phagwara, India

Kamini Solanki
Parul Institute of Computer Application
Parul University
Vadodara, India

Varsha Vimal Sood
Department of Electronics and Communication Engineering
Chandigarh University
Mohali, India

Priya Trivedi
AISECT College of Professional Studies
Indore, India

Hassan Usaman
Department of Bio Engineering
Integral University
Lucknow, India

Vaishnavi Vajirkar
All India Shri Shivaji Memorial Society Institute of Information Technology
Pune, India

Aysegül Yanık
Ankara University
Ankara, Turkey

Mertkan Yanık
Ankara University
Ankara, Turkey

Mustafa Zor
Department of Computer Engineering
Ankara University
Ankara, Turkey

1 Machine Learning–Based Early Fire Detection System Using a Low-Cost Drone

Ayşegül Yanık, Mertkan Yanık,
Mehmet Serdar Güzel, and Gazi Erkan Bostancı

CONTENTS

1.1 INTRODUCTION

The use of drones is becoming popular in many fields [3, 21, 27]. This chapter aims to bring a new and improved perspective based on visual indications relating to early detection systems for forest fires. At this time, when the number of systems [1] developed by utilizing unmanned aerial technology is increasing day by day, unmanned aerial vehicles can be used to achieve the targets of minimizing the destruction of our forests, which are the lungs of the world, and optimizing the use of workforce and time resources. It is proposed that a system based on the detection of smoke images by unmanned aerial vehicles can provide a great benefit in reducing the error rate in fire detection. The microprocessor in the system has been trained with deep learning methods and has been given the ability to recognize smoke images, which are the earliest diagnostic sign of fire. The most fundamental problem in the common algorithms used in fire detection is the high level of false alarms and the overlook rate. Confirming the result obtained from the detection and defining additional proof will increase the reliability of the system as well as the accuracy. Since the unmanned aerial vehicle provides mobile vision [2, 28], the point of view can be controlled by

DOI: 10.1201/9781003221333-1

FIGURE 1.1 Three-dimensional model of the drone and the actual developed drone.

the ground station, which can manipulate it for the sake of the accuracy of the result. An application developed in line with the subject of the chapter was implemented in a simulation environment, and the advantages of an early fire detection system and the results of the analysis are discussed in the conclusion of the chapter.

1.1.1 Motivation

Once the signs of fire are inspected in a real-time process [24, 28], they have obvious distinguishing features in terms of motion, color spectrum, and textural structure. In this way, fire can be easily separated from other natural components in the sky and forest by means of various filtering, edge detection, and color recognition tools. However, it is too late to intervene when the fire and flame become visible. For this reason, based on smoke data, which is the earliest sign of fire, a convincing fire diagnosis will pave the way for a faster reaction. Nevertheless, the results obtained from image or video input by image processing algorithms [7, 23, 29, 30] in today's developed applications contain a large number of errors. In the event of a fire, no signal is given, or the system goes into an alarm state when there is no fire. These conditions should be minimized for early and on-site intervention. At this stage, a control system patrolling with unmanned aerial vehicles will function in terms of identifying fire and providing a clearer view [12, 13]. The mounted microprocessor, which is powered by aerial vehicle internally, processes data to detect smoke patterns by using Tensorflow Library before sending results to the control station. The main objective of this study is to minimize the rate of false reports and omissions and to optimize the process. The overall design of the drone is illustrated in Figure 1.1. Also, the bottom and top body designs of the drone are shown in Figures 1.2 and 1.3.

1.2 MATERIALS AND METHODS

Forest fire is a natural disaster, and the resulting havoc can be significantly reduced by early detection and early intervention. Despite the best efforts of firefighters, massive spread of fire is unavoidable in some situations, such as traffic, late or false alarms, or where the fire is in a difficult area to reach. In the case of forest fires, the location and

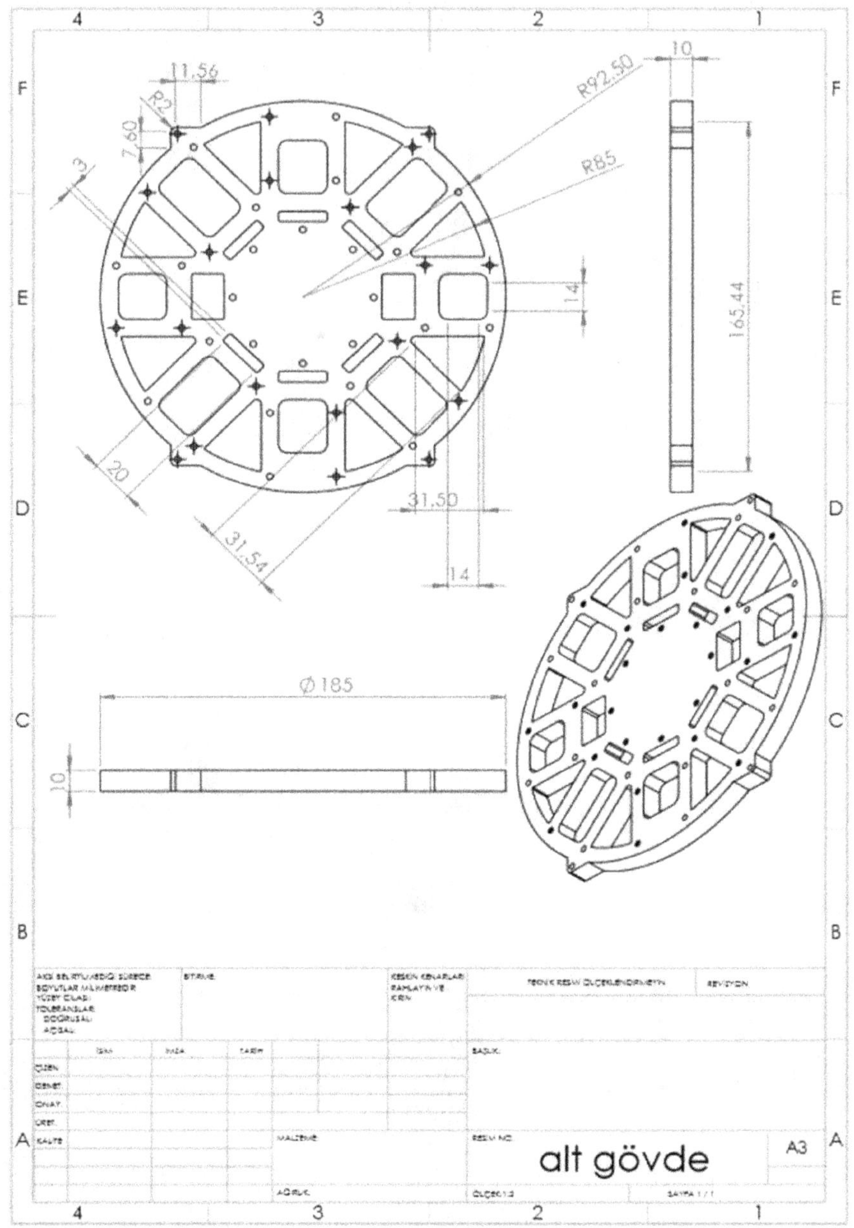

FIGURE 1.2 Bottom body design of the developed drone.

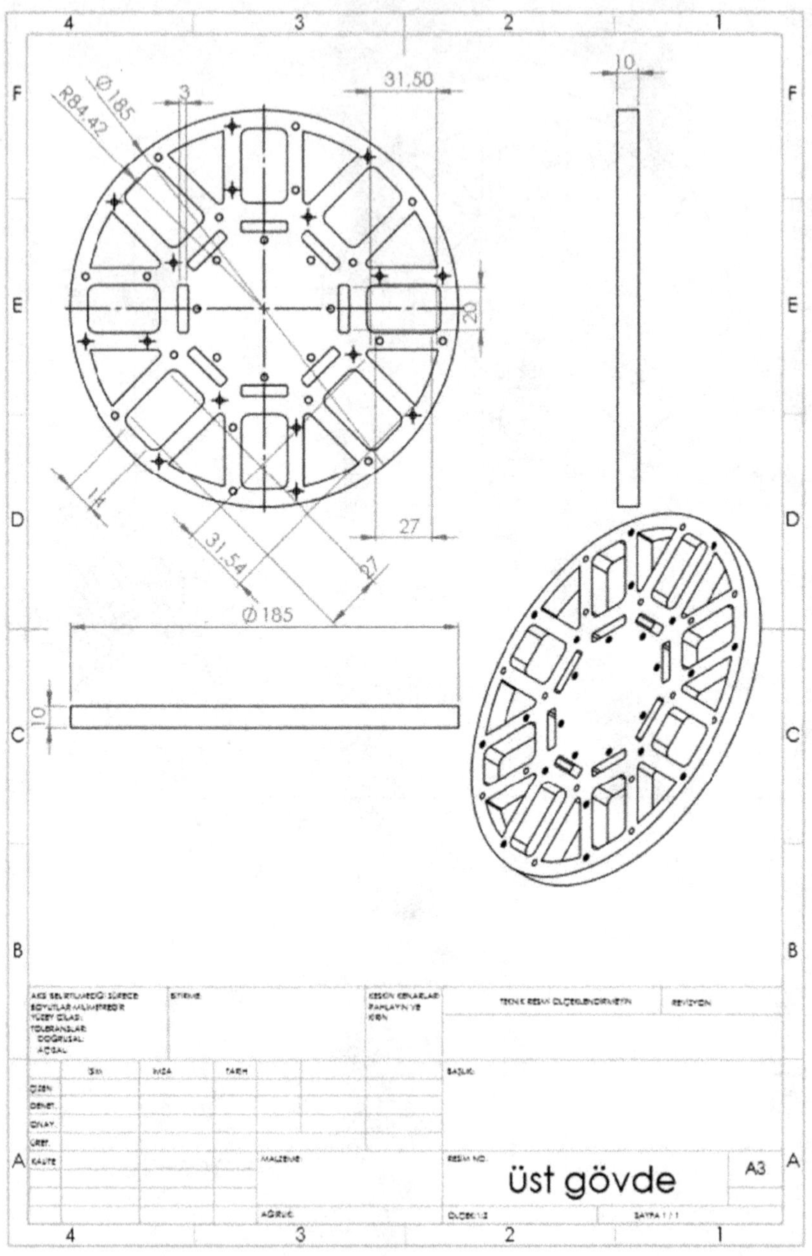

FIGURE 1.3 Top body design of the developed drone.

the time required to get there are the most important obstacles to overcome, because forest fires have the fastest spreading speed and are the most noticeable type of fire [3, 18]. Due to the fact that forest fires occur frequently, the green areas in our world are becoming smaller each day. Forest fires have intensified in some regions. Helicopters are involved in firefighting as well as water tankers and pumper vehicles, enabling faster results to be obtained using land vehicles [21]. The objective of this study is to minimize the error in the detection of forest fires by the ability of unmanned aerial vehicles to view while in motion. The Raspberry Pi device (see Figures 1.4 and 1.5) is a powerful and high-capacity device used in many robotic and Internet of Things (IoT) projects [14], offering the programmer a mobile production environment. The first step in making the Raspberry Pi minicomputer usable with unmanned aerial vehicles will be to connect the device to the system, which can be fed from the internal battery of the unmanned aerial vehicle. Other possible methods, such as feeding the device with a power bank to be fixed on the unmanned aerial vehicle, are not desirable, since this would increase the weight of the vehicle.

The current and voltage values of the electricity required from the lithium polymer battery of the Raspberry Pi are high enough to avoid the need to feed the device from the vehicle with trip wiring. For Raspberry, a regulator is placed between the power distributor and the inlet slot to overcome this obstacle. A converter battery adjuster circuit is used for this, whereby the black pin ends of the battery eliminator circuit (BEC) connect to the ground end of the power distributor, the red pin end of the circuit to the voltage end of the distributor, and the triple end of the rechargeable circuit to the Raspberry Pi 2.4 and 6 function pin inputs. When designing a frame for the *four-rotor rotary engine* unmanned aerial vehicle to be used in the project, a perforated model made of carbon fiber material was preferred because of the aerodynamic effect and the center of gravity distribution. When looking at the drone designs, the

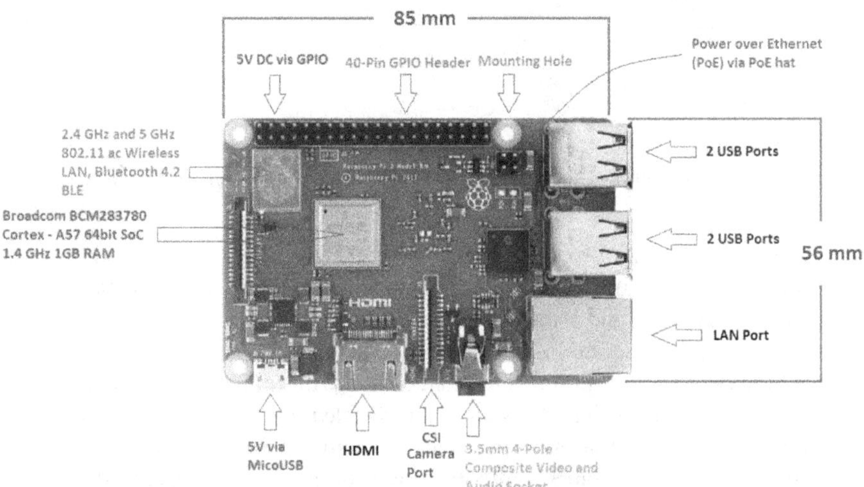

FIGURE 1.4 Three-dimensional configuration of Raspberry Pi device.

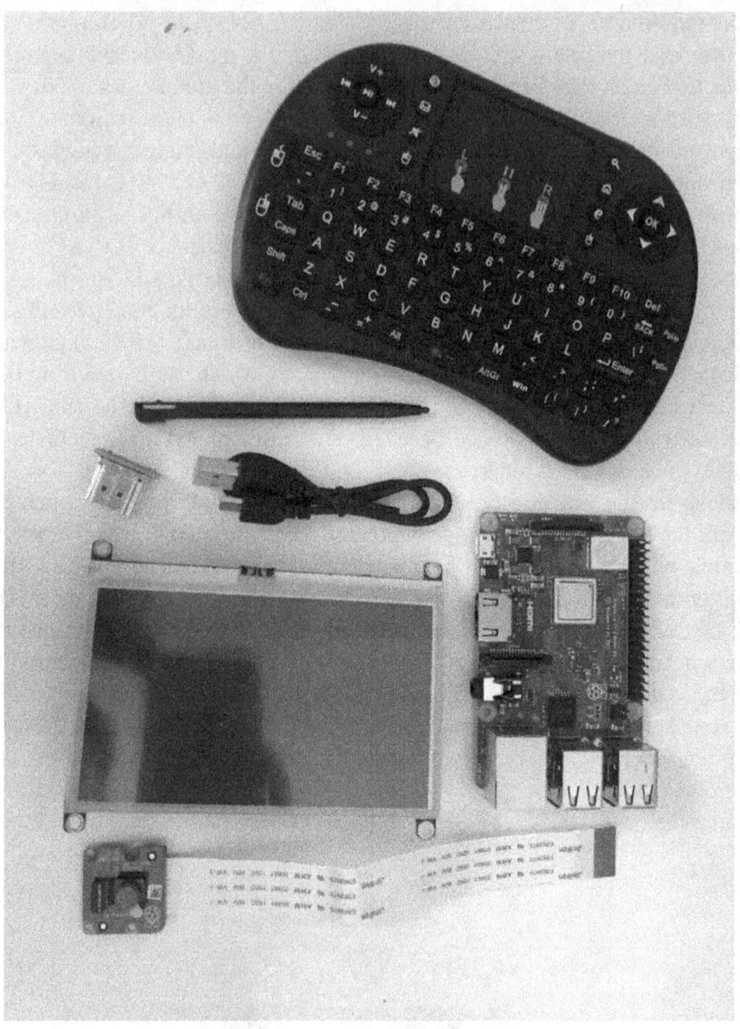

FIGURE 1.5 Three-dimensional Raspberry Pi device and other devices used in this project.

reason for choosing the H-type frame is evident: the potential of drones with this type of frame to perform more robust rolling while undertaking load-bearing tasks due to the reliable mass base and arms extending away from the frame. H-type drones can recover more easily after taking sharp bends, and it can be predicted that the battery placed in the center of the hull will last longer, as it will suffer less damage in possible drops. However, in the original design of the vehicle used in the project, it is planned to place electronic components in the sandwich model on the cordless body.

Thus, since the electronics and battery will be located in a sheltered compartment during trials and flight training, the second of the two main differences between the X- and H-type cases is irrelevant. The fastest movement to be made during the fulfillment of the project requests will be movement to the left and right. Considering this requirement, the most obvious difference that makes the H frame reasonable when the X and Y frames are compared will also lose its importance. Generally, a type X frame is used in racing drones. Considering that the fast speed of the vehicle will provide wider control, the advantages of using the X-type drone in racing vehicles played a role in the choice of the X-type body in the design of the project vehicle. These advantages are that the distribution of the center of gravity is more balanced, the mechanisms involved in drone moves are in a symmetrical co-operating principle, and the pilot can be more easily orientated to the vehicle control commands. In the project vehicle, the distance of the engines from the body/center arm length was designed as 30 cm. F*our-rotor rotary engine* unmanned aerial vehicles have a well-established place with distinct advantages among lightweight multicopters. The most important advantage is that the unstable drone design minimizes the balance handicap due to the motor/wing symmetry compared with fixed wings. Each engine to be added will add weight to the design, but at the same time, since the increase in engine power will directly increase the handling, acceleration capacity (torque), and performance of the drone, the four-engine design will balance this contrast well, producing the most appropriate results. The electronic speed control unit, engine, and propeller components make up the drone propulsion system. Speed control in brushless motors can be achieved with constant torque, since there is no friction, and the motor does not arc or dust but warms up and works at high efficiency. In the design of the vehicle, it was found appropriate to use four 770 kV short shaft Avroto brushless motors. In each engine, four propellers made of high-strength plastic/nylon or carbon fiber/nylon material will be used, taking into account the estimated weight of the project vehicle and the risks that may arise in possible collision. Instead of a three-blade propeller, a two-blade propeller was selected with an inclination angle of 4.5″ (11.43 cm) and propeller diameter of 10″ (25.4 cm). This selection was based on equivalent racing drones. When calculating the thrust value of the unmanned aircraft system, d symbolizes the propeller diameter in inches and is included in the formula as 10″ (25.4 cm), which is a known variable. The pitch value represents the value of the road traveled in a propeller lap in inches, and $V0$ represents the forward propeller speed in m/s. In the forward dynamic propulsion equation, the test data of the Avroto M2814-11S engines used in the vehicle were used when calculating the thrust value [11]. The fact that the vehicle has a unique design is one of the main objectives and achievements of the project. In order to carry the sub-systems of the vehicle (Raspberry Pi, camera module, landing gear, battery, four electronic speed controls, etc.) in a balanced and sheltered way, the model formed by the parallel placement of two empty plates, called the sandwich model, has been found appropriate. Taking into account the aerodynamic elements, the sandwich model of the vehicle is supported by the perforated body design (see Figure 1.1) [3, 21]. After the vehicle was created, flight attempts were made in two ways, by mounting the battery above and below the case. The working principle of the load-carrying mechanism in unmanned aerial vehicles with rotary wings is based on the downward force of the air produced by the thrust units. Thus, any design in which propellers interact negatively with each other or another object affects the stability and performance of the vehicle.

After this was observed in the test results, the batteries were placed under the body. During flight tests, there were no losses except for the propeller.

1.2.1 Software Design

The Raspbian operating system, which can run many programs smoothly and quickly, is installed on the device. OpenCV [17] is suitable for the use of image processing [9] and is also installed on the device [14]. TensorFlow library [16], a leading machine learning library, is also installed to deal with object detection problems. It is possible to install and write programs with standard Linux commands through the terminal command line after the installation of the operating system on the microcontroller. The Python 3.6 language is utilized to bring artificial learning [14, 22] skills into the tool. In the model used to distinguish the smoke image from the background elements, several post-processing steps are carried out involving background separation, color and edge modeling, and data normalization. With the help of built-in methods, functions, and defined values in the libraries used, the steps to be taken during the calling and use of the model after the training phase are carried out in a cost-free manner and in order to prevent the processor from being exhausted. The Materials and Methods section covers the installation and start-up steps of the application, the identification and training of the model, and the process of installing the model on a newly purchased Raspberry. As mentioned earlier, Raspberry Pi is supported by OpenCV and TensorFlow libraries, integrated into a custom-made drone so as to detect forest fires. TensorFlow is a library used to train the system for the object to be distinguished in object recognition applications [10, 11]. In the TensorFlow model, the distinguishing properties of the object are extracted using a data type called a tensor. In terms of the TensorFlow library, a tensor is a multidimensional data structure consisting of primitive (integer, float) data contained within a multidimensional array. It serves to keep the difference and similarity mathematically determined when certain regions of the images in the stream are specified to contain the same object. The TensorFlow diagram aims to identify the factors that are effective in the most accurate classification of test images and to produce a property map for the objects in the model, with improvements and training phases. When the object recognition algorithm is running, a series of computational operations, which are made up of nodes and can be represented by a flow diagram, are executed when the object to be distinguished is identified in the model. Generally, each node takes a tensor, calculates it, and outputs a tensor. In learning models, the aim is to obtain different and improved outputs with the same inputs. In order to achieve this, tensors in the flow are updated, and iteration is maintained by using TensorFlow variables, operations, placeholders, and constants. Improvement updates in machine learning cannot be done manually; an application programming interface called "loss function", which will minimize the instantaneous value of the function used to measure the difference between the estimated success value of the model and the success of the correlation value, is fed to that object when a similar relationship is encountered in the region containing the object in the other pictures. The training pictures in the batch, the specific iteration unit, are finally compared with the test data, and the model is updated and optimized according to the difference between the classification and the class to which it actually belongs. The statistics and results of the mathematical evaluations that the system is trained to make in order to give fast and low error rates are given in this section. The ssdlite mobilenet model, a sub-model of the supervised deep learning school called Model Zoo, which contains pre-educated models in the TensorFlow library, is taken as an example for training in this study [6, 8]. There is a speed–precision swap that creates a kind of balance between sub-models. The criterion to be considered when choosing a model is the purpose for which the model will be used, because in programming, there is always a trade-off balance that makes it necessary to waive certain things while improving others. In the real-time object recognition application, the speed of the model is in the foreground, while the accuracy rate remains in the background. Essentially, the models, trained with the COCO [8, 27] training set, were investigated, and the ssdlite mobilenet model, most likely to give the best results when working on real-time smoke data on Raspberry, was adapted for this study. An example of smoke detection within the developed software system is given in Figure 1.6.

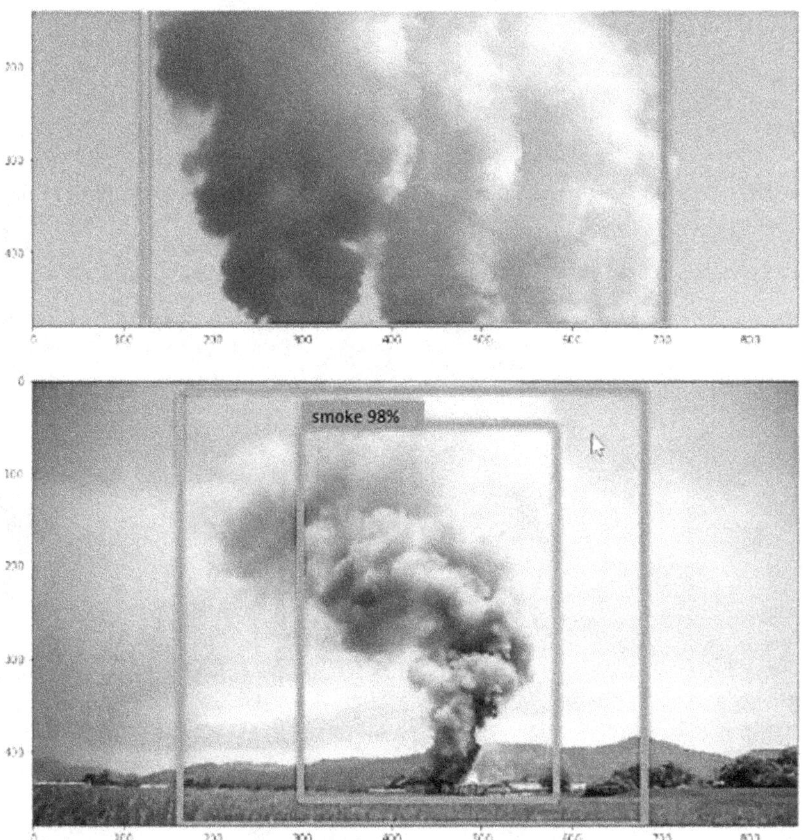

FIGURE 1.6 Example of smoke detection with 98% probability.

For the training process, 280 images representing smoke are used; most of those were obtained from Google, whereas some of them were generated in physical environments. Consequently, those data are employed in the training phase of the model. Change in the loss functions is illustrated in Figure 1.7. Figure 1.8 illustrates images

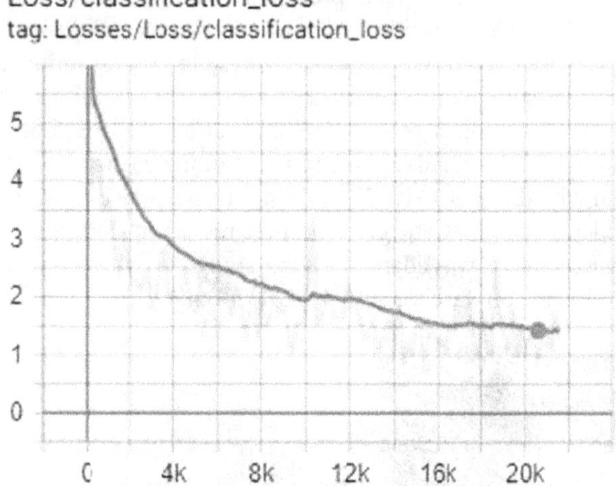

FIGURE 1.7 Loss function variation (y-axis) during the iterations (x-axis) (90% of data for training).

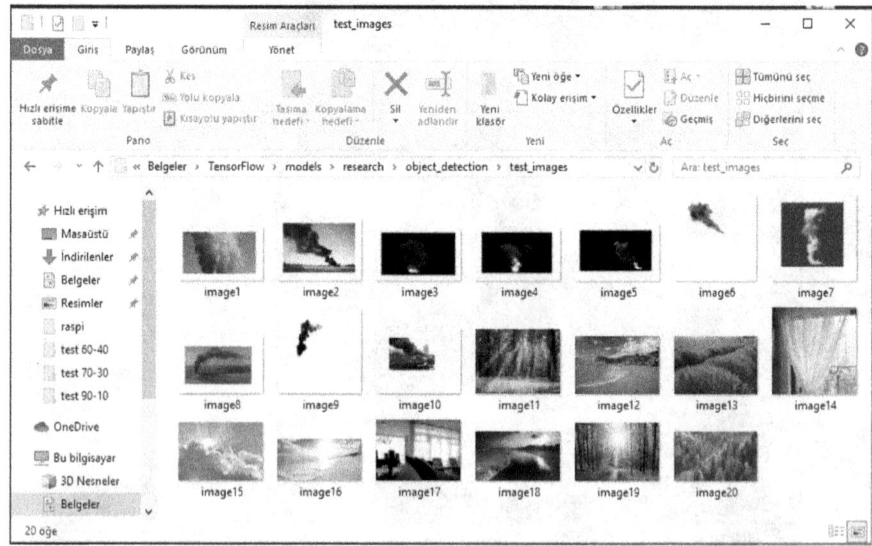

FIGURE 1.8 Dataset obtained from Google and corresponding overall accuracy of the dataset.

FIGURE 1.9 Example data generated by the author using the drone.

obtained from datasets, whereas Figure 1.9 illustrates an example of data generated by the authors using the developed drone.

1.3 RESULTS

When evaluating the loss graphs, it would not be correct to call the model with the lowest loss value that we obtained the most successful model. It shows only that this model can be trained more quickly than other models. The training ended in different models and different iterations in each model. For instance, Figure 1.7 illustrates a model employing 90% of data for training. Comparing the results with the point where each model reaches the optimum level, comparing the results, and comparing the loss values at the end of the training can be considered as a relatively accurate comparison parameter. There are two versions of TensorFlow running on the central processing unit (CPU) and the graphics processing unit (GPU). The model is trained in two ways. The same model working on the same data was able to reach the point where the training process carried out on the CPU took 10 hours, 32 minutes, and 54 seconds, and on the GPU, 1 hour, 39 minutes, and 47 seconds. In this case, it was observed that the model was trained 6.34 times faster compared with the CPU when the deep learning algorithms developed on visual data were run on the GPU. A number of analysis procedures were carried out on four models run on the same control group to choose between the different percentages of training and test data dropped and the four different models trained. The success of the models in different respects was estimated, the models were compared with each other, and one of them was chosen for use. Four models are defined based on the distribution of the training and test data (Table 1.1).

There are some standard evaluation parameters used to measure the overall performance of deep learning models [26]. These are total accuracy, positive interpretation

TABLE 1.1
Definition of Models Used in the Experimental Part

Model Name	Training Data (%)	Test Data (%)
M1	60	40
M2	70	30
M3	80	20
M4	90	10

power, negative interpretation power, sensitivity, F-rating, and specificity. When calculating these values, DP: true-positive, DN: true-negative, YP: false-positive, and FN: false-negative are used. The true-positive expression refers to the number of operations in the sample space in which the model contains smoke for visual data that actually contain smoke. The value of true-negative indicates the number of visual data that the model classifies as smoke-containing. On the one hand, the false-positive does not actually contain smoke; on the other hand, the false-negative characterizes the number of visual data classified by the model as though they actually contain smoke, when in fact they do not. High true-positive and true-negative values show that the model produces realistic results. Models with a large number of false-positives are likely to trigger the fire department in the absence of fire, and models with a large number of false-negatives can lead to neglect because the fire is not seen. Therefore, conducting a numerical analysis before deciding on the model to be used is of great importance for the efficiency of the project. For all four models, the total accuracy, positive interpretation power, negative interpretation power, sensitivity, F-evaluation, and specificity values were calculated based on the results of the estimation values table prepared using the values recorded in the output table. Total accuracy is the most intuitive and basic of the evaluation criteria. These equations are given here:

$$\text{accuracy} = \frac{\text{TP} + \text{TN}}{\text{TP} + \text{FP} + \text{TN} + \text{FN}}$$

$$\text{precision} = \frac{\text{TP}}{\text{TP} + \text{FP}}$$

$$\text{recall} = \frac{\text{TP}}{\text{TP} + \text{FN}}$$

$$\text{Fmeasure} = \text{precision} \frac{\text{recall}}{\text{precision} + \text{recall}}$$

A number of analysis procedures were carried out on four models run on the same control group to choose between the different percentages of training and test data

TABLE 1.2
Performance Evaluation of Training Models Used in the Experiment

Model	M_1	M_2	M_3	M_4
Accuracy	0.55	0.55	0.85	0.70
Precision	0.70	0.05	0.70	0.70
Recall	0.40	0.30	0.80	0.60
FScore	0.51	0.38	0.75	0.64

FIGURE 1.10 Some test results obtained from the physical environment.

dropped and the four different models trained. Estimation success of the models in different respects was measured, models were compared with each other, and one of them was chosen. The preferred model is the third model (M3), trained with 70% training and 30% test data. Table 1.2 demonstrates the overall results of the corresponding physical experiments. According to these, Model 3 is determined as the best model for all evaluation criteria. For the physical experiment, the M3 model was loaded on Raspberry Pi, and test studies were carried out. In order to carry out the flight test in accordance with the rules, a remote and sheltered area near the Kızılcahamam Soguksu National Park was chosen, and the test flight and related shots were performed in daylight. As a smoke source, brazed gel was ignited in a footed enamel brazier, and some plastic waste material was added for the brushing and bulking of smoke (see Figure 1.9). Figures 1.10 and 1.11 illustrate the experimental results, validating both the performance of the drone and its smoke recognition

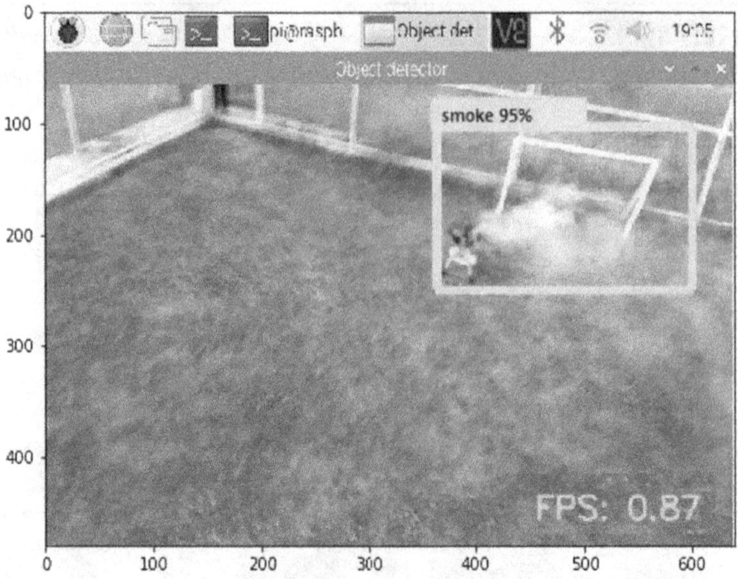

FIGURE 1.11 Some other test results obtained from the physical environment.

capacity. During the test flight of the system (see Figures 1.9, 1.10, and 1.11), images of smoke and the vehicle were taken both from inside Raspberry (see Figures 1.9, 1.10, and 1.11) and from outside. Since the vehicle does not have a vision system (FPV [first person view]) that can help the pilot to control the vehicle more easily, it has been very difficult to center the smoke in the images taken from the vehicle. For this reason, it is recommended to use a visual support mechanism while driving in similar studies [4]. In addition, due to vibration, the images obtained are blurry. However, even though the similarity rate was not successful in each smoke input, the smoke recognition model worked flawlessly, negligible localization errors were observed, and the test flight was successfully completed.

1.4 CONCLUSIONS

This chapter introduces a new low-cost drone equipped with image processing and object detection abilities for smoke and fire recognition tasks in forests. The statistics and results of the mathematical evaluations that the system is trained to use in order to give fast and low error rates are given. The ssdlite mobilenet model, a sub-model of the supervised deep learning model called Model Zoo, which contains pre-educated models in the TensorFlow library, is taken as an example for education in this [25] project. There is a speed–precision trade-off that creates a kind of balance between sub-models. The criterion to be considered when selecting a model is the purpose for which the model will be used, because in programming, there is always a trade-off balance that makes it necessary to waive certain things while improving others. In the real-time object recognition application, the speed of the model is in the foreground, and the accuracy rate remains a minor preference. In the project, the models that were trained with the COCO training set were investigated, and the ssdlite mobilenet model, which was most likely to give the best results when working on real-time smoke data on Raspberry, was modified. When evaluating the suitability of the model used, it should be taken into consideration that each model cannot work with the same speed and accuracy rate in each dataset. A model based on the dimension of colors cannot successfully distinguish colored objects from the background when these objects are very close to the background color. The preferred model successfully performed its task. Based on these results, it was considered appropriate to use the drone and the early fire detection system for the detection of forest fires. In order to improve the system, it is possible to increase the maximum flight time of the unmanned aircraft so that the area that can be controlled in a single flight can be expanded. In this eco-friendly project, the system can be supplied from convertible energy sources such as solar energy instead of a lithium polymer battery. In addition, in cases where smoke is detected by the unmanned aircraft, it is possible to estimate how far the smoke is from the vehicle by means of monocular depth detection algorithms, and this improvement enables the location of the fire to be reported to the ground station. In addition to the smoke recognition application, the same model can be trained to detect flame and fire visuals [5, 15, 19, 20], and a downward camera can be added to the unmanned aerial vehicle to perform a downward fire/flame scan with the vehicle suspended above the smoke detection zone. This will enhance the accuracy of the project outcome. The authors hope that the study will shed light on similar scientific studies.

ACKNOWLEDGMENTS

Part of this study was published in the MSc thesis [25] entitled "Visual Based Early Fire Detection System with Unmanned Aerial Vehicles, 2019" by the first author.

CONFLICTS OF INTEREST

The authors declare no conflict of interest.

REFERENCES

1. Akyurek, S., Yılmaz, M. A., & Taskıran, M. (2012). Insansız Hava Aracları(Muharebe Alanında ve Terorle Mucadelede Devrimsel Donusum. Technical Report No: 53, Ankara, Turkey.
2. Boroujeni, N. S. (2019). Monocular vision system for unmanned aerial vehicles. PhD Thesis, Carleton University, Canada.
3. Cetinkaya, O. T., et al. (2019). A fuzzy rule based visual human tracking system for drones. In 4th International Conference on Computer Science and Engineering (UBMK), pp. 1–6. Ankara, Turkey.
4. Chen, T., Wu, P., & Chiou, Y. (2014). An early fire-detection method based on image processing. International Conference on Image Processing (ICIP04), Singapore. Meyer,

J., Du, F., & Clarke, W. (2009). Design considerations for long endurance unmanned aerial vehicles. *Aerial Vehicles.* doi:10.5772/6482.

5. Frizzi, S., Kaabi, R., Bouchouicha, M., Ginoux, J., Moreau, E. & Fnaiech, F. (2016). Convolutional neural network for video fire and smoke detection, *IECON 2016 - 42nd Annual Conference of the IEEE Industrial Electronics Society*, pp. 877–882, doi: 10.1109/IECON.2016.7793196. Florence, Italy.
6. Gad, A. F. (2018). Tensorflow recognition application. *Practical Computer Vision Applications Using Deep Learning with CNNs*, *6*, 229–294. doi:10.1007/978-1-4842-4167-7.
7. Lin, G., et al. (2017). Smoke detection in video sequences based on dynamic texture using volume local binary patterns. *KSII Transactions on Internet and Information Systems*, *11*. doi:10.3837/tiis.2017.11.019.
9. Jain, N., Yerragolla, S., Guha T., & Mohana, T. (2019) Performance Analysis of Object Detection and Tracking Algorithms for Traffic Surveillance Applications using Neural Networks, 3rd International conference on I-SMAC (IoT in Social, Mobile, Analytics and Cloud) (I-SMAC), 2019, pp. 690–696, doi: 10.1109/I-SMAC47947.2019.9032502. Pallamad, India.
10. Joseph, A., & Geetha, P. (2020). Facial emotion detection using modified eyemap–mouthmap algorithm on an enhanced image and classification with tensorflow. *Vis Comput* 36, 529–539. https://doi.org/10.1007/s00371-019-01628-3.
11. Krishnadas, A., & Nithin, S. (2021). A comparative study of machine learning and deep learning algorithms for recognizing facial emotions, 2nd International Conference on Electronics and Sustainable Communication Systems (ICESC), pp. 1506–1512, doi: 10.1109/ICESC51422.2021.9532745. Coimbatore, India.
12. Lin, G., et al. (2017). A novel fire detection approach based on CNN-SVM using tensorflow. *Intelligent Computing Methodologies Lecture Notes in Computer Science*, 682–693.
13. Lin, G., Zhang, Y., Zhang, Q., Jia, Y., Xu, G. & Wang, J. (2017). Smoke detection in video sequences based on dynamic texture using volume local binary patterns, KSII Transactions on Internet and Information Systems, vol. 11, no. 11, pp. 5522–5536, 2017. DOI: 10.3837/tiis.2017.11.019.
14. Maksimovic, M., Vujovic, V., Davidović, N., Milosevic, V., & Perisic, B. (2014). Raspberry Pi as Internet of Things hardware: Performances and Constraints.
15. Patel, P., & Tiwari, S. (2012). Flame detection using image processing techniques. *International Journal of Computer Applications*, *58*(18), 13–16. doi:10.5120/9381-3817.
16. Pattanayak, S. (2017). Introduction to deep-learning concepts and tensorflow. *Pro Deep Learning with Tensorflow*, 89–152. 1st Edition, OReilly. doi:10.1007/978-1-4842-3096-1_2.
17. Santana, P., Gomes, P., & Barata, J. (2012). A vision-based system for early fire detection. 2012 IEEE International Conference on Systems, Man, and Cybernetics (SMC), Seoul, Korea (South). doi:10.1109/icsmc.2012.6377815
18. Shaqura, M., & Shamma, J. S. (2017). An automated quadcopter CAD based design and modeling platform using solidworks API and smart dynamic assembly. In Proceedings of the 14th International Conference on Informatics in Control, Automation and Robotics. doi:10.5220/0006438601220131.
19. Suleyman, G., Ugur, B., & Enis, A. (2019, February 05). Deep convolutional generative adversarial networks based flame detection in video. Retrieved from https://arxiv.org/abs/1902.01824.
20. Toreyin, B., Dedeoglu, Y., & Cetin, A. (2005). Flame detection in video using hidden Markov models. In IEEE International Conference on Image Processing 2005. doi:10.1109/icip.2005.1530284.

21. Unal, A., Bostanci, G. E., Sertalp, E., Guzel, M. S., & Kanwal, N. (2018). Geolocation based augmented reality application for cultural heritage using drones. 2nd International Symposium on Multidisciplinary Studies and Innovative Technologies (ISMSIT), pp. 1–4. Ankara, Turkey.
22. Verghese, V. R., & Verghese, D. C. (2021). Autofhm: A Python Library for Automated Machine Learning, 3rd International Conference on Inventive Research in Computing Applications (ICIRCA), pp. 860–867, doi: 10.1109/ICIRCA51532.2021.9544859. Coimbatore, India.
23. Wang, T., Liu, Y., & Xie, Z. (2011). Flutter analysis based video smoke detection. *Journal of Electronics & Information Technology, 33*(5), 1024–1029. doi:10.3724/sp.j.1146.2010.00912.
24. Xu, G., Zhang, Y., Zhang, Q., Lin, G., & Wang, J. (2017) Deep domain adaptation based video smoke detection using synthetic smoke images, Fire Safety Journal, Volume 93, Pages 53–59, ISSN 0379-7112. https://doi.org/10.1016/j.firesaf.2017.08.004.
25. Yanik, A. (2019). Visual based early fire detection system with unmanned aerial Vehicles, MSc thesis, Ankara University, Turkey.
26. Yilmaz, A. A., Guzel, M. S., Bostanci, G.E., & Askerzade, I. (2021) A novel action recognition framework based on deep-learning and genetic algorithms. IEEE Access, 8, 631-644.w
27. Yilmaz, A. A., Guzel, M. S., & Askerbeyli, I. (2018). A vehicle detection approach using deep learning methodologies. arXiv:1804.00429.
28. Zhang, F. (2015). *Obstacle Detection Using Monocular Camera for Low Flying Unmanned Aerial Vehicle.* MSc Thesis, Carleton University, Canada. doi:10.22215/etd/2015-10769.
29. Zhang, Q., Lin, G., Zhang, Y., Xu, G., & Wang, J. (2018). Wildland forest fire smoke detection based on faster R-CNN using synthetic smoke images. *Procedia Engineering, 211*, 441–446. doi:10.1016/j.proeng.2017.12.034.
30. Zhou, B., Song, Y., & Yu, M. (2015). Video fire smoke detection based on static features of smoke. *Journal of Advanced Computing.*

2 Computer Vision

Practical Approach to Facial Detection Techniques for Security Applications

Shivam Gupta

CONTENTS

2.1 INTRODUCTION

Figure 2.1 shows two friends conversing with each other regarding the features of biometrics and shows how the advancement in the field of biometrics has attracted them and made them enthusiastic about studying, using, and understanding biometrics.

To reaffirm the above statement, let us have a glimpse of some of the statistics and surveys, as shown in Figure 2.2, conducted by many reputed authorities regarding the usage, applications, and acceptability of biometrics by people worldwide.

From these facts and figures, it is clearly concluded that evolution in the field of biometric features of human beings has initiated a more secure and reliable pathway for many different applications. Now, let us have a clearer understanding of the term "biometrics" [1].

DOI: 10.1201/9781003221333-2

VIKRAM: "Hey Dipak! How are you doing?"

DIPAK: "Good. How about you?"

VIKRAM: "I am fine too. It looks like you got a new phone."

DIPAK: "Yes. I bought it recently."

VIKRAM: "If you bought it then it must be having something special."

DIPAK: "Yes, you are right. It is the latest smartphone with the latest technologies used to secure the phone and its data with the help of different biometric features like face, fingerprint and iris recognition."

VIKRAM: "Do you think it is worth buying this phone for some useless biometric locks?"

DIPAK: "Yes, of course. The biometrics are the most reliable, secured and unique way of securing your data from leaking and stealing. They are highly precised and very much confidential since no one else can access your system using your biometrics."

VIKRAM: "Wow! Seems to be pretty much cool. I think now I am going to gather some more knowledge about the biometrics like you."

DIPAK: "Best of luck for that. See you then. Bye."

VIKRAM: "Thanks. Bye."

FIGURE 2.1 A conversation between two friends on biometric locks.

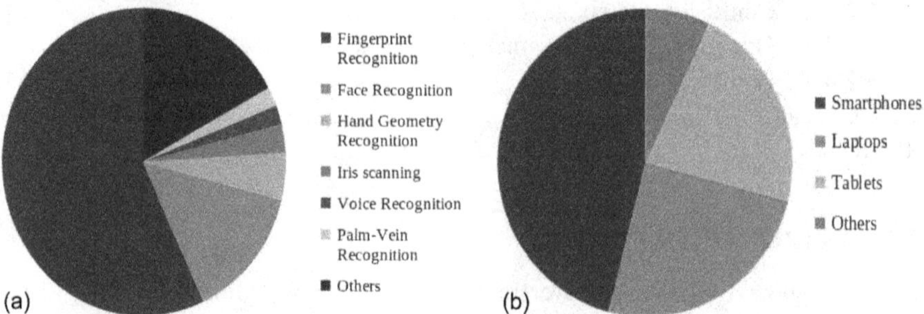

FIGURE 2.2 Statistics regarding (a) different types of biometric locks used by people and (b) different types of electronic gadgets using biometrics as security lock.

Biometrics is defined as the calculations and analysis performed to study the different physiological or behavioural characteristics of human beings, which they either possess from birth or adopt afterwards. Figure 2.3 shows the classification of different biometrics that have been proved efficient and effective so far and are widely used in different fields for authentication and other purposes, as shown in Figure 2.4. Developments in modern-day technologies, powerful computing units,

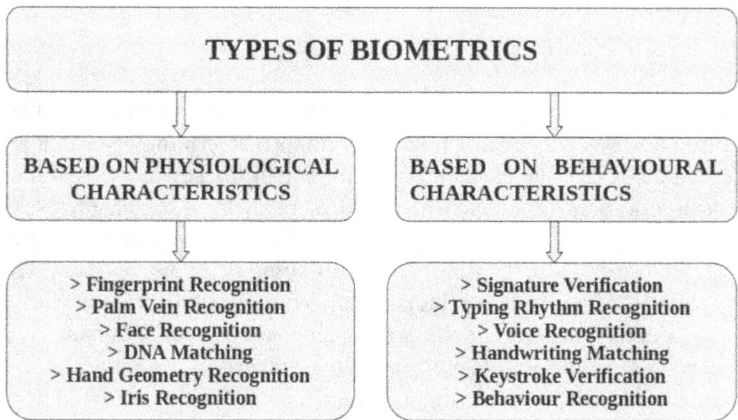

FIGURE 2.3 Different types of biometric studies based on different human body characteristics.

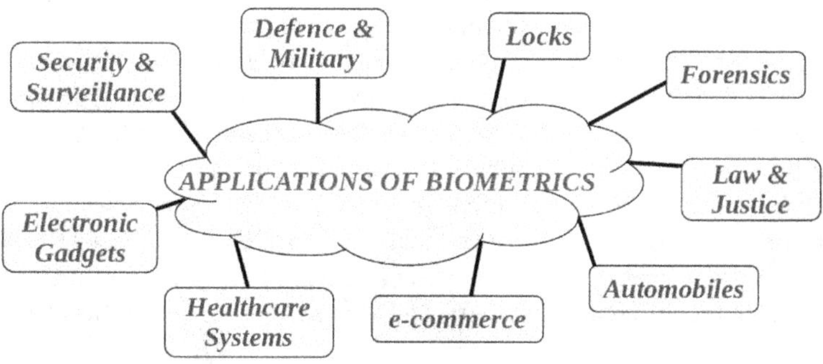

FIGURE 2.4 Some of the real-world fields affected by the use of biometrics.

and large datasets have made it possible to develop many different complex algorithms and techniques so as to detect, record, decode, analyse, recognise, and store the biometric characteristics of different people with exceptionally high accuracy, recall, and precision values. Researchers and scientists have been constantly researching in depth the different biometric characteristics of the human body and developing different techniques in order to study the unique patterns of the brain as well as the heart of different people. The most important features of biometrics that make them extremely useful are [2, 3]:

- Highly reliable
- Non-transferable
- Unique
- Universal
- Widely acceptable

- Confidential
- High-performance

Now, before diving in depth into the different facial detection techniques, let us familiarise ourselves with some of the other important terminologies that are likely to appear in the rest of the chapter. Faces are one of the key features of the human body and a vital area of study in the field of biometrics, which involves exploring, analysing, and comparing face patterns, landmarks, and keypoints of different people so as to record the similarities and dissimilarities among them. Computer vision is one of the branches of artificial intelligence, which focuses on working with images and videos to enable computers to understand them and automate a variety of tasks that require human assistance. One of the subparts of computer vision is facial detection, which deals with studying and recognising the patterns and keypoints of different human faces and then using them to identify and extract the other known or unknown faces present in the input images and videos.

As the title of the chapter suggests, this chapter primarily focuses on the implementation of the different facial detection techniques and algorithms. It uses a step-by-step approach to showcase the implementation, working, and results of different facial detection techniques. The main aim of this chapter is to provide readers with a glimpse of the capabilities of computer vision using a hands-on approach. This chapter can be very beneficial for readers who want to learn about the implementation of the different facial detection techniques with less theory and more practical results [4].

The primary contents of the chapter include the introduction to facial detection with a brief distinction from facial recognition, a brief but clear knowledge about the working of facial detection techniques, resultant output images of the dataset comprising random test images, a well-structured comparative analysis of all the techniques used based on the different parameters, and others. The features of this chapter are:

- Less theory and a more practical approach.
- Well-structured and fully commented Python codes.
- Real output results of the images in the test dataset.
- Links to all the models, datasets, codes, and results discussed in the chapter.
- In-depth analysis of all the techniques and algorithms used in the chapter [5].

Now, let us move ahead to another section of the chapter, which gives a clearer understanding of facial detection.

2.2 FACIAL DETECTION

Facial detection, as the name suggests, is a part of computer vision that focuses on detecting all the human faces present in an image or video with a further step of extracting them and bounding them inside the boxes. The facial detection techniques are either used exclusively or as an initial layer of facial recognition. The term "facial detection" can sometimes be confused with facial recognition, but that is not correct at all. Facial recognition is a step ahead of the facial detection process;

TABLE 2.1
Distinction between Facial Detection and Facial Recognition

Parameters	Facial Detection	Facial Recognition
Definition	Facial detection refers to the detection of the presence of human faces in input images or videos.	Facial recognition refers to detecting the human faces present in the images or videos as well as recognising their identity by matching from the predefined database.
Database	They require no predefined database to detect human faces.	They require a predefined database to recognise the detected human faces.
Model Examples	Haar cascades classifiers, MTCNN.	Local Binary Pattern Histogram (LBPH) algorithm, Siamese networks.
Applications	Face filters, Augmented Reality, security and surveillance.	Authentication, identification purposes, forensics, e-commerce.

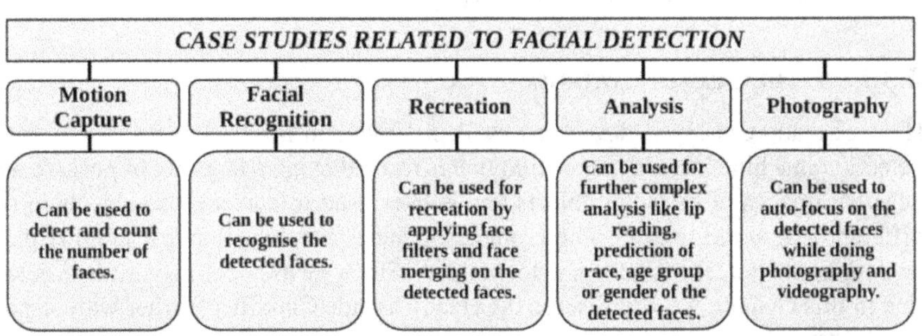

FIGURE 2.5 Different case studies of facial detection in real-world scenarios.

it involves labelling the detected faces by extracting them and comparing their facial keypoints with a set of other images of human faces stored in a predefined database [6] (Table 2.1).

Facial detection techniques were, obviously, developed earlier than facial recognition techniques and serve efficiently in a variety of real-world applications with the task of detecting human faces in images and videos. Some of the case studies where facial detection plays a primary role are in Kak et al. and Chen et al. [7, 8] (Figure 2.5).

Now, let us have a look at some of the facial detection techniques used in the chapter for detection as well as evaluation purposes.

2.3 FACIAL DETECTION TECHNIQUES

In this section, we will discuss some of the facial detection techniques:

1. Haar Cascade Classifier
2. Maximum-Margin CNN Object Detector model (MMOD) Face Detector

3. Histogram of Oriented Gradients (HOG) Face Detector
4. Multi-Task Cascaded Convolutional Neural Network (MTCNN) Face Detector
5. Deep Neural Networks (DNN) Face Detector

Before diving deep into the working of these detectors, here are some of the prerequisites that must be taken into account in order to use these models:

- Programming language used – Python
- Operating system used – Ubuntu
- Integrated Development Environment used – Pycharm Community Edition
- Libraries used – OpenCV, numpy, dlib, mtcnn, os, time [9]
- GitHub link – https://github.com/hmsgupta3062/Chapter-on-Facial-Detection-Techniques
- Google drive link – https://drive.google.com/drive/folders/1WTHFh9OTl7Rzk8mOp7FWs260ZFfwHYZ? usp=sharing

2.3.1 Haar Cascade Classifier

Haar Cascade Classifier, developed by Paul Viola and Michael Jones, is an object detection machine learning algorithm that is trained using a large set of positive and negative images of different objects and is later used to detect those objects in the different real-world images. These objects include face detection, car number plate detection, eye detection, person detection, etc. Since we are dealing with face detection in this chapter, we will discuss the Haar Cascade Classifier further with respect to face detection.

The Haar Cascade algorithm, in brief, is as follows:

1. A large number of positive (image containing faces) images and negative (image not containing faces) images are used to train the classifier.
2. Each and every Haar feature (some of them are shown in Figure 2.6) is applied to all the training images.

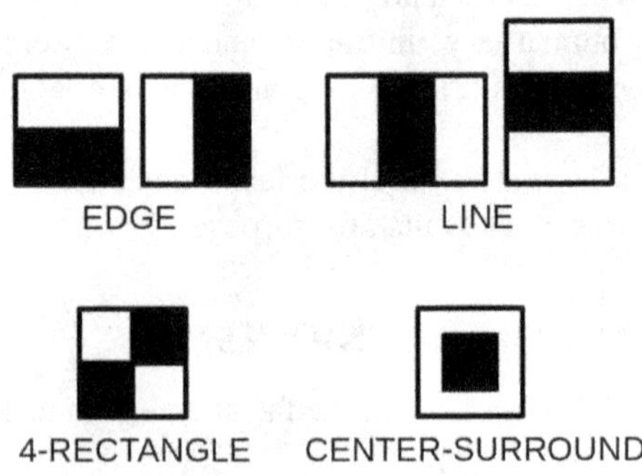

FIGURE 2.6 Some of the Haar features.

3. Firstly, the images are initialised with equal weights, which subsequently increase in the case of misclassified images. The process is repeated, and every time a new set of error rate and feature weights is calculated and that feature is selected if the required error rate, accuracy, or number of features is achieved [10].
4. With the help of the weak classifiers obtained, a final classifier made of 6000 features is computed using the weighted sum of the weak classifiers. This method is known as Adaboost.
5. Instead of applying all the 6000 features to the input image, a cascade of the classifiers is used, which involves dividing all the 6000 features into 38 different stages and passing on the image to the next stage if and only if it passes the current stage (see Figure 2.7).

The pros of the Haar Cascade Classifier include real-time performance on a central processing unit (CPU), simple model architecture, and being invariable to different scaling conditions. But, its major drawbacks are many false predictions, resulting in a decrease in precision and recall values, and inefficiency against images with obstructions and different perspectives [11, 12].

Syntax =>

- Haar Cascade Classifier – cv2.CascadeClassifier(self, *args, **kwargs)

Parameters =>

- Haar Cascade Classifier – *args = path to the Haar Cascade Classifier model.
- – **kwargs = None.

Returns =>

- Haar Cascade Classifier – Object of type 'cv2.CascadeClassifier'.

2.3.2 MMOD Face Detector

DLIB is one of the famous open-source toolkits written in the C++ language and is available for users to implement machine learning and data analytics techniques with ease in real-world applications. It also contains easy-to-use Python bindings for Python developers to interact directly with the dlib methods.

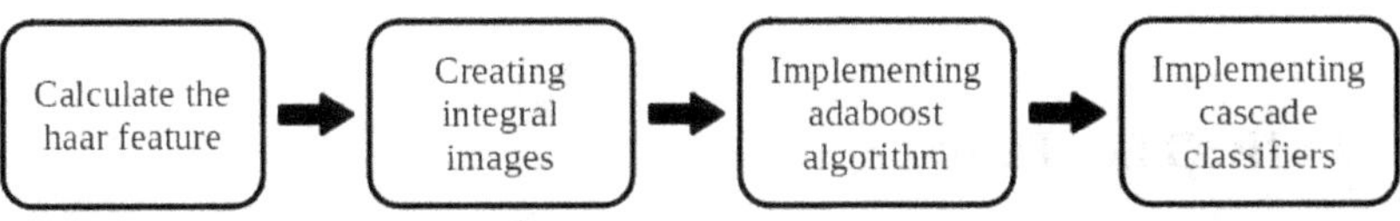

FIGURE 2.7 A brief block diagram of the steps required to train a Haar classifier.

In this subsection, before diving into the practical implementation of the DLIB and its Convolutional Neural Network (CNN)-based approach to detect faces, we are going to get to know something about them. The CNN-based face detector method of dlib called MMOD is a highly efficient and robust facial detection method, which is capable of detecting human faces irrespective of the lighting conditions, perspectives, and obstacles. The main benefit of this facial detection technique is in applications where the main priority is to accurately and precisely detect the maximum number of human faces present in the image with improper conditions [13].

This model uses a CNN trained on a dataset of 7220 images from various sources and has a very easy and simple training process. It can efficiently detect human faces with different orientations, obstacles, and lighting conditions which have the shape of more than 80 × 80 (width × height). After detecting the faces, the detector returns an object of type "_dlib_pybind11.mmod_rectangles", which contains the coordinates of the detected faces. But, this method fails to serve applications that have time constraints, thus preventing it from being implementable in real-time applications. To use this technique in real time, it is often advised to run it on a powerful graphics processing unit (GPU) rather than a CPU [14].

Besides detecting faces in the images or the videos and bounding them inside the boxes, DLIB comes with an additional method, which can be used for the localisation of facial keypoints on the detected faces and is known as "shape predictor". The shape predictor technique takes the input image and the detected face and localises the various keypoints or landmarks, such as nose, mouth, eyes, eyebrows, etc. (as shown in Figure 2.8), on the detected face, better known as face ROI (Region Of Interest). The DLIB facial landmark detector returns a set of 68 facial landmark coordinates based on the iBUG 300-W dataset.

Syntax =>

- MMOD Face Detector – dlib.cnn_face_detection_model_v1(self, filename)
- DLIB Shape Predictor – dlib.shape_predictor(self, *args, **kwargs)

Parameters =>

- MMOD Face Detector – filename = path to the mmod face detector model.
- DLIB Shape Predictor – *args = path to the DLIB shape predictor model.
  **kwargs = None.

Returns =>

- MMOD Face Detector – Object of type '_dlib_pybind11.cnn_face_detection_model_v1'.
- DLIB Shape Predictor – Object of type '_dlib_pybind11.shape_predictor'. [15]

2.3.3 HOG Face Detector

As previously discussed in subsection 2.3.2, DLIB is an open-source C++ toolkit used to apply machine learning and data analytics techniques in real-world tasks.

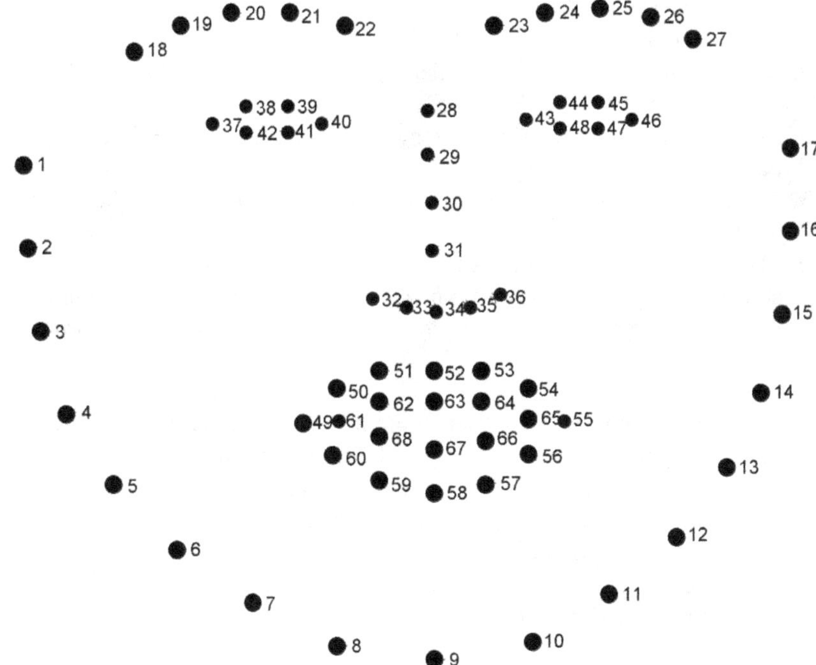

FIGURE 2.8 The 68 facial keypoints predicted by the shape predictor.

In subsection 2.3.2, we have already discussed one of the famous dlib face detector models: the MMOD face detector model. Although it is a very efficient, robust, and accurate face detector model and performs exceptionally well irrespective of the lighting conditions, obstacles, and orientations, due to its time complexity on CPUs, it cannot be used for real-time applications. Because of this, there is another DLIB method that is used for facial detection, known as HOG + Linear Support Vector Machine (SVM) face detector. This detector is based on HOG and linear SVM.

The HOG-based DLIB frontal face detector is computationally efficient and is thus capable of detecting faces present in the image or video in real time. It can perform well even on CPUs with very low time complexity as compared with the MMOD face detector.

The HOG feature descriptors have proven very effective to detect and localise different objects. They are simple, less powerful, but faster and efficient alternatives to the DNNs. They capture the most valuable information, i.e., outline information, shape, and appearance of objects, from the images. The algorithm, in brief, to calculate the HOG feature descriptors is as follows:

1. Preprocessing the image for correcting image aspect ratio, gamma correction, etc.
2. Computing the gradient values of the image horizontally and vertically using the following kernels:

Horizontal gradient kernel = [−1,0,1]
Vertical gradient kernel = [−1,0,1]

3. Dividing the image into the smaller regions called cells that are connected to each other.
4. Next, we find the edge orientation by calculating the magnitude and direction of the gradient of edges.
5. The edge gradients and orientations values of all the pixels in the cell are used to generate the histogram of that cell.
6. The histogram of every cell is normalised with the help of pixel intensities across the larger image region, known as blocks, to improve the accuracy.
7. Finally, all the resultant histograms are concatenated together to form the HOG descriptor.

The limitations of the HOG face detector are that it is not as accurate and robust as the MMOD face detector. It is inefficient in detecting faces in images with improper orientations and perspectives, bad lighting conditions, or obstacles. But in some cases, it can perform well as compared with the MMOD face detector [16, 17].

Along with the HOG frontal face detector, we can also apply the dlib shape predictor method on the input so as to mark the 68 key facial landmarks on the detected faces in the images, as shown in Figure 2.8.

Syntax =>

- HOG Face Detector – dlib.get_frontal_face_detector()
- DLIB Shape Predictor – dlib.shape_predictor(self, *args, **kwargs)

Parameters =>

- HOG Face Detector – None.
- DLIB Shape Predictor – *args = path to the DLIB shape predictor model.
  – **kwargs = None.

Returns =>

- HOG Face Detector – Object of type '_dlib_pybind11.fhog_object_detector'.
- DLIB Shape Predictor – Object of type '_dlib_pybind11.shape_predictor' [18].

2.3.4 MTCNN Face Detector

MTCNN is a complex CNN consisting of three different CNNs placed in a cascaded structure, which means that the output of the previous layer serves as input to the next layer, as shown in Figure 2.9. These CNNs are termed:

- Proposal Network (P-Net) – responsible for detecting possible facial regions through a shallow CNN.
- Refine Network (R-Net) – responsible for refining the facial regions and producing bounding boxes for the actual facial regions.

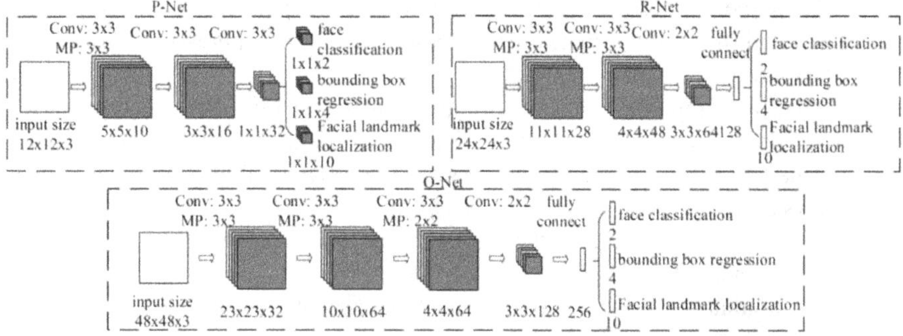

FIGURE 2.9 Diagram of cascaded architecture of the MTCNN face detector model.

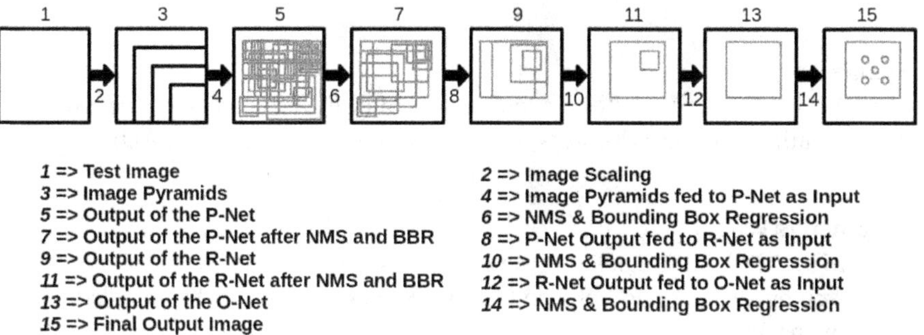

FIGURE 2.10 Pipeline of MTCNN face detector model on an input image.

- Output Network (O-Net) – responsible for filtering the detected faces and detecting the facial landmarks of the resultant faces [19].

The MTCNN face detector model is described briefly in the following, with the overall pipeline of the MTCNN model depicted in Figure 2.10.

- *Stage 1:* Creating multiple pyramids of the input image to feed in the P-Net. Gathering output and deleting low confidence bounding boxes. Transforming 12 × 12 kernel coordinates to the original image's coordinates. Applying Non-Maximum Suppression for kernels in each of the scaled images as well as for all the kernels. Transforming the bounding box coordinates to original image coordinates and reshaping them to squares.
- *Stage 2:* Pad out of the bounded boxes and supply the scaled images to R-Net. Gather the output and delete low confidence bounding boxes. Apply Non-Maximum Suppression for all the boxes. Transforming the bounding box coordinates to original image coordinates and reshaping them to squares.
- *Stage 3:* Pad out of the bounded boxes and supply the scaled images to O-Net. Gather the output and delete low confidence bounding boxes. Transforming

the bounding box coordinates as well as facial keypoints coordinated to the original image coordinates. Apply Non-Maximum Suppression for all the boxes.

- *Stage 4:* Compiling all the results and returning an object of data type 'dictionary' (or a list of dictionaries in the case of multiple images), which contains three keys for each and every face detected in the input image. The keys are:
  1. 'box': which contains the top left coordinates, height and width of the bounding boxes.
  2. 'confidence': which represents the probability of confidence of prediction.
  3. 'keypoints': which contains the coordinates for key face landmarks like left and right eyes, nose, and left and right mouth corners for the detected face [20, 21].

Syntax =>

- MTCNN Face Detector – mtcnn.mtcnn.MTCNN(self, weights_file=None, min_face_size=20, steps_threshold=None, scale_factor=0.709)

Parameters =>

- MTCNN Face Detector – weights_file = path of the file containing the weights of P-Net, R-Net, and O-Net. Default are the weights bundled with the package.
  - min_face_size = minimum face size to detect.
  - steps_threshold = step threshold value.
  - scale_factor = scaling factor.

Returns =>

- MTCNN Face Detector – Object of type 'mtcnn.mtcnn.MTCNN'.

2.3.5 DNN Face Detector

OpenCV contains a pre-trained Deep Learning–based facial detection method. This method was trained in the Caffe framework, which was created using the RESNET-10 model architecture and is based on the Single Shot Detector (SSD) framework. It was trained using a large dataset. The DNN face detector is real time, highly accurate, faster, easy to implement, and robust against images with bad resolution, improper lighting conditions, perspective, facial expressions, and face position [22] (Figure 2.11).

Syntax =>

- DNN Face Detector – cv2.dnn.readNetFromCaffe(prototxt, caffeModel=None)

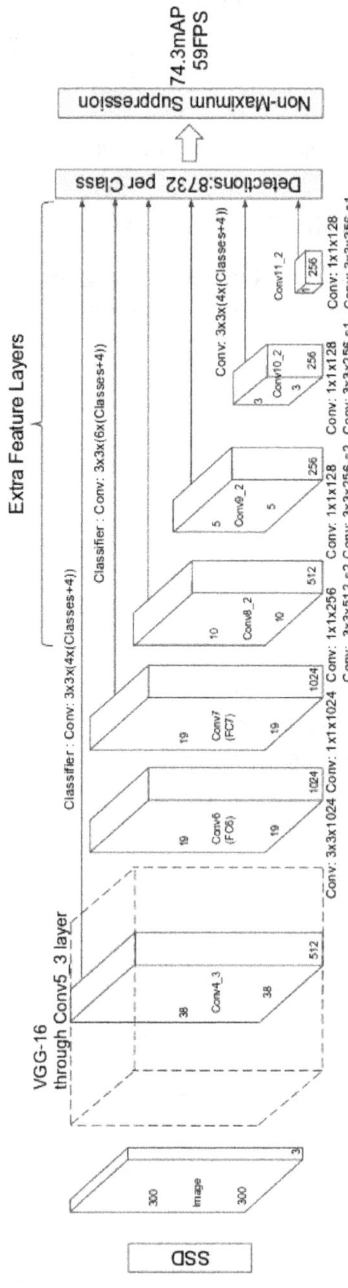

FIGURE 2.11 SSD-based RESNET face detector model.

Parameters =>

- DNN Face Detector – prototxt = path of RESNET model's prototxt file (contains model architecture i.e. layers).
    - caffeModel = path to RESNET model's caffemodel file (has weights of the model layers).

Returns =>

- DNN Face Detector – Object of type 'cv2.dnn_Net'.

2.4 RESULTS AND COMPARATIVE ANALYSIS

This section contains the well-structured and descriptive results of all the models. The results are represented with the help of visualisations and tables, and the section also contains a deep comparative analysis of the models based on different parameters. The models were tested on a predefined dataset of different images, which consists of open-source images from different sources. The description of the dataset is given in Table 2.2 [23].

The annotations used in the upcoming tables and visualisations are as follows:

- Gxx – group_xx.jpg
- Pxx – person_xx.jpg
- Txx – thing_xx.jpg
- HAAR – Haar Cascade Classifier
- HOG – HOG Face Detector
- MMOD – MMOD Face Detector
- MTCNN – MTCNN Face Detector
- DNN – DNN Face Detector
- TP – True Positive
- FP – False Positive
- FN – False Negative
- TN – True Negative
- TFC – Total Face Count
- TT – Total Time Taken
- ACC – Accuracy
- PRE – Precision
- REC – Recall

TABLE 2.2
Description of the Test Dataset Images

Categories	Quantity	File Name
Group of People	16	group_xx.jpg
Single Person	10	person_xx.jpg
Miscellaneous	4	thing_xx.jpg

The different evaluation parameters that were used to evaluate the performance of the models are:

- Confusion Matrix: It is defined as the table that is used to calculate the performance of the model based on the predicted as well as the actual label values.
- Accuracy: It is defined as the ratio of the correctly predicted values to the total predicted values [24].

$$\text{Accuracy} = \frac{\text{TP} + \text{TN}}{\text{TP} + \text{FP} + \text{FN} + \text{TN}}$$

- Precision: It is defined as the ratio of the correctly predicted true values to the total true predicted values.

$$\text{Precision} = \frac{\text{TP}}{\text{TP} + \text{FP}}$$

- Recall: It is defined as the ratio of the correctly predicted positive values to the total values that are positive.

$$\text{Recall} = \frac{\text{TP}}{\text{TP} + \text{FN}}$$

- Time Complexity: It is defined as the time taken by the algorithm to run.
- True Positive: An outcome is said to be true positive when its actual label as well as the predicted label is "true" or "1".
- False Positive: An outcome is said to be false positive when its actual label is "false" or "0" but the predicted label is "true" or "1".
- False Negative: An outcome is said to be false negative when its actual label is "true" or "1" but the predicted label is "false" or "0".
- True Negative: An outcome is said to be true negative when its actual label as well as the predicted label is "false" or "0" [25, 26].

Figure 2.12 depicts the confusion matrix that correctly represents the way to evaluate the different facial detection methods on the test dataset images.

2.4.1 Count of Detected Faces and Their Time Analysis

The individual count of the faces detected as well as the time complexity of all the models on the test dataset images is given in Table 2.3.

Figure 2.13 shows the visualisation of the number of faces detected by all the models on each and every image of the test dataset.

Figure 2.14 shows the visualisation of the time taken by all methods to detect the faces individually in the images of the test dataset.

CONFUSION MATRIX

		PREDICTED	
		POSITIVE (Model has detected a face)	**NEGATIVE** (Model hasn't detected face)
ACTUAL	**POSITIVE** (The face is present)	**TRUE POSITIVE (TP)** (The face is present & The model has detected the face)	**FALSE NEGATIVE (FN)** (The face is present but Model has not detected the face)
	NEGATIVE (The face is not present)	**FALSE POSITIVE (FP)** (The face is not present but The model has detected a face)	**TRUE NEGATIVE (TN)** (The face is not present & Model has not detected the face)

$$Accuracy=\frac{TP+TN}{TP+FP+FN+TN} \qquad Precision=\frac{TP}{TP+FP} \qquad Recall=\frac{TP}{TP+FN}$$

FIGURE 2.12 Confusion matrix on the basis of which the models are evaluated.

2.4.2 Confusion Matrix

The values of confusion matrix of all the models on the complete test dataset are shown in Table 2.4.

The visualisation of the confusion matrix of all the models on the complete test dataset is shown in Figure 2.15.

2.4.3 Other Evaluation Parameters

The evaluation scores of the models on parameters like accuracy, precision, recall, and total time complexity on the complete test dataset images are shown in Table 2.5.

The visualisation of the models on the other evaluation parameters is depicted in Figure 2.16.

Figure 2.17 represents another visualisation for the total time complexity of all the models in seconds over the complete test dataset.

2.4.4 Comparative Analysis

In subsections 2.4.1, 2.4.2, and 2.4.3, we have evaluated the different facial detection techniques on our custom test dataset created with open-source images. The evaluation parameters used were count of the faces detected, time taken individually on the test dataset images, confusion matrix, accuracy, precision, recall, and time complexity on the complete test dataset.

From Table 2.3, Table 2.4, and Table 2.5, it can be clearly concluded in Table 2.6.

All the models were tested and evaluated for the predefined set of parameters and values. The results are likely to change due to the following properties:

- Variation in image size, shape, and other geometrical properties.
- Variation in the values of the parameters used at the time of detection.

TABLE 2.3
Number of Faces Detected and Time Complexity of the Models on the Test Images

Test Image Name	Count of Faces Detected					Time Complexity (Seconds)				
	HAAR	HOG	MMOD	MTCNN	DNN	HAAR	HOG	MMOD	MTCNN	DNN
G01	12/13	13/13	13/13	13/13	13/13	0.15117	0.10541	2.67807	2.61640	0.05798
G02	28/35	34/35	27/35	33/35	35/35	0.03178	0.16698	4.01863	0.45739	0.04069
G03	14/16	14/16	16/16	14/16	16/16	0.03627	0.19946	5.05686	0.39982	0.05852
G04	18/20	19/20	21/20	19/20	18/20	0.07468	0.48900	11.1954	0.83345	0.05013
G05	10/10	9/10	10/10	9/10	10/10	0.03601	0.31221	6.19209	0.43481	0.04916
G06	25/31	30/31	31/31	30/31	31/31	0.06017	0.30349	6.73304	0.52714	0.06421
G07	6/7	7/7	7/7	6/7	7/7	0.03470	0.32266	6.26088	0.47568	0.05785
G08	5/5	5/5	5/5	5/5	5/5	0.03576	0.21558	4.86598	0.42793	0.06431
G09	8/14	14/14	14/14	14/14	12/14	0.03583	0.29771	5.74384	0.49048	0.05870
G10	6/15	13/15	14/15	11/15	13/15	0.04462	0.23160	5.27126	0.48976	0.05930
G11	3/3	3/3	3/3	3/3	3/3	0.05611	0.60504	16.2468	0.62445	0.05087
G12	10/13	7/13	4/13	13/13	6/13	0.03839	0.21719	4.34697	0.47493	0.05431
G13	5/5	5/5	5/5	5/5	5/5	0.08905	0.59659	15.2128	0.74361	0.05781
G14	4/5	0/5	0/5	5/5	5/5	0.01104	0.09857	1.78286	0.30194	0.06414
G15	7/7	7/7	7/7	7/7	7/7	0.04166	0.22759	5.22349	0.47025	0.05836
G16	5/4	4/4	4/4	4/4	4/4	0.06756	0.55860	12.9640	0.57009	0.05225
P01	1/1	1/1	1/1	1/1	1/1	0.00937	0.12803	2.95222	0.43838	0.05738
P02	1/1	1/1	1/1	1/1	2/1	0.02627	0.22179	5.04210	0.36855	0.05369
P03	1/1	1/1	1/1	1/1	1/1	0.02280	0.41709	8.94276	0.46381	0.04041
P04	1/1	1/1	1/1	1/1	1/1	0.03727	0.26923	5.56944	0.42175	0.04227
P05	2/1	1/1	1/1	1/1	1/1	0.02866	0.50981	10.3219	0.57749	0.06255
P06	1/1	1/1	1/1	1/1	1/1	0.02975	0.38595	8.51864	0.50248	0.05689
P07	1/1	1/1	1/1	1/1	2/1	0.03050	0.43154	8.22227	0.50290	0.05355
P08	1/1	1/1	1/1	1/1	6/1	0.02064	0.14797	3.48941	0.40856	0.07616
P09	1/1	1/1	1/1	1/1	2/1	0.02049	0.33586	6.83404	0.42601	0.05490
P10	1/1	1/1	1/1	1/1	1/1	0.04922	0.35773	8.23268	0.46582	0.05535
T01	0/0	0/0	0/0	0/0	0/0	0.03248	0.41778	8.47432	0.49894	0.07322
T02	0/0	0/0	0/0	0/0	0/0	0.02291	0.13698	3.22676	0.38254	0.07001
T03	0/0	0/0	0/0	0/0	1/0	0.02209	0.24773	5.47660	0.37967	0.05783
T04	0/0	0/0	0/0	0/0	0/0	0.01522	0.13909	3.32630	0.35607	0.06928

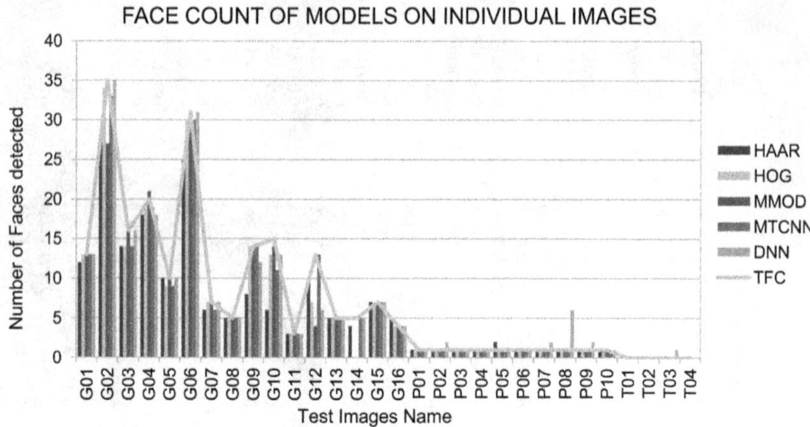

FIGURE 2.13 Number of faces detected in the test images individually by the model.

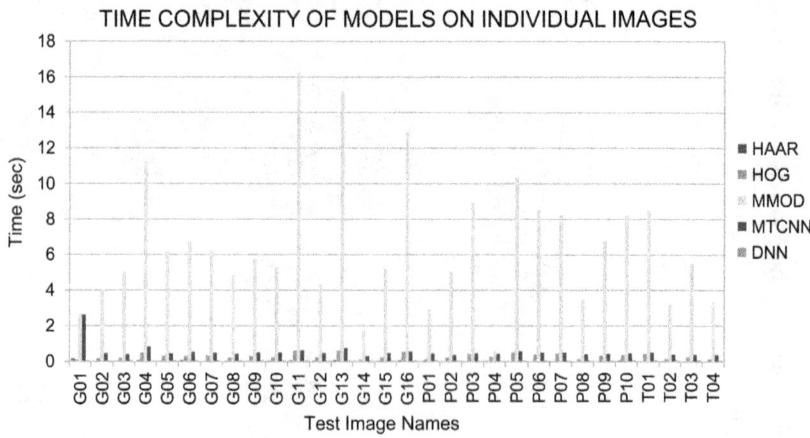

FIGURE 2.14 Time complexity of the models on each test image.

- Difference in the lighting, scaling, orientation, translation. and perspective conditions.
- Using any methods, like image smoothening, image sharpening, or others, for preprocessing images [27, 28].

2.5 CONCLUSION

As we have seen in the chapter, there are plenty of real-life applications where the use of biometrics has proved to be very productive and effective. As we proceeded further with discovering the efficiency of biometrics, we dived deeper into one of the key biometric studies, known as facial detection, and discussed its advancements and

TABLE 2.4
Confusion Matrix of the Models on the Test Dataset

			Predicted									
			True					False				
		MODELS	**HAAR**	**HOG**	**MMOD**	**MTCNN**	**DNN**	**HAAR**	**HOG**	**MMOD**	**MTCNN**	**DNN**
Actual	**True**	HAAR	173					40				
		HOG		194					19			
		MMOD			190					23		
		MTCNN				200					13	
		DNN					200					13
	False	HAAR	4					4				
		HOG		0					4			
		MMOD			1					4		
		MTCNN				1					4	
		DNN					9					3

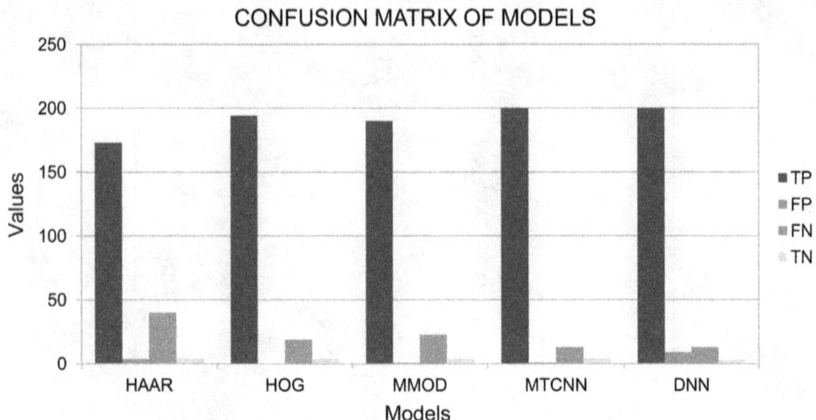

FIGURE 2.15 Confusion matrix of models on the test dataset.

TABLE 2.5
Scores of the Models for Other Evaluation Parameters

		Evaluation Parameters			
		ACC	**PRE**	**REC**	**TT (sec)**
Models	**HAAR**	0.800904	0.977401	0.812206	1.212611
	HOG	0.912442	1.000000	0.910798	9.094422
	MMOD	0.889908	0.994764	0.892018	202.4227
	MTCNN	0.935779	0.995024	0.938967	16.53127
	DNN	0.902222	0.956937	0.938967	1.722235

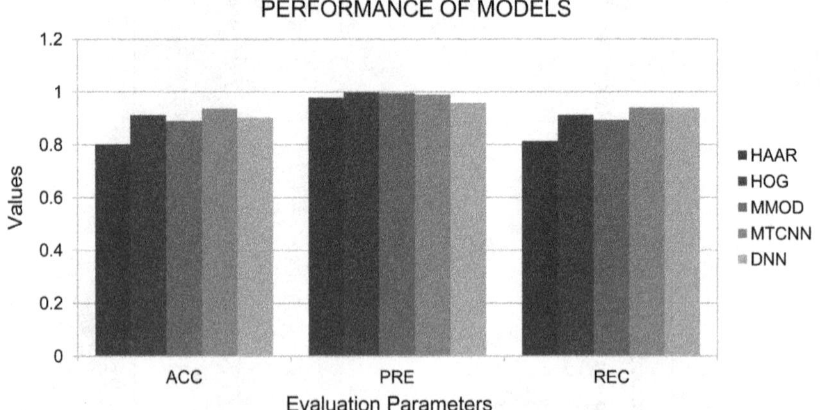

FIGURE 2.16 Scores of the models computed for other evaluation parameters.

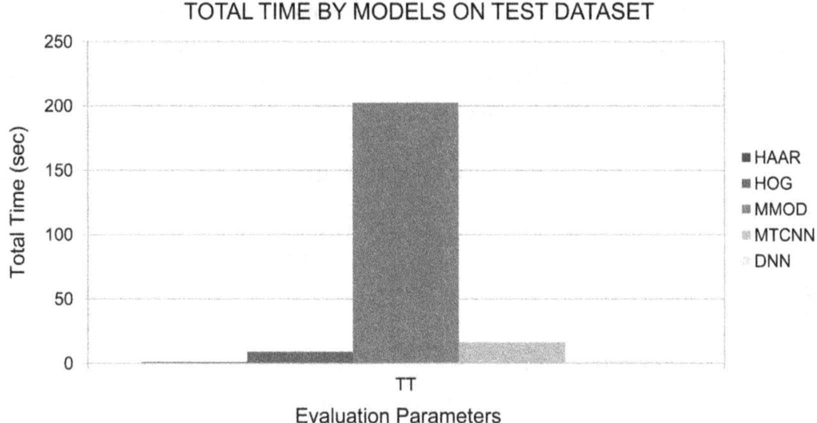

FIGURE 2.17 Total time taken by the models on the complete test dataset.

TABLE 2.6
Conclusion of Models Derived Using Different Evaluation Parameters

Parameters	Model with LOWEST Value	Value	Model with HIGHEST Value	Value
TT (sec)	**HAAR**	**1.212611**	MMOD	202.4227
TP	HAAR	173	**MTCNN, DNN**	**200**
FP	**HOG**	**0**	DNN	9
FN	**MTCNN, DNN**	**13**	HAAR	40
TN	DNN	3	**HAAR, HOG, MMOD, MTCNN**	**4**
ACC	HAAR	0.800904	**MTCNN**	**0.935779**
PRE	DNN	0.956937	**HOG**	**1.000000**
REC	HAAR	0.812206	**MTCNN, DNN**	**0.938967**

Note: The model name and the value in bold indicate the best performer for a particular parameter.

implementation in different case studies. We also discussed some of the well-known facial detection techniques in practice with a little theory before hands-on.

After defining, using, testing, and evaluating the five models, we can conclude that every facial detection technique has its own advantages and disadvantages, which are based on the parameters and inputs they are evaluated on. With the help of Table 2.6, we can reach the conclusion that based on the different needs of the applications, like time requirements, space available, or efficiency requirements, the models can be selected and tested on a series of parameter values so as to obtain the best performance and results from the model. For example, a time-constrained application can opt for the Haar Cascade Classifier, whereas a scenario that requires efficiency can approach any complex model, like the MTCNN face detector. We can even train our own facial detectors using facial datasets, machine learning algorithms, neural networks, and other resources based on the user's and the application's requirements [29, 30].

Also, we have seen that the facial detection techniques are not limited to those discussed in the chapter; other algorithms can also be used to detect faces, and these may have their own pros and cons. After detection of faces, the output can either be used solely or as an input to complex techniques like facial recognition, Augmented Reality, and others [31].

REFERENCES

1. Q. Cao, L. Shen, W. Xie, O. M. Parkhi and A. Zisserman, "Vggface2: A Dataset for Recognising Faces Across Pose and Age", in 2018 13th IEEE International Conference on Automatic Face & Gesture Recognition (FG 2018), Jodhpur, India, May 2018, 67–74.
2. S. Zhang, X. Wang, A. Liu, C. Zhao, J. Wan, S. Escalera, H. Shi, Z. Wang and S. Z. Li, "A Dataset and Benchmark for Large-Scale Multi-Modal Face Anti-Spoofing", in 2019 IEEE/CVF Conference on Computer Vision and Pattern Recognition (CVPR), Long Beach, CA, USA, April 2019, 919–928.
3. Y. Shi and A. K. Jain, "Probabilistic Face Embeddings", in *Proceedings of the IEEE/CVF International Conference on Computer Vision*, Seoul, Korea, August 2019, 6902–6911.
4. H. Zhao, X.-J. Liang and P. Yang, "Research on Face Recognition Based on Embedded System", *Mathematical Problems in Engineering*, Volume 2013, September 2013, 1–6.
5. P. Santemiz, L. J. Spreeuwers and R. N. J. Veldhuis, "Side-View Face Recognition", *Researchgate*, January 2010.
6. K. Grolinger and A. M. Ghosh, "Deep Learning: Edge-Cloud Data Analytics for Iot", Electrical and Computer Engineering Publications, 2019.
7. S. F. Kak, F. M. Mustafa and P. Valente, "A Review of Person Recognition Based on Face Model", *Eurasian Journal of Science & Engineering*, Volume 4, Issue 1 (Special Issue), September 2018, 157–168.
8. Y. Chen, X. Zhu and S. Gong, "Person Re-Identification by Deep Learning Multi-Scale Representations", in 2017 IEEE International Conference on Computer Vision Workshops (ICCVW), Venice, Italy, 2017, 2590–2600.
9. B. Babenko, M.-H. Yang and S. Belongie, "Visual Tracking With Online Multiple Instance Learning", in 2009 IEEE Conference on Computer Vision and Pattern Recognition, Miami, FL, USA, June 2009, 983–990.
10. O. Dürr, D. Browarnik and R. Axthelm, "Deep Learning on a Raspberry Pi for Real Time Face Recognition", *Researchgate*, January 2015, 1–4.
11. B. Fröba and A. Ernst, "Face Detection With the Modified Census Transform", Sixth IEEE International Conference on Automatic Face and Gesture Recognition, 2004. Proceedings, Seoul, Korea, 2004, 1–6.
12. P. Dattani, "Face Detection Based on Image Processing Using Raspberry Pi 4", *International Research Journal of Engineering and Technology*, Volume 07, Issue 02, February 2020, 120–123.
13. L. K. Gunnemeda, S. C. Gadde, H. Guduru, M. B. Devarapalli and S. K. Peketi, "Iot Based Smart Surveillance System", *International Journal of Advance Research and Development*, Volume 3, Issue 2, 2018, 166–171.
14. Y. Jin, J. Kim, B. Kim, R. Mallipeddi and M. Lee, "Smart Cane: Face Recognition System for Blind", in 3rd International Conference on Human-Agent Interaction, Daegu, Korea, October 2015, 145–148.

15. F. Ahmad, A. Najam and Z. Ahmed, "Image-Based Face Detection and Recognition: 'State of the Art'", *International Journal of Computer Science Issues*, Volume 9, Issue 6, February 2013, 1–4.
16. S Suthagar, A. S. Ponmalar, Benita, Banupriya and Beulah, "Smart Surveillance Camera Using Raspberry Pi 2 and Opencv", *International Journal of Advanced Research Trends in Engineering and Technology*, Volume 3, Special Issue 19, April 2016, 177–181.
17. M. Lal, K. Kumar, R. H. Arain, A. Maitlo, S. A. Ruk and H. Shaikh, "Study of Face Recognition Techniques: A Survey", *International Journal of Advanced Computer Science and Applications*, Volume 9, Issue 6, 2018, 42–49.
18. K. Enriquez, "Faster Face Detection Using Convolutional Neural Networks & The Viola-Jones Algorithm", May 2018.
19. P. Kb and J. Manikandan, "Design and Evaluation of a Real-Time Face Recognition System Using Convolutional Neural Networks", in Third International Conference on Computing and Network Communications (CoCoNet'19), Trivandrum, Volume 171, 2020, 1651–1659.
20. K. Zhang, Z. Zhang, Z. Li and Y. Qiao, "Joint Face Detection and Alignment Using Multi-Task Cascaded Convolutional Networks", *IEEE Signal Processing Letters*, Volume 23, Issue 10, April 2016, 1–5.
21. J. Du, "High-Precision Portrait Classification Based on Mtcnn and Its Application on Similarity Judgement", *Journal of Physics: Conference Series*, Volume 1518, 2020, 1–9.
22. M. Coşkun, A. Uçar, Ö. Yıldırım and Y. Demir, "Face Recognition Based on Convolutional Neural Network", in 2017 International Conference on Modern Electrical and Energy Systems (MEES), November 2017, 376–379.
23. S. Khan, M. H. Javed, E. Ahmed, S. A. Shah and S. U. Ali, "Facial Recognition Using Convolutional Neural Networks and Implementation on Smart Glasses", in 2019 International Conference on Information Science and Communication Technology (ICISCT), Karachi, Pakistan, March 2019, 1–6.
24. J. Wang and Z. Li, "Research on Face Recognition Based on CNN", in 2nd International Symposium on Resource Exploration and Environmental Science, 2018, 1–5.
25. N. S. Zabidi, N. M. Norowi and R. W. O. K. Rahmat, "A Survey of User Preferences on Biometric Authentication for Smartphones", *International Journal Of Engineering & Technology*, Volume 7, 2018, 491–495.
26. G. Hemalatha and C. P. Sumathi, "A Study of Techniques for Facial Detection and Expression Classification", *International Journal Of Computer Science & Engineering Survey (IJCSES)*, Volume 5, Issue 2, April 2014, 27–37.
27. A. Kumar, A. Kaur and M. Kumar, "Face Detection Techniques: A Review", *Artificial Intelligence Review*, Volume 52, Issue 2, August 2019, 927–948.
28. M.-Y. Shieh and H. Tsung-Min, "Fast Facial Detection by Depth Map Analysis", *Mathematical Problems in Engineering*, Volume 2013, October 2013, 1–10.
29. Y. Ji, S. Wang, Y. Zhao, J. Wei and Y. Lu, "Fatigue State Detection Based on Multi-Index Fusion and State Recognition Network", *IEEE Access*, Volume 7, May 2019, 64136–64147.
30. M. Sivaram, V. Porkodi, A. S. Mohammed and V. Manikandan, "Detection of Accurate Facial Detection Using Hybrid Deep Convolutional Recurrent Neural Network", *Ictact Journal on Soft Computing: Special Issue On Artificial Intelligence And Deep Learning*, Volume 09, Issue 02, January 2019, 1844–1850.
31. M. Sharma, J. Anuradha, H. K. Manne and G. S. C. Kashyap, "Facial Detection Using Deep Learning", *IOP Conference Series: Materials Science And Engineering*, Volume 263, Issue 4, 2017, 1–9.

3 Image Segmentation
Classification and Implementation Techniques

Hina J. Chokshi and Abhishek Agrawal

CONTENTS

DOI: 10.1201/9781003221333-3

3.1 INTRODUCTION

Image segmentation implies that splitting a picture into several distinct portions must meet certain criteria. In various requirements, the goal is for the areas to represent different portions of the image, similar to how crops, town regions, and forests appear on satellite television for computer images. In three-dimensional business item imaging, the areas are also diagrammatically represented as line segments and spherical arc segments in various analytic approaches. Regions can also be defined by a border made up of pixels, or by segmentation toward a specific way using a circle, ellipse, or polygon. We can still partition the picture into foreground areas of interest and historical past portions to be ignored as long as the attention-grabbing areas don't cover the entire picture [1] (Figure 3.1).

There are two basic objectives of image segmentation. The main goal is to decompose the image into components for analysis. In some circumstances, the environment may be sufficiently properly maintained so that the segmentation method

FIGURE 3.1 Football match image (a) and segmentation into regions (b). Each defined region is a set of connected pixels having the same color.

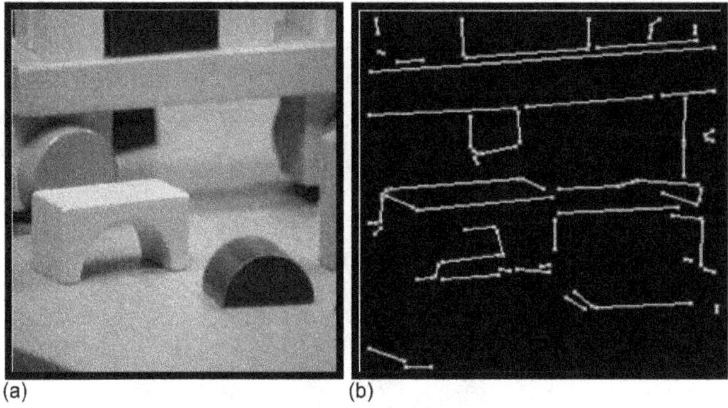

FIGURE 3.2 Blocks image (a) and extracted set of straight line segments (b).

faithfully extracts all of the elements that need to be evaluated further. For example, of splitting a person's face from a color video clip [2]. In advanced cases, such as extracting an entire avenue community from a greyscale aerial picture, the segmentation may well be very difficult and would possibly need computer code with advanced domain-constructing knowledge (Figure 3.2).

Streak segments have been removed by the Object Recognition Toolkit (ORT) package. The question of whether or not segmentation can be carried out for numerous distinct domains using enormous bottom-up techniques that do not require any particular region information is a crucial one. This toolkit offers segmentation techniques that may be used with a unique domain. The prospects of one segmentation tool working well for all subjects look dim. Knowledge has shown that a system vision package implementer must be able to select from a toolkit of strategies and maybe customize a solution based on the application's statistics. This approach, like standard cluster algorithms and line and spherical arc detectors, employs unique segmentation algorithms. The segmentation of a colored image of a football match into regions of near-consistent color is shown in Figure 3.1. The avenue segments extracted from an image of toy blocks are represented in Figure 3.2. In each scenario, keep in mind that the results are far from realistic. However, segmentation may offer useful input for higher-degree system-controlled processing; for example, it could provide gaming enthusiasts with a useful resource.

3.2 HOW IMAGE SEGMENTATION WORKS

Using a mask or a labeled photo, image segmentation involves dividing a photo into sections of photo elements, most likely in the form of a diagram. By breaking a photo into segments, it is possible to concentrate solely on the most important parts of the image rather than the complete image. A famous technique is to look for abrupt discontinuities in pixel values, which commonly mean edges that outline an area. Another commonplace region approach is to discover similarities inside the regions of a photo. Some techniques that involve this approach are area growing, clustering, and thresholding [3]. Several possible photo segmentation procedures have been improved during the course of the year by utilizing domain-specific data to successfully tackle segmentation challenges in certain software programme areas. Therefore, we will start with one of the most completely clustering-based processes in image segmentation: K-means clustering.

3.3 APPLICATIONS OF DIGITAL IMAGE PROCESSING

Some of the fundamental fields wherein virtual photograph processing is broadly used are as follows:

- Image sharpening and restoration
- Medical field
- Remote sensing
- Transmission and encoding
- Machine/Robot vision
- Color processing
- Pattern recognition
- Video processing
- Microscopic imaging
- Others

3.3.1 Image Sharpening and Restoration

Image sharpening and recovery refers to photos taken with a digital camera in order to improve the quality of the image or accomplish the desired result. It refers to what

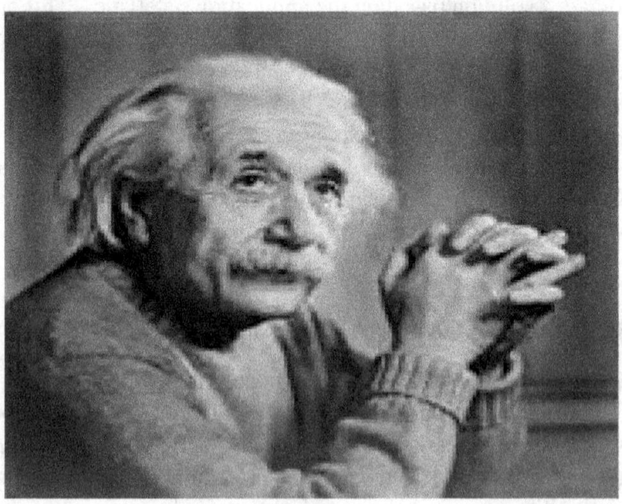

FIGURE 3.3 Original image.

Photoshop on occasion does. Zooming, blurring, sharpening, grayscale to color conversion, police painting edges and other ways around [4], picture retrieval, and image reputation are all examples of this. Common examples are:

Original image (Figure 3.3)
Zoomed image (Figure 3.4)
Blurred image (Figure 3.5)
Sharp image (Figure 3.6)
Edges (Figure 3.7)

FIGURE 3.4 Zoomed image.

FIGURE 3.5 Blurred image.

FIGURE 3.6 Sharp image.

FIGURE 3.7 Edge represented image.

3.3.2 Medical Field

The common applications of digital image processing (DIP) in the medical field are:

- Positron emission tomography scan
- Gamma ray imaging
- UV imaging
- Medical computed tomography
- X-ray imaging

3.3.2.1 Ultraviolet Imaging

In the domain of remote sensing the globe is scanned using the satellite cameras to analyze the information concerning it. One specific use of numerical picture technique software in the field of remote sensing is to detect infrastructure damage resulting from an earthquake. Despite the severe destruction in the center, it takes a long time to understand the damage in this case. The earthquake's affected area is so large that estimating the damage with human sight alone is impossible, or it is significantly stressful and time-consuming technique [5]. As a result, numerical image processing is currently the approach of choice. A photograph of the setup area is taken from a high point on the ground (Figure 3.8).

Then, it is analyzed to identify the numerous sorts of harm caused by the earthquake [6]. The key steps in the evaluation are:

- Extraction of edges
- Analysis and smoothing of various types of edges

3.3.2.2 Transmission and Encoding

When transmission was first introduced, an image was transmitted from London through a submarine cable. Figure 3.9 shows the photograph that was sent.

The photograph took 3 hours to travel from one locality to the other. Now, we can share video feed or CCTV pictures from one place to another in seconds [4]. This suggests that this discipline, too, has received a lot of attention: not just goal transmission, but also encoding. Several cameras, both high and low bandwidth, have been created to encode images for distribution via the internet and other means.

FIGURE 3.8 UV imaging.

FIGURE 3.9 Transmitted photograph.

FIGURE 3.10 Obstacle detection.

3.3.2.3 Machine/Robot Vision

Among the numerous challenges that robotics faces currently, one of the most vital is to improve machine vision and assemble robots that are able to see things, verify them, determine obstacles, etc. Considerable work has been contributed on this subject, including portable laptop vision [4].

3.3.2.4 Obstacle Detection

Another application zone of image segmentation is obstacle detection in a photograph. It can easily detect the obstacles in a given area (Figure 3.10).

3.3.2.5 Line Follower Robot

Most robots presently paint by following the street and so are known as line follower robots. This helps a robot to move further on its dedicated path and complete

FIGURE 3.11 Line follower robot.

its allocated tasks. This has however been performed following the image path (Figure 3.11).

3.3.2.6 Color Processing

Color processing entails the manipulation of coloured images and the utilization of extremely specific colored regions. It might be an RGB colour model, a YCbCr colour model, or an HSV colour model. It also entails locating the colored photos' transmission, storage, and mystery writing [7].

3.3.2.7 Pattern Recognition

Photographic style and a variety of other domains in alternative gadget mastering are included in pattern recognition (a department of synthetic intelligence). Photograph processing is used in sample [8] to feature the items in a photo, and gadget mastering is used to educate the gadget for the extrude. Pattern reputation is used in mobile computer-assisted diagnosis, handwriting reputation, photo reputation, and many more applications.

3.3.2.8 Video Processing

A video is nothing more than the proper representation of images. A video's good quality is determined by the number of frames/images per minute, as well as the standard set by all of us [9]. Video processing consists of noise reduction, detail improvement, motion detection, frame fee conversion, quantitative relation conversion, colored house conversion, etc.

3.4 REQUIREMENT FOR IMAGE SEGMENTATION

Image segmentation has a wide range of applications in a variety of industries and might be a key element in creative and discerning PCs. Image segmentation is used in a number of applications, including:

3.4.1 Face Recognition

The first-class face technology present in your iPhone and advanced protection systems uses photo segmentation to identify your face. It must be able to select the various features of your face with great care, ensuring that no undesirable party has access to your phone or gadget [10].

3.4.2 Number Plate Identification

To impose fines and facilitate searches, many internet site visitors' lights and cameras utilize large number of plate identification. The use of number plate identification technology allows users of a website to familiarize themselves with an automobile's ownership information. It divides the number plate and its statistics from the rest of the components using innovative method of photograph segmentation. This technology has simplified the fining method drastically for governments.

3.4.3 Image-Based Search

When using Google Search with a scanned photograph, it appears that the segmentation process uses photo segmentation strategies to detect the contents of the photograph and compare their conclusions to relevant photographs in order to produce results.

3.4.4 Medical Imaging

In specific medical sectors, image segmentation strategies are carried out to identify cancer cells, maintain tissue volumes, run digital surgical treatment simulations, and do intra-surgical treatment navigation. Image segmentation has numerous packages within the medical sector. Image segmentation has uses in manufacturing, agriculture, safety, and lots of risk sectors. The use of photo segmentation algorithms will increase as our innovative and discerning generation of computers becomes considerably superior. For example, some manufacturers have begun to redefine image segmentation strategies to find faulty products. The rule of thumb may also completely capture the desired components from the image of the object and identify them as faulty or optimal in this case [11]. This gadget reduces the risk of human errors and makes checking much more affordable for the organization. Common implementations of image segmentation are in Python, C, C++, and MATLAB.

3.5 TYPES OF IMAGE SEGMENTATION

Image segmentation is an upcoming research area consisting of various techniques for segmentation. On the basis of different features, characteristics, and parameters, the image segmentation techniques are classified as shown in the following subsections.

3.5.1 Approach-Based Classification

Image segmentation is performed by identifying an object first. To segment an object, it must first be identified; a set of guidelines will not suffice, as it can only classify the sorts of components [12]. All photo segmentation tasks are initiated with object identification only. The performance of the algorithms and how good they are at identifying and finding objects while gathering information on the relevant pixels are used to segment an image [13]. Different methods are used to perform this task, as follows.

3.5.1.1 Region-Based Approach (Similarity Detection)

The approach consists of locating similar pixels based on a selected threshold, area growing, area spreading, and area merging. Different machine learning algorithms and clustering algorithms use this method to find the unknown characteristics and features of an image [14, 15]. Various classification algorithms use this method for feature detection and image segmentation.

3.5.1.2 Boundary-Based Approach (Discontinuity Detection)

This method is a good option for location-based methodology for object detection. In this situation, the pixels within the boundary that have similarity are identified, as opposed to location-based methodology, which locates pixels based on similarity functions [16]. They simply recognize on the edge of numerous pixels and separate them from the rest of the images using techniques such as edge detection and line detection, as well as different equivalent algorithms for the same.

3.5.2 Technique-Based Classification

Different image segmentation techniques use their own unique techniques to locate an image and segment it. Depending on the type of image to be segmented and the type of data to be extracted from it, corresponding techniques are selected. The image segmentation is furthermore classified as shown in the following.

3.5.2.1 Structural Techniques

The structural facts associated with the image used for processing are needed through the algorithms. The information covered is the pixels of the image, histograms, pixel density, and distributions together with the image distribution and one-of-a-kind relevant records. Then you'll need structural records on the area where you want to lower the picture's size [17]. You'll need the records so that your set of rules can find the area. The algorithms we use for these implementations look at the area-based complete technique.

3.5.2.2 Stochastic Techniques

These strategies need the discrete pixel value of an image in preference to the form of the whole picture or the photograph represented. Because of this, to carry out the image segmentation process on multiple images requires a large number of facts [18].

Machine learning algorithms including the K-means clustering and artificial neural network (ANN) algorithms are used.

3.5.2.3 Combined/Hybrid Techniques

As the name "hybrid" implies, these cross-breed algorithms utilize both primary and stochastic methods [19]. To part a picture, these hybrid approach algorithms are combined to separate the required insights of the perfect picture as well as the discrete pixels. As a result, they make use of the fundamental concepts of the ideal spot and discrete pixel.

3.6 IMAGE SEGMENTATION TECHNIQUES

For segmenting an image, several techniques can be applied:

1. Thresholding Segmentation
2. Edge-based Segmentation
3. Region-based Segmentation
4. Watershed Segmentation
5. Clustering-based Segmentation Algorithms
6. Neural Networks for Segmentation

All these techniques are briefly explained in the following subsections.

3.6.1 Thresholding Segmentation

The threshold segmentation method is said to be one of the most powerful methods. For an image, the pixels are divided using its threshold value. The technique enables separating an object from its background. In this technique each pixel value of an image is compared with its specified threshold. If the image has very little noise then this method of segmentation is considered as more appropriate. The thresholding approach accurately turns a grayscale image into a binary image with the added benefit of segmentation (required and now no longer favored sections) [20]. The thresholding segmentation may be categorized into many unique categories.

3.6.1.1 Simple Thresholding

In a number of the thresholding methods, the image's substance is replaced with two colors, white and black. If the intensity value of the image element is in smaller quantity, then it's replaced with a black pixel; if it is high, then it is replaced with the white color [21]. This method is acceptable for beginners in image segmentation, as it is said to be clean thresholding.

3.6.1.2 Otsu's Binarization

To accomplish clean thresholding, a threshold value is chosen and picture segmentation is performed. How can you be sure, though, that the seemingly correct value you chose was the correct one? While determining truly one-of-a-kind values and selecting one is the clean process for this, it isn't the most cost-effective. For applying Otsu binarization, an image with a graphical illustration of a chart having a foreground peak and a background peak is chosen. The approximate value of the center of those peaks as your threshold value can be calculated using the Otsu binarization pattern [21]. If the image is a bimodal image, this technique can be used to determine the issues from the image's chart. Essentially, this method is used to scan documents or remove supplementary material from a file. But, it's going to have many limitations. Such pictures are often used for footage and aren't bimodal.

3.6.1.3 Adaptive Thresholding

Thresholding based on one constant value is not considered to be an efficient method for an image. Really one-of-a-type photos have one-of-a-type background conditions that have an effect on their properties. For an image, one has to find some constant threshold value for image segmentation. Also, one can find one of the similar types of threshold value for numerous sections of an image. This method works properly with footage that has numerous lighting conditions [21].

3.6.2 Edge-Based Segmentation

Edge-based segmentation is one of the most implemented methodologies in the field of segmentation in image processing. It focuses on the principal edges of various devices in a photograph. This is a key stage since it allows you to learn about the various devices available within the image, as the edges carry a wealth of information that you can use [22]. Edge detection is the recommended technique, as it permits removal of unwanted and inessential data from the image. It effectively makes an image of smaller size. Algorithms for part-based segmentation are fully segmented. Installation edges throughout a picture are in step with the variations in texture, contrast, gray level, color, saturation, and one-of-a-type properties. The image can be enhanced using the effects available with the available resources, as well as combining all of the connecting sides into part chains that more closely fit the image borders [23]. There are many edge segmentation techniques, and based on their characteristics, they are divided into two main categories, as follows

3.6.2.1 Search-Based Edge Detection

The search-based edge detection method locates the edges by finding the edge strength, using the gradient magnitude, and then finding the local directional maxima of the gradient magnitude using the gradient direction [24].

3.6.2.2 Zero Crossing–Based Edge Detection

This technique searches for average global gradient retrieved from the image. Generally, you'd be told to pre-approach the image to remove unnecessary noise and assemble it, which is a lot less arduous than having to stare at the edges [25]. Canny, Prewitt, Deriche, and Roberts pass are the most commonly used methods. They make it a lot easier to reconnect with discontinuities and resolve the issues. The image must be segmental, and the close edges must all cluster appropriately into a binary image, to achieve minimum partial segmentation [26].

3.6.3 Region-Based Segmentation

The image is divided based on similar capabilities. This section shows a set of pixels outside of a portion of the input image, which may be a small segment.. After locating the Starting(Seed) point [27–28], a set of rules for region-based segmentation should be forced to either add extra pixels to them or cut them back so that they'll combine with unique seed points. We have a tendency to be about to categorize location-based entirely absolutely segmentation into subsequent groups using those two strategies.

3.6.3.1 Region Growing

In this segmentation technique, the pixels are grouped together based on distinct similarity values. For implementation, it begins with a small group of pixels and then forms a large group. A location growth formula must be compelled to choose the image's primary ingredient, compare it to the neighboring pixels, and begin building the situation with the accessible resource of finding suites to the seed point. The set of guidelines will recognize every different seed pixel which does not belong to any present region as long as the current regions pixel does not increase anymore. Location growing algorithms are used to avoid such an error. This algorithm increases a handful of regions at specific time intervals.

3.6.3.2 Region Splitting and Merging

Techniques for region splitting and merging are used for performing these two functionalities on an image. The image is first split into regions based on similar features and attributes and then, is merged with the adjacent parts [29]. The collection of policies in location splitting considers the entire image, whereas the set of policies in location growth focuses on a single spot. This method employs a divide-and-conquer strategy, in which an image is separated into a group of comparable attributes and then suited to the predefined conditions. The recursive rules that perform this challenge are also known as split-merge algorithms [30].

3.6.4 Watershed Segmentation

Watershed segmentation refers to one form of region-based method. This method separates basins from each other. This technique decomposes an image completely and assigns a pixel either to a region or to a watershed. It is highly suitable for medical image segmentation.

A watershed represents a set of regulations that manipulates an image as if it had been a geography map. It specializes in shaping basins, which are the polar opposite of ridges, and floods the basins with markers until they reach the watershed lines browsing the ridges, rather than employing pixel ridges. Compared with the ridges, the basins have hundreds of markers, through which the image is segmented into a number of regions in step with the peak pixel. This technique converts an image or photo into its corresponding geographic map. Using the gray value of the pixels, this method creates a mirror topography image. This leads to representing a landscape with valleys and ridges in a three-dimensional (3-D) image [31]. This technique displays the regions of a 3-D image as "catchment basins". This technique is highly preferred in medical image segmentation, such as magnetic resonance imaging (MRI), clinical imaging, and many more.

3.6.5 Clustering-Based Segmentation Algorithms

Clustering-based segmentation algorithms are used to discover the hidden statistics inside an image that may not be visible in a normal case. A cluster is a set of rules that divides the image into different groups of pixels, known as clusters, which have comparable features [32]. It differentiates the data factors into clusters and represents them as special clusters. Popular algorithms used with this technique are ok-approach, fuzzy c-approach (FCM), ok-approach, and progressed ok-approach algorithms. Because of its efficiency, the ok-approach clustering algorithm is preferred for picture segmentation [33]. The fuzzy c-approach algorithm categorizes factors into several clusters based on their range of membership. A few clustering algorithms based on their functionality are mentioned in the following subsections.

3.6.5.1 K-Means Clustering

The K-means clustering algorithm works on a different set of rules by gaining information from the image. The image is classified on the basis of the selected range of clusters. The clustering technique begins with dividing the image into ok pixels that constitute ok cluster centroids [34]. Each item is assigned to the organization that supports the distance between them and the centroid on this basis.

3.6.5.2 Fuzzy C Means

The Fuzzy C Means algorithm groups the pixels of an image into different clusters. It may also be the case that a pixel may belong to a single cluster. Moreover, pixels may have many stages of similarities with each cluster [35]. The set of rules applied by this approach is efficient for detecting the accuracy of image segmentation. Cluster algorithms are best suited for photograph segmentation needs.

3.6.6 Neural Networks for Segmentation

Neural networks play an important role in image segmentation. Artificial intelligence is used to examine an image and identify the objects, human faces, text data, and other required information. Convolutional neural networks (CNNs) are said to be preferred for image segmentation, as they can extract the different image features and statistics at a much faster rate and with highest accuracy [36]. Recently, the experts at Facebook AI analysis (FAIR) created a Mask R-CNN, which is a deep gaining knowledge of layout that created pixel-clever masks for individual image elements. It falls into the category of an improved model of the faster R-CNN item detection architecture. For each element present in the image, the quicker R-CNN employs two distinct fact devices. Other additional information associated with an image can be extracted using the Mask R-CNN algorithm. This algorithm first generates the function map of the image. Then, the device applies the area idea network (RPN) at the function maps and generates the element proposals with their objectless scores. As soon as it is completed, the pooling layer is applied to the suggestions, allowing them to be conveyed in a uniform manner [37]. Finally, the proposals to the corresponding layer for class are passed by the device, and the output is generated.

3.7 IMPLEMENTATION AND PRE-REQUISITES

Different image processing and segmentation techniques are commonly used in different computer programming languages like C++, Java, and matplotlib. The reason for integration is its flexibility to support modularity; it can be easily implemented and used in different technological tool analyzing techniques, deep analyzing techniques, and expertise stacks. The added feature of Python makes it feasible for implementing unique methods in SimpleITK, Mahotas, image scikit-photograph, matplotlib, OpenCV, and many more. Python libraries are a difficult method of implementation, as it requires basic understanding of Pandas and Python language. To have greater control over the various libraries that might be used in this approach, one must have good command and understanding of it and also of the tool analyzing it for performing segmentation in images. The different pre-processing steps related to the image segmentation can be easily done using the smart statistics of linear algebra, differential equations, and calculus. Expertise in neural networks, and also in CNNs, is important for the ANN implementation of photograph processing and segmentation. Python is enriched with a major set of libraries for removing noise, differentiating contours, and creating histograms.

3.8 FUTURE SCOPE

In the field of DIP, image segmentation will open enormous opportunities for researchers in the near future. With the advancement of different segmentation techniques, there will be unlimited numbers of robots in the world transforming technology completely digitally. Verbal commands, waiting for the desires of government agencies, translating languages, analyzing clinical conditions, completing surgery, reprogramming defects in human DNA, and automating all modes of transportation will be among the advancements in picture processing and synthetic intelligence.

With the increasing power and sophistication of new computers, the concept of computation will expand beyond these limitations, and in the future, image technique generation will improve, as well as the human sensory system. Long-term trends in afar sensing may lean toward stepped forward sensors that record a stable picture in several spectral channels. Graphics expertise is becoming vital in image processing applications. The long-term satellite image processing packages for computers are based completely on the imaging levels ranging from planetary exploration to police research packages. Large-scale homogenized cell arrays of true circuits are being used to carry out image segmentation and to explain sample-forming phenomena. The cell neural community is a programmable set of closely connected neural networks that has evolved into a paradigm for future imaging methods. This device's applications include semiconducting cloth retina, sample generation, and other areas.

3.9 CONCLUSION

Image segmentation is said to be a basic step to analyze an image. Processing an image, classifying an image, and understanding the information extracted from it is termed image segmentation. It separates a given image into homogeneous sections, simplifying and making the image more comprehensible. The chapter discusses the different digital image segmentation methods and the techniques used for processing an image, such as edge detection, thresholding, clustering, and region growing. An analysis of the results of the segmentation techniques, and which methods give better results, is provided. Discussions related to different image segmentation techniques are presented. The implementation of image segmentation in various fields, along with its benefits and limitations, is also discussed. Different factors like image content, the texture homogeneity of an image, and spatial characteristics affect the process of image segmentation. Thus, image segmentation methods, techniques, and algorithms have been recognized as one of the most promising and significant fields of modern research in each and every aspect of the industry 4.0 revolution.

REFERENCES

1. V. Gurusamy, S. Kannan, and G. Nalini. "Review on image segmentation techniques." October 2014, p. 273127438.
2. Y. J. Zhang. "A survey on evaluation methods for image segmentation." *PR*, 29(8): 1335–1346, 1996.
3. S. K. Dewangan "Importance & applications of digital image processing." *International Journal of Computer Science & Engineering Technology*, 7(7), Jul 2016: 316–320.
4. A. Bindhu and K. K. Thanammal "Analytical study on digital image processing applications." 7(6), June 2020.
5. S. Li, et al. "Pixel-level image fusion: A survey of the state of the art." *Information Fusion*, 33, 100–112, 2017.
6. M. A. Viergever, et al. "A survey of medical image registration: Under review." 140–144, 2016.
7. R. Ramanath, W. E. Snyder, Y. Yoo, and M. S. Drew. "Color image processing pipeline." *IEEE Signal Processing Magazine*, 22(1), 34–43, Jan. 2005.
8. P. Devijver and J. Kittler *Pattern Recognition: A Statistical Approach*. Prentice Hall, 1982.
9. C. Eckes and J. C. Vorbrüggen "Combining data-driven and model-based cues for segmentation of video sequences." In: Proceedings of the World Congress on Neural Networks, pp. 868–875, 1996.
10. J. Savitha and A.V. Senthil Kumar. "Image segmentation based face recognition using enhanced SPCA-KNN method." *International Journal of Computer Science and Information Technologies*, 6(3), 2115–2120, 2015.

11. G. Litjens, et al. "A survey on deep learning in medical image analysis." *Medical Image Analysis*, 42, 60–88, 2017.
12. F. J. Estrada and A. D. Jepson. "Quantitative evaluation of a novel image segmentation algorithm." 2005 IEEE Computer Society Conference on Computer Vision and Pattern Recognition (CVPR'05). Vol. 2. IEEE, 2005.
13. S. V. Kasmir Raja, A. Shaik, and A. Kadir. "Moving towards region-based image segmentation techniques: A study." *Journal of Theoretical and Applied Information Technology*, 5(1), 2009.
14. M. R. Khokher, A. Ghafoor, and A. M. Siddiqui. "Image segmentation using multi-level graph cuts and graph development using fuzzy rule-based system." *IET Image Processing*, 7(3): 201–211, 2012.
15. V. Dey, Y. Zhang, and M. Zhong. "A review on image segmentation techniques with Remote sensing perspective." In ISPRS, Vienna, Austria, July 2010, Vol. XXXVIII.
16. S. Inderpal and K. Dinesh. "A Review on Different Image Segmentation Techniques." *IJAR*, 4(April), 2014.
17. Y. Y. Boykov and M.-P. Jolly. "Interactive graph cuts for optimal boundary & region segmentation of objects in N-D images." In Proceedings of 8th IEEE International Conference On Computer Vision, Vancouver, Canada, July 2001, Vol. I, P.105.
18. K. Haris, S. N. Efstratiadis, N. Maglaveras, and A. K. Katsaggelos. "Hybrid image segmentation using watersheds and fast region merging." *IEEE Transactions on Image Processing*, 7(12), December 1998.
19. Y.-H. Wang. *Tutorial: Image Segmentation*. National Taiwan University, Taipei, Taiwan, ROC.
20. P. Singh. "A new approach to image segmentation." *International Journal of Advanced Research in Computer Science and Software Engineering*, 3(4), April 2013.
21. S. S. Al-amri and N. V. Kalyankar. "Image segmentation by using threshold techniques." *Journal of Computing*, 2(5), May 2010.
22. M. Sarif, M. Raza, and S. Mohsin, "Face recognition using edge information and DCT." *Sindhu University Research Journal*, 43(2), 209–214, 2011.
23. S. Lakshmi and D. V. Sankaranarayanan. "A study of edge detection techniques for segmentation computing approaches." IJCA Special Issue on *"Computer Aided Soft Computing Techniques for Imaging and Biomedical Applications" CASCT*, 2010.
24. M. Sharif, S. Mohsin, M. Y. Javed, and M. A. Ali. "Single image face recognition using laplacian of Gaussian and discrete cosine transforms." International Arab Journal of Information Technology, 9(6), 562–570, 2012.
25. B. Sumengen and B. Manjunath. "Multi-scale edge detection and image segmentation." In Proceedings of the European Signal Processing Conference, 2005.
26. X. Yu and J. Yla-Jaaski. "A new algorithm for image segmentation based on region growing and edge detection." In Proceedings of the IEEE International Sympoisum on Circuits and Systems, 1991, pp. 516–519.
27. H. G. Kaganami and Z. Beij. "Region-based segmentatio versus edge detection." 2009 Fifth International Conference *on* Intelligent Information Hiding *and* Multimedia Signal Processing, 1217–1221, 2009.
28. I. Karoui, R. Fablet, J. Boucher, and J. Augustin. "Unsupervised region-based image segmentation using texture statistics and levelset methods." In Proceedings of the WISP IEEE International Symposium on Intelligent Signal Processing, 2007, pp. 1–5.
29. Y. M. Zhou, S. Y. Jiang, and M. L. Yin. "A region-based image segmentation method with mean-shift clustering algorithm." In Proceedings of the 5th International Conference on Fuzzy Systems and Knowledge Discovery, 2008, pp. 366–370.

30. C. Cigla and A. A. Alatan. "Region-based image segmentation via graph cuts." In Proceedings of the 15th IEEE International Conference on Image Processing, 2008, pp. 2272–2275.
31. M. Bai and R. Urtasun. "Deep watershed transform for instance segmentation." In *Proceedings of the IEEE Conference on Computer Vision and Pattern Recognition*, 2017.
32. J. Wang and X. Su. "An improved K-means clustering algorithm." In Proceedings of the IEEE 3rd International Conference on Communication Software and Networks, 2011, pp. 425–429.
33. X. Zheng, et al. "Image segmentation based on adaptive K-means algorithm." *EURASIP Journal on Image and Video Processing*, 1(2018), 1–10.
34. N. A. Mohamed and M. N. Ahmed. "Modified fuzzy c – mean in medical image segmentation." In *Proceedings of the IEEE International Conference on Acoustics,* Speech, *and Signal Processing*, 6, 3429–3432, 1999.
35. N. MongHien, N. ThanhBinh, and N. Q. Viet. "Edge detection based on Fuzzy C means in medical image processing system." In 2017 International Conference on System Science and Engineering (ICSSE), Ho Chi Minh City, 2017, pp. 12–15, doi: 10.1109/ICSSE.2017.8030827.
36. E. Saravana Kumar and K. Vengatesan "Trust based resource selection with optimization technique." *Cluster Computing*, 22, 207–213, 2019 (Scopus).
37. S. Prabu, V. Balamurugan, and K. Vengatesan. "Design of cognitive image filters for suppression of noise level in medical images." *Measurement*, 141, 296–301, 2019.

4 Image Processing with IoT for Patient Monitoring

Hina J. Chokshi and Maya Mehta

CONTENTS

DOI: 10.1201/9781003221333-4

4.1 INTRODUCTION

Image Processing combined with the Internet of Things (IoT) has opened the gates for monitoring the health of patients. Internet of Medical Things (IoMT) in the healthcare domain gives the perfect roadmap to utilize medical resources. As the global population increases day by day, it is difficult to reach each and every person and provide medical services. Patients who are suffering from critical illness require to be monitored on a regular basis, 24/7, so that proper medical assistance can be provided to them in case of emergency. When the illness is spread across the world, the patient to staff ratio becomes difficult to manage. The major challenge is to provide specialized and efficient medical attention to patients. In hospital, patients admitted to the intensive care unit (ICU) are observed and monitored on a constant basis, whereas patients admitted to special rooms and general wards are observed twice a day. The situation is often worse for patients admitted to public hospitals. The regular way of treating patients is for a doctor to personally and physically monitor a patient, mining the patient's health records for documentation of mobility events or physical observation of the patient [1]. This methodology consumes a lot of time and effort, and at times, it becomes difficult for healthcare workers and doctors to give special attention simultaneously to a large number of patients. Therefore, the need arises for an efficient and effective IoT-enabled patient monitoring system to provide efficient and consistent patient monitoring.

In this situation, IoMT plays a significant role by making use of things that are always lying around the patient to collect real-time data, such as a wheelchair, oxygen pump, smart watch, or smart phone. Equipped with sensors, these can become smart and intelligent devices that can generate data according to the situation. All these devices are connected using communication protocols, and by enabling the Internet using a smartphone, data about the patient can be centralized. Here, data can be images or reports. Centralized data help medical professionals to track, monitor, and provide superlative care to their patients. Images taken by smart devices or stored in a central database can be processed and used by medical professionals. Image processing helps professionals to analyze the patient's situation and give prescriptive analysis. These data can be used by other non-governmental organizations or institutes for monitoring civil crimes, etc. In the era of healthcare 4.0 and the pandemic, the situation arises where tracking and monitoring should be processed without physically examining the patient unless the worst condition arises.

This chapter basically focuses on the application areas of IoMT devices and their limitation and benefits. It discusses how different devices can be used for medical care and also standardizing the presentation of data. All devices use different methods of data presentation, so it is difficult to process data. Using a standard interface for capturing patients' details, IoMT devices shall easily enable the processing of patients' records. The chapter also covers the layer architecture of data storage and communication. Layer architecture gives remote data access and quick reports to medical professionals and also gives a detailed history of the patient using a centralized database. A healthy workforce is an asset to any country. The use of IoMT could result in a 50% reduction in disease, saving medical resources and making IoMT accessible to people on a low income.

4.2 IoT IN THE MEDICAL DOMAIN

With the advancement of IoT applications, traditional medical applications are turning into smart healthcare applications. Human health is defined as a state of complete physical, mental, and social well-being by the World Health Organization (WHO) [2]. An increase in life expectancy of individuals is observed through the advancements in technological and medical research, coupled with increased awareness about health and hygiene [3]. With the help of smart healthcare solutions, a person can easily monitor their own body parameters and be health conscious. Remote access to such devices enables the utilization of the available resources to the maximum level. IoT is considered the backbone of healthcare applications. The computing capacity of the real-world components is increased by using fog-based and cloud-based solutions. IoT elements, known in brief as "things", are fitted with sensors that have the capacity to wirelessly connect with the Internet as well as with each other. Figure 4.1 represents the traditional way of treating patients versus the smart way of treating and monitoring patients.

Using cloud-based solutions, IoT helps real-world electronic components to perform advanced computing. IoT enables these real-world components to be self-sufficient and provides them with the benefit of remote access and faster data sharing [4]. Sensors have the tendency to convert a physical quantity, such as temperature, heart rate, blood pressure level, sugar level, etc., into numerical values, as they are simple electronic devices [5]. Data collected by smart devices can be, for example, numeric values, images, string literals, etc. The different electronic devices available, like smartphones, iPads, tablets, and smart watches, have multiple inbuilt sensors, which record an individual's different body parameters. These body parameters are then stored in the form of numerical values in the cloud through the IoT devices

FIGURE 4.1 Traditional patient monitoring versus smart patient monitoring using IoT. (Image courtesy of Creative Commons, pixabay.com.)

or components, from which meaningful information can be retrieved. Due to these enhanced inbuilt functionalities, IoT devices are used in every sector of smart city construction: medical sector, transport, construction, energy conservation, CCTV monitoring, agriculture, and many more [6]. Several benefits offered by the IoT make it suitable for application in the medical sector as well. Research on smart electronic systems, when combined with medical science, has led to the increased use and application development of various IoT-based healthcare solutions. Figure 4.2 shows the representation of IoT in various medical sectors. In all, the use of IoT has led to decreasing the distance between patient and doctor and providing an affordable healthcare solution.

4.2.1 Data Communication between Different Layers in IoT

Data representation refers to data and images collected from the lower layer to centralized storage. The Fog layer is responsible for remotely processing the data stored in the warehouse.

Figure 4.3 shows the whole architecture.

4.2.1.1 Internet of Healthcare Things (IoHT) Network Layer

The IoHT is the collection of things that are always lying around the patient and constantly monitoring the patient's health. This is very important to provide real-time information about the patient. These healthcare devices are fitted with sensors, which make the devices smart and intelligent. This layer connects all devices with each other and makes an internal network as described (Figure 4.4). Using Internet

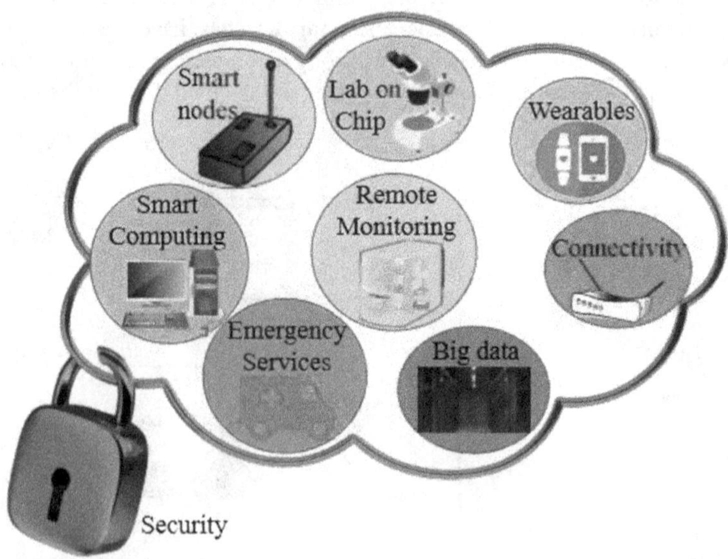

FIGURE 4.2 Representation of IoT in the medical sector.

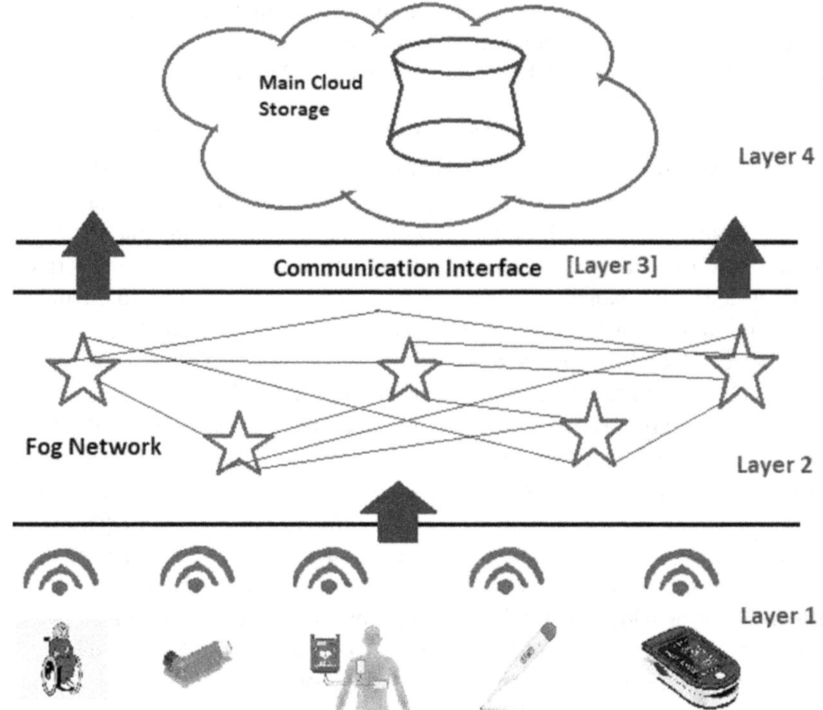

FIGURE 4.3 Layer architecture of IoT application.

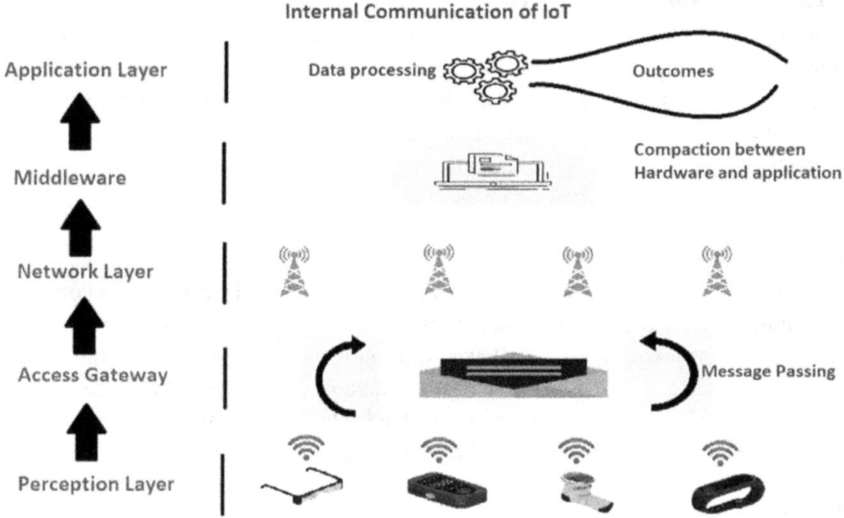

FIGURE 4.4 Communication of IoT network. (Al-Qaseemi, S.A, et al., IoT architecture challenges and issues: Lack of standardization. In *Future Technologies Conference*, 978-1-5090-4171-8/16. IEEE, 2016.)

data access, this layer forwards data to the fog computing layer. The main challenge is to create a secure and authenticated system. Researchers have introduced an "IoT based healthcare system which is optimal and secure" [29].

4.2.1.1.1 Presentation of Data and Communication in IoT Layer

As the IoT network is based on different devices, standardization of data presentation is a big constraint on analyzing data [12,17,18]. Recommendation ITU-T H.860 ("Multimedia e-health data exchange services: Data schema and supporting services") specifies a common health schema applicable to a wide range of health systems [19]. Normalized data gives perfect analysis power to the machine. Data communication between internal networks of IoT is based on an original three layers of architecture and an updated five layers of flexible architecture [18], which can be used for communication. Figure 4.4 shows five-layer architectural communication [20]. In Figure 4.4, researchers proposed five different layers with different functionalities, where the perception layer consists of physical devices and is also called the sensing layer due to the devices being fitted with sensors. These devices collect data and pass it through the network layer. Between the network layer and the perception layer, an access gateway is used to manage internal communication between smart devices in a particular environment. The network layer provides the facility of wired or wireless connection for communication. Middleware provides flexible compaction between hardware devices and software applications. Data received by application layer is processed and can be either sent back to smart applications of the perception layer or to a remote server, i.e. a fog remote node.

4.2.1.2 Fog Computing Layer

The fog layer is a network of a remote server that quickly accesses remotely available data, processes it, and sends a report about the patient to professionals. This layer provides distributed services, which makes services like patient monitoring easier and faster. Many diseases can be visualized and analyzed by keeping a check on pictorial representation, processing, and segmentation. All these activities are performed by the fog computing layer.

4.2.1.3 Communication Interface

The communication interface provides a platform for both layers, i.e. fog computing and cloud layer. All data processed by fog remote server is sent to the upper layer and medical professionals. Upper layer communication is performed by the communication interface. The IoT network collects data in different formats, which is very difficult to analyze and use; in this layer, the system can normalize data and make the database consistent. Figure 4.5 shows data standardization.

4.2.1.4 Cloud Layer

This layer provides centralized storage for data. Sometimes, professionals need the medical history of the patient, based on which treatment can commence, and it will be very helpful for patient monitoring. Doctors should be familiar with some

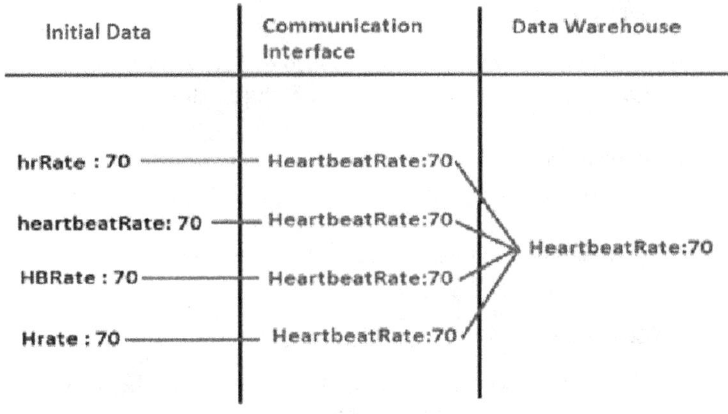

FIGURE 4.5 Data standardization.

symptoms such as spasms or high blood pressure, in case the patient forgets to inform them about these symptoms; then, due to centralized data, doctors can easily access the information and monitor the patient. Hassanalieragh et al. [26], Tyagi et al. [27], and Plamenka et al. [9] explain the necessity of the IoT in the medical imaging field to provide optimized healthcare services. Remote monitoring of patients suffering from any disease gives suggestions for precautions to be taken.

4.3 APPLICATION AREAS OF MEDICAL IOT

Different IoT-based smart healthcare solutions help in organizing drugs efficiently, maintaining digital health records efficiently, and offering healthcare assistance through different telephone services. A few application areas of medical IoT are listed here.

4.3.1 Patient Monitoring and Tracking

IoT enables real-time patient monitoring and tracking. Critically ill people generally need to be cared for in isolation, so patients should not visit any place without assistance. A real-time patient monitoring and tracking system overcomes this problem and allows the patient to change their environment. This technology is also useful for the elderly and many other people.

4.3.2 IoT for Big Data

Big data is another emerging technology where gathering accurate and precise data is very important. IoT provides a platform to generate real-time data on a situation. To gain benefits from big data and analysis, data needs to be accurate and precise. Here, security and privacy are the main challenges, but researchers are constantly finding solutions to overcome this problem.

4.3.3 IoT Wearable Devices

Many new wearable devices have been introduced by IoT, enabling a comfortable life for patients. Devices like wristbands, necklaces, shirts, shoes, handbags, caps, and smartwatches provide details about the person's health monitoring. Hearables enable a deaf person to hear, ingestible sensors monitor the medication in our body and act as alarms for any disease, Moodables help to improve the person's mood throughout the day. Healthcare charting is used by doctors to verbally record patients' data through voice commands, and makes data readily accessible 24/7.

4.3.4 Emergency Services

IoT devices reduce the emergency room wait time, enabling the doctor to be contacted and the required medication to be obtained on time.

4.3.5 Smart Computing

IoT devices covering the next-generation computing paradigms support wireless networking with high data transfer rates and autonomous decision-making capabilities.

4.3.6 Smart Nodes

IoT smart nodes represent how IoT is represented on its end nodes. They are basically the sensor nodes fixed to different hardware components of the application. The main challenge is how these smart nodes minimize power consumption, disseminate the data by the most effective routes, and by self-adaptation capabilities sense change in the environment.

4.4 IMAGE PROCESSING IN MEDICAL IOT

Image processing in the medical domain enhances work efficiency. Today, types of sensors and high-definition cameras are available that give internal and external images of the human body. These images can be further processed and analyzed using image segmentation, enhancement, edge detection, etc. Processed images provide a broader view for treatment and diagnosis and make medical services more accurate. Researchers have found some areas of the medical domain where image segmentation promotes accessible services. Figure 4.6 describes the medical domain of image processing.

4.4.1 Remote Patient Monitoring

This area works efficiently if the situation requires elderly people or patients to be moved to their homes. Professionals can easily access their present plight by installing a remote sensing camera into their local environment [10]. Here, developers can use fog computing for remote image processing.

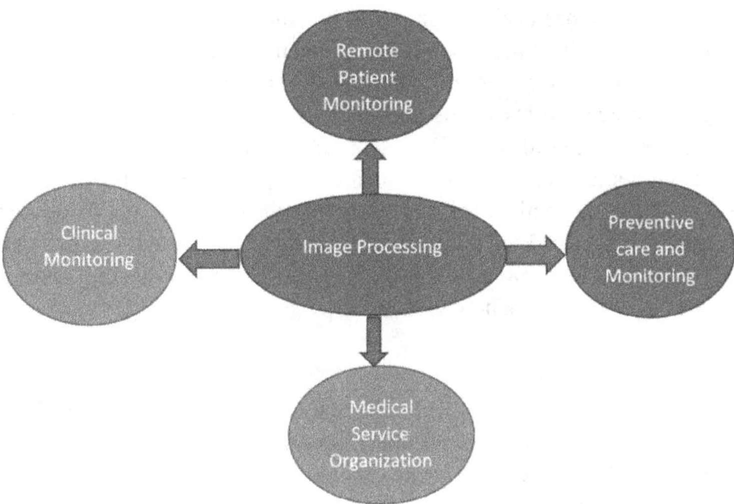

FIGURE 4.6 Image processing in medical applications.

4.4.2 Preventive Care and Monitoring

Lethal diseases include AIDS, malignant cancer, heart diseases, and brain tumor. Patients suffering from these diseases cannot be hospitalized in the early stages, but professionals can give preventive care and remote monitoring during the onset of sudden symptoms using cameras.

4.4.3 Clinical Monitoring

Hospital wards or room environments can be converted into IoT-based networks so that real-time reports about patients can be generated on time.

4.4.4 Medical Service Organization

Medical service organizations make use of monitoring systems for providing superlative care to their patients.

4.4.5 Different Applications Equipped with Image Processing

V. Akkala et al. [21] introduced an IoT-based remote ultrasound imaging application for kidney defects and their treatment. The system architecture works on wireless communication protocols and uses radio waves for communication over the Internet from the IoT network to centralized cloud storage, where the server is able to process data and produce valuable information. In Iuliana et al. [22], smart medical equipment fitted with medical imaging is used to generate expanded and enhanced medical images that are available in real time. Tao Liu et al. [23] proposed an IoT-based application for healthcare services for patient monitoring and to manage medical activities in real time. Figure 4.6 shows the importance of IoT. In constantly reporting to medical professionals and institutions about patient details, IoT plays a key role. Borovska et al. [24,25] present key causes and effects of IoT on medical area using home telephony. Tolga Soyata et al., [26], Tyagi et al. [27], and Amit Agarwal et al. [28] have all presented optimized access to IoT and its utilization for patient monitoring and taking precautionary action. The pandemic situation has also resulted in the need for remote patient monitoring. Taking all this into consideration, researchers have proposed a new system.

4.4.5.1 Proposed System

Whenever a person is suffering from a disease, the patient must visit the hospital and go through the initial medical check-up and tests. Each disease has its own specified tests, which give the overall situation of the patient. If the patient is at a critical stage, then it is advisable to admit them. The initial tests consist of urine tests, sodium and potassium tests, blood tests, hormone tests, and disease-specific tests, and the data is stored on a centralized cloud database. Various indications are obtained by biosensors [7]. After this process, when the patient is admitted, it is difficult to monitor and visit the patient frequently. Generally, doctors can visit a patient two or three times a day. The health of the patient is crucial, so regular monitoring is required. To overcome this problem, image processing using IoMT gives the optimal solution. Constant measurement of blood pressure, temperature, sugar level, pulse rate, and electrocardiogram and electromyogram [8,9] gives actual parameter values for patient health. Using a high-definition camera, the system constantly observes and takes pictures of the patient, which can be useful for further treatment.

4.4.5.2 System Description

Figure 4.7 demonstrates the actual scenario of the environment. Here, researchers utilize IoT technology along with image processing. Using this application, professionals will be updated with the disease symptoms, and treatment can be started without wasting time. With the help of basic symptoms collected by a smart electronic device, noticeable changes in health can be seen. After getting the basic symptoms, doctors can easily prescribe initial tests for the patient without wasting time. As directed by the doctor, tests are performed by various departments, and all departments upload the patient's report to a centralized database and fog server, which combines and processes all data and employs the Random Forest algorithm to predict disease, with all possible predictions forwarded to professionals. Once the disease is recognized, medical care can be provided to the patient. Once the patient commences medical services, the application constantly monitors the patient, giving facial expression data and other normal parameter data in real time. This system is divided into three stages: first, notifying changes in health parameters; second, initial testing conducted by a different department of the medical domain; and third,

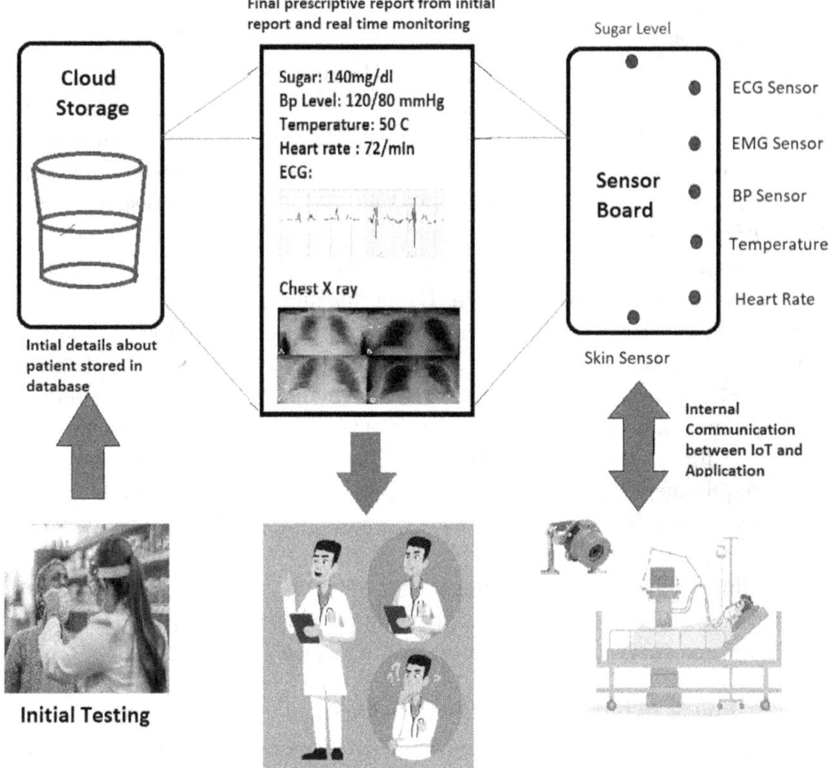

FIGURE 4.7 Complete working scenario for patient monitoring using image processing.

after commencement of initial treatment, constantly monitoring the patient in the particular environment.

4.4.5.3 Communication System

An IoT-based intelligent network gives real-time data regarding the patient's condition. This is the first layer of initial communication. These devices are always situated around the patient, are used to capture images and gather medical information, and are responsible for delivering data to the remote server. This network uses a communication protocol like zigbee or zwave and has access to the Internet, through which data is sent to the remote fog server. The research by Tuli et al. [11] targeted the requirement of fog computing systems for the development of a framework named Health Fog for deep learning along with the real-time application of the analysis of cardiac diseases. The study was successful in the development of the system architecture of the Health Fog model associated with the IoT system. These types of model for various diseases can be used for the initial data requirements. The communication system collects real-time data, which is further utilized by the upper layer to recognize diseases.

4.4.5.4 Disease Recognition

Many types of diseases can be easily recognized by visualization, like skin diseases and other diseases that mostly affect human body temperature or heartbeat rate or sugar level, which are normal parameters. These parameters can be easily monitored using wearable devices [12], and if any fluctuation is found, initial testing can be conducted, which gives the actual scenario in the early stages of illness. Here, researchers combine cloud computing [13] and IoMT [14–16] for disease detection. This technique allows prevention and cure.

The author of this chapter has proposed a new algorithm named Random Forest for identifying the class of algorithm. Using machine learning, a supervised algorithm system can be made intelligent to identify the domain of any disease. This algorithm uses a training data set to create a random forest tree for inputted data. In the human body, mostly, disease does not have any physical appearance, but using a high-definition cameras system can identify where the patient has pain with behavioral input [10]. The Random Forest algorithm works by training the machines as per the requirement. According to each input, a Random Forest tree is generated, which gives prediction of disease. Table 4.1 and Figure 4.8 describe the data set and Random Forest tree of one domain.

According to data gathered by the IoT network, then the patient may have bacterial pneumonia, Covid 19 pneumonia, or viral pneumonia if all parameters fall into the relevant category. A Random Forest tree is formed as a result of this.

4.4.5.5 Image Acquisition and Pre-Processing

The human body may suffer from many types of diseases. Skin diseases [13] are easy to capture and process by image processing methodology, but other many diseases cannot be directly captured and processed by image processing. This problem can be solved as shown in Figure 4.4. Initially, all tests are conducted by professionals, and the results are stored in a centralized data store. After disease detection, the patient can be monitored remotely or in a medical organization where the system constantly monitors the patient's behavior [10] and normal parameters with the help of the IoT

TABLE 4.1
Parameters for Coronavirus Disease

Coronavirus Symptoms	Percentage	Related Test	Outcomes
Fever	83–99%	RTPCR	Virus presence
Cough	60–80%	Chest X-ray	Effect on lungs
Fatigue	45–70%	CT scan	Detailing of lungs
Anorexia	40–80%	Blood test	For blood indices
Shortness of breath	30–40%		
Myalgia	11–35%		
Incubation period	2–14 days		

RTPCR: reverse transcription–polymerase chain reaction.

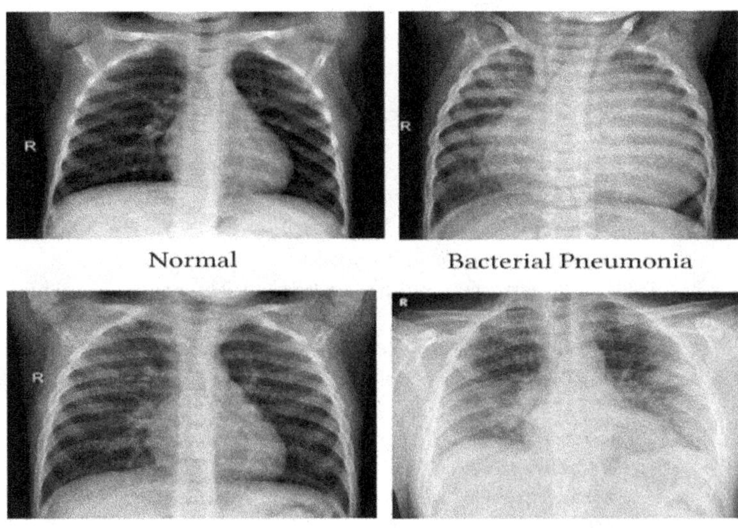

FIGURE 4.8 Possible outcomes from Table 4.1 parameters and chest X-ray. (a) Normal; (b) bacterial pneumonia; (c) viral pneumonia; (d) Covid-19 pneumonia.

network. With this aim, some images have been taken from the Internet and pre-processed so that they can be used as a training data set. Pre-processing requires image enhancement and sharpened image segmentation. Images are converted into 256 × 256 size. For disease detection, the system needs to acquire data and images from different sources, like an X-ray machine [33], magnetic resonance imaging (MRI) [31], computed tomography (CT scan) [32], and an ultrasound machine [30]. The researcher has described the entire working for analyzing images captured from the medical equipment. In data, images are very important, as they describe and visualize the causes of disease. Currently, the medical domain has many resources for collecting images, but the main benefits can be optimized when data is processed in real time.

4.4.5.6 Image Segmentation

After acquiring images from different sources, the system performs image segmentation techniques to identify the affected region of the organ using MatLab function rgb2gray(Image), which gives types of images like Roberts, sobel, prewitt, canny, approxcanny, zerocross, and log. All these image types are based on edge detection methodology. Table 4.2 shows all techniques. The projected outcome is presented to the doctor after segmented images based on various approaches are compared to the test data set. After segmentation, the image can be enhanced so that an improved-quality image gives a better result.

4.4.5.7 Feature Extraction

After segmentation of the image, features are extracted from it. Processing the image enables the machine to extract unique features from it. The application has already

TABLE 4.2
Image Segmentation Techniques

Method	Description
Sobel	Using the Sobel approximation to the derivative, it finds edges at those points where the maximum gradient of the image is found.
Prewitt	Using the Prewitt approximation to the derivative, it finds edges at those points where the maximum gradient of the image is found.
Roberts	Using the Roberts approximation to the derivative, it finds edges at those points where the maximum gradient of the image is found.
Canny	This technique detects the edges by tracing the local maxima of the gradient of the image. The edge function calculates the gradient. Using the derivative of a Gaussian filter, this
	technique uses two thresholds for detecting strong and weak edges. It also includes weak edges in the output if they are connected to strong edges. As two thresholds are used, this technique is less preferable compared with the other techniques and more likely to detect true weak edges.
Log	After filtering the image with a Laplacian of Gaussian (LoG) filter, it finds edges by looking for zerocrossings.
Zerocross	After filtering the image, it finds edges by looking for zerocrossings.
Approxcanny	Using the approximate version of the Canny edge detection algorithm, it finds the edges. The technique also provides faster execution time at the expense of less precise detection. It normalizes the floating point images in a range of [0, 1].

been fed with a training data set of disease symptoms, and based on this, the inputted pre-processed image is used to identify a particular disease. The main disadvantages of the traditional method are wastage of time and manual image analysis by a professional, which is not affordable when it comes to someone's health. This application is much faster at extracting features and can update professionals within a second.

4.4.5.8 Advantages of Proposed Application

4.4.5.8.1 High-Precision Diagnostics

The application gives high-quality images by enhancing the method. Very high-resolution machine vision cameras offer high sensitivity and image quality, which enables the system to detect the smallest details, whether inspecting retina, skin, or other tissues for diseases. High-sensitivity cameras can be used for diagnostic systems, such as retina imaging devices.

4.4.5.8.2 Reveal the Invisible

Industry 4.0 is achieving new heights in industries where image processing enables the invisible to be revealed. Sometimes, it is possible that humans make mistakes and cannot predict some areas of images, whereas machines can accurately identify similar clusters. The camera captured images will play an important role. Using infrared imaging at various wavelengths makes visualization possible. Long-wave infrared (LWIR) cameras make it possible to see what conventional cameras cannot detect. For example, thermal imaging can be used to detect inflamed tissue or insufficient blood circulation.

4.4.5.8.3 Assist Medical Trainees

Image processing and prediction assists medical trainees to analyse and review disease. Medical trainees are learning professionals and need assistance at all times.

4.4.5.8.4 Speed and Precision

Image processing and the proposed system increase the speed and precision of services. Health is very important, and more lives can be saved by applying patient monitoring in real time.

4.4.5.9 Challenges of Application

This application has the following challenges in analyzing the working scenario and the nature of human disease:

a. A high-definition camera is needed to identify small changes in physical appearance.
b. Because of initial parameter fluctuations, the doctor can be assisted in real time but cannot accurately identify the disease type.
c. Some diseases require initial testing for accurate identification.

4.5 BENEFITS AND LIMITATIONS OF IOT

IoT is applied to multiple domains of multiple industries. IoT in the healthcare industry has converted the medical process into a digital approach. The wireless network created byIoT is used to trace and monitor the patient's health constantly in real time. Using IoT with an image processing system makes medical services more reliable and accurate, and the most important area is emergency management. The role of IoT in the medical field is to give professionals the opportunity for rapid monitoring of the disease fluctuation and recovery progress and to provide advice and precautions in the meantime [21] Professionals can easily access the patient's medical history. The IoT in the medical field scales up a smart and healthy environment [23]. IoT-integrated technologies enable professionals to diagnose the patient efficiently in the meantime [25]. However, despite the importance of IoT and image processing in the medical domain, a few drawbacks have been observed to cause problems in real-time monitoring. Large-scale wireless networking and device mobility management, data with ACID qualities (Atomicity, Consistency, Integrity, and Durability), data compression, data security, and privacy are the other side of the coin. These drawbacks need to be overcome using different techniques and changing network infrastructure.

4.6 FUTURE SCOPE

IoT is a large domain of research and work, where researchers can apply multiple technologies and get multiple benefits. Image processing using IoT gives enhanced capability to the medical industry, where real-time monitoring will be more accurate and optimized. Researchers are trying to fill the gap between the current scenario and future capabilities. Though it is difficult to predict any diseases in the human body, the Random Forest algorithm gives a possible prediction from the given input. In future, these application capabilities can be enhanced using machine learning algorithms. Different patterns can be fed to the machine to learn how and which diseases can be recognized accurately. As many patterns are available, machines will be able to accurately identify disease, which will be one of the great achievements in patient health monitoring systems. If disease is known at an early stage, then prevention and cure can be achieved.

4.7 CONCLUSION

The results of the research show scope for improving the quality of life through smart healthcare solutions. Detection of a patient's disease in the early stages by using image processing and analyzing the health records helps doctors to improve the patient's treatment as well as increase the ratio of healthy patients. To monitor and classify the patient's disease, IoT is essential for sending the images and giving feedback. The proposed system not only monitors the patient but also gives predictions about the disease in real time. Generally, people ignore the early symptoms of any disease and try different home remedies instead of taking the symptoms into consideration. Sometimes, they may recover by using home remedies, but many times, this may be lethal. In these lethal situations, the proposed application plays a vital role in early diagnosis and improves patient monitoring.

REFERENCES

1. S. Yeung et al. "A computer vision system for deep learning-based detection of patient mobilization activities in the ICU." *NPJ Digital Medicine*2, no. 1 (Dec. 2019): 1–5. Doi: 10.1038/s41746-019-0087-z.
2. World Health Organization. http://www.who.int/about/mission/en/.
3. S. Majumder, E. Aghayi, M. Noferesti, H. M. Tehran, T. Mondal, Z. Pang, and M. J. "Smart homes for elderly healthcare – Recent advances and research challenges." *Sensors* 17, no. 11 (2017): 2496.
4. S. P. Mohanty, E. Kougianos, and P. Guturu. "SBPG: Secure better portable graphics for trustworthy media communication in the IoT." *IEEE Access* 6 (2018): 5939–5953.
5. S. P. Mohanty and E. Kougianos. "Biosensors: A tutorial review." *IEEE Potentials* 25, no. 2 (2006): 35–40. Deen, Smart Home for Elderly Healthcare – Recent Advances and Research Challenges, Sensors (Basel) 11 (2017).
6. E. Kougianos, S. P. Mohanty, G. Coelho, U. Albalawi, and P. Sundaravadivel. "Design of a high-performance system for secure image communication in the internet of things." *IEEE Access* 4 (2016): 1222–1242.

7. K. Georgakopoulou, C. Spathis, N. Petrellis, and A. Birbas. "A capacitive to digital converter with automatic range adaptation." *IEEE Transactions on Instrumentation and Measurement* 65, no. 2 (2016): 336–345.
8. N. Petrellis, I. Kosmadakis, M. Vardakas, F. Gioulekas, M. Birbas, and A. Lalos. "Compressing and filtering medical data in a low cost health monitoring system." In Proceedings of the 21st PCI 2017, Larissa, Greece, 28–30 September 2017. [Google Scholar]
9. Y. Zhang, L. Sun, H. Song, and X. Cao. "Ubiquitous WSN for healthcare: Recent advances and future prospects. *IEEE Internet Things Journal*1 (2014):311–318. [Google Scholar] [CrossRef].
10. IEEE. *Scored2020_Patient Monitoring System using Computer Vision for Emotional Recognition and Vital Signs Detection.*
11. S. Tuli, N. Basumatary, S. S. Gill, M. Kahani, R. C. Arya, G. S. Wander, and R. Buyya. "HealthFog: An ensemble deep learning based smart healthcare system for automatic diagnosis of heart diseases in integrated IoT and fog computing environments."*Future Generation Computer Systems*104 (2020): 187–200. [CrossRef].
12. P. Kumari, M. López-Benítez, G. MyoungLee, and T.-S. Kim. "Wearable internet of things: From human activity tracking to clinical integration." IEEE publication978-1-5090-2809-2/17.
13. L.-s. Wei, Q. Gan, and T. Ji. "Skin disease recognition method based on image color and texture features." *Hindawi Computational and Mathematical Methods in Medicine*2018: 8145713, 10 pages. Doi: 10.1155/2018/8145713.
14. M. Saranya, R. Preethi, M. Rupasri, and S. Veena. "A survey on health monitoring system by using IOT." *IJRASET*6 no. III (March 2018). ISSN: 2321–9653, IC Value: 45.98.
15. M. M. Islam, A. Rahaman, and M. R. Islam. "Development of smart healthcare monitoring system in IoT environment." *SN Computer Science* 1 (2020): 1–11.
16. A. Chandy, "A review on IoT based medical imaging technology for healthcare applications." *Journal of Innovative Image Processing*1, no. 1 (2019): 51–60. Doi: https://doi.org/10.36548/jiip.2019.1.006
17. J. Saleem, B. Adebisi, R. Ande, M. Hammoudeh, and U. Raza. "IoT standardisation – challenges, perspectives and solution." In Proceedings of ICFNDS' 18, Amman, Jordan, 9 pages, June 26–27, 2018. Doi: 10.1145/3231053.3231103.
18. S. A. Al-Qaseemi, H. A. Almulhim, M. F. Almulhim, and S. R. Chaudhry. "IoT architecture challenges and issues: Lack of standardization." In Future Technologies Conference, 978-1-5090-4171-8/16. IEEE, 2016.
19. L. DeNardis. "E-health standards and interoperability." In *ITU-T Technology Watch Report*, Apr 2012.
20. G. B. Rehm, S. H. Woo, X. L. Chen, B. T. Kuhn, I. C. Puch, N. R. Anderson, J. Y. Adams, and C.-N. Chuah. "Leveraging IoTs and machine learning for patient diagnosis and ventilation management in the intensive care unit." *IEEE Pervasive Computing* 19, no.3 (2020): 68–78. doi: 10.1109/mprv.2020.2986767 IEEE 2020.
21. K. D. Krishna, V. Akkala, R. Bharath, P. Rajalakshmi, A. M. Mohammed, S. N. Merchant, and U. B. Desai. "Computer aided abnormality detection for kidney on FPGA based IoT enabled portable ultrasound imaging system." *IRBM* 37, no. 4 (2016): 189–197.
22. I. Chiuchisan. "An approach to the Verilog-based system for medical image enhancement." In 2015 E-Health and Bioengineering Conference (EHB), pp. 1–4. IEEE, 2015.
23. D. Lu, and T. Liu. "The application of IOT in medical system." In 2011 IEEE International Symposium on IT in Medicine and Education, vol. 1, pp. 272–275. IEEE, 2011.

24. S. S. Al-Majeed, I. S. Al-Mejibli, and J. Karam. "Home telehealth by internet of things (IoT)." In 2015 IEEE 28th Canadian Conference on Electrical and Computer Engineering (CCECE), pp. 609–613. IEEE, 2015.
25. P. Borovska, D. Ivanova, and V. Kadurin. "Experimental framework for the investigations in internet of medical imaging things ecosystem."*QED*17 (2017): 20–21.
26. M. Hassanalieragh, A. Page, T. Soyata, G. Sharma, M. Aktas, G. Mateos, B. Kantarci, and S. Andreescu. "Health monitoring and management using Internet-of-Things (IoT) sensing with cloud-based processing: Opportunities and challenges." In 2015 IEEE International Conference on Services Computing, pp. 285–292. IEEE, 2015.
27. S. Tyagi, A. Agarwal, and P. Maheshwari. "A conceptual framework for IoT based healthcare system using cloud computing." In 2016 6th International ConferenceCloud System and Big Data Engineering (Confluence), pp. 503–507. IEEE, 2016.
28. P. Borovska, D. Ivanova, and I. Draganov. "Internet of medical imaging things and analytics in support of precision medicine for the case study of thyroid cancer early diagnostics." *Serdica Journal of Computing, Bulgarian Academy of Sciences, Institute of Mathematics and Informatics* 12, 1–2 (2018): 047p–064p.
29. SEA. *A Secure and Efficient Authentication and Authorization Architecture for IoT-Based Healthcare Using Smart Gateways.* SEA.
30. S. Nikolov, and J. ArendtJensen. "Virtual ultrasound sources in highresolution ultrasound imaging." In *Medical Imaging* 2002*: Ultrasonic Imaging and Signal Processing*, vol. 4687, pp. 395–405. International Society for Optics and Photonics, 2002.
31. V. S. Khoo, D. P. Dearnaley, D. J. Finnigan, A. Padhani, S. F. Tanner, and M. O. Leach. "Magnetic resonance imaging (MRI): Considerations and applications in radiotherapy treatment planning." *Radiotherapy and Oncology*42, no. 1 (1997): 1–15.
32. S. J. Zinreich, D. W. Kennedy, A. E. Rosenbaum, B. W. Gayler, A. J. Kumar, and H. Stammberger. "Paranasal sinuses: CT imaging requirements for endoscopic surgery." *Radiology*163, no. 3 (1987): 769–775.
33. J. C. Kieffer, A. Krol, Z. Jiang, C. C. Chamberlain, E. Scalzetti, and Z. Ichalalene. "Future of laser-based X-ray sources for medical imaging." *Applied Physics B*74, no. 1 (2002): s75–s81.

5 Theory, Practical Concepts, Strategies and Methods for Emotion Recognition

Varsha K. Patil, Vijaya Pawar,
Vaishnavi Vajirkar, Vedita Kharabe,
Nimisha Gutte, and Mustafa Sameer

CONTENTS

DOI: 10.1201/9781003221333-5

5.1 INTRODUCTION

5.1.1 Human Behavior and Emotions

From childhood, every person's behavioral pattern is ruled by emotions. The reasons for a particular action and their counter effects produce behavioral patterns. These behavioral patterns drive the emotions and again, create causes for counter emotions and behavioral patterns. Thus, one can say that human behavior and emotions are both counter effects and complementary. Humans are different from other living creatures on earth on the basis of emotions and intelligence. But, intelligence can be influenced by emotions. Cognitive thinking on a particular problem has multiple dimensions, like paying attention, linking problem solving with learning, and adapting skills. Cognitive acts like thinking, paying attention, grasping, memorizing, and solving complex real life or mathematical problems are influenced by the up-and-down tides of emotions. Capturing faces and automatically identifying emotions can lead to understanding people's behavioral traits and emotional patterns. This can provide the big picture of some people's tendency to avoid problems.

Emotions are reactions to a mindful psychological feedback accompanied by biological and behavioral variations in the human body. Fighting or avoiding problem are two basic human behavioral tendencies, which are highly relevant. Modern progress in machine learning and image processing to detect and understand human affective states by capturing and analyzing facial expressions is called human behavior detection. Human brain research has proved that the behavior of human beings is mainly determined by emotions. Rational behavior and emotions can be in harmony; for example, when dangerous stimuli are present, fear is present, and sometimes emotions like fear cause greater effort. On the other hand, emotions can also be on the opposite side to rational behavior; e.g., shame and lying to cover up a situation are counterparts of emotions. Emotions influence the world that we are experiencing at any instant because of options chosen by individuals on the basis of emotions (Majkowski et al., 2017). Thus, to predict human behavior, there is no better way than emotion recognition.

Mostly, in order to perform human emotion and behavior recognition, global rather than local features are taken into consideration for digital image processing. There are many other approaches, like human behavior analysis through the dataset of electroencephalography (EEG) and electrocardiography (ECG) or through building various convolutional neural network (CNN) and recurrent neural network (RNN) models based on actions and movements. In this chapter, we have proposed human behavior detection using emotion recognition. We have studied, performed, and analyzed various techniques of emotion recognition. We have demonstrated the practical implementation of visual analysis of human face sentiments with the help of digital image processing algorithms by utilizing image processing libraries. These methods, explained here, are useful to predict human behavior. Happiness, anger, disgust, sadness, surprise, and fear are the six basic recognized emotions, as shown in Figure 5.1.

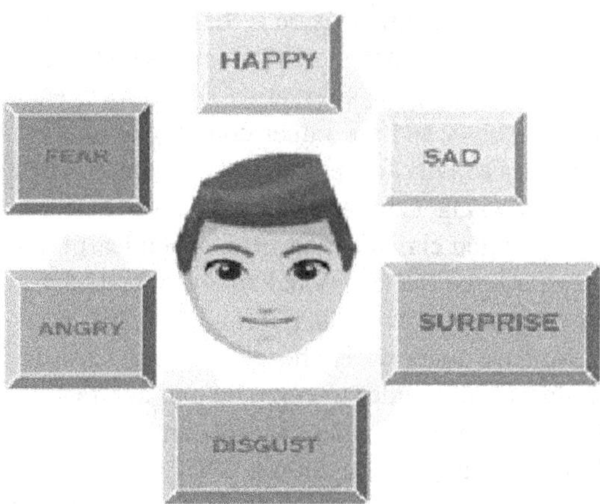

FIGURE 5.1 Human emotions.

To distinguish these emotions automatically by a facial expression method, we need to merge the science of facial expressions, artificial intelligence, and psychological rules.

Describing emotional occurrences in terms of behavioral principles will lead to a more comprehensive science of emotion that has the potential to go beyond our first intuitions and theoretical concepts (Burghardt, 2019). From the various orders of emotion, one after another, a complete behavioral analysis can be done. Prediction of human behavior using facial emotion analysis provides a huge contribution in situations where the person in question is moving and we need to track his/her face and monitor the activity constantly.

5.2 EMOTION RECOGNITION AND ITS TYPES

Facial expressions are nonverbal communication. Along with nonverbal communication, speech is considered in emotion recognition (Sugur et al., 2019). It is an act of acknowledging subtle or implicit as well as clear-cut, explicit sentiments and mental states from facial expressions using technological advancements such as image processing, machine learning, and computer vision. Facial landmarks show emotions (Byun & Lee, 2020). Emotion recognition is a healthy exchange of expressions in relationships. Currently, it is an important topic of investigation related to human development, emotional welfare, and social adjustment. Emotion recognition has a fundamental role in the understanding of compassion. It is suitable for the prediction of anti-social behavior. Along with intelligence quotient (IQ), today, emotional quotient (EQ) is crucial. Hence, to model EQ, emotion recognition is important. The goal of facial emotion recognition is to recognize facial emotion states without error or human intervention (Azizan & Fatimaah, 2020). The purpose of emotion recognition is to understand human behavior characteristics with pattern recognition applied to emotion recognition. This field is a crucial field of study in order to achieve efficient human–computer interaction. Human vision is sharp, trained and tested by the human brain. Computer vision is designed to match human vision by analyzing digital image inputs, similar to what the human observation sense does. Human emotions are a mixture of blends and colors and are rather complex to analyze. Detecting human emotions is a day-to-day activity for human intelligence, but detecting a mixture of emotions on the face by an algorithm is a difficult task to perform. In recent years, in most pattern recognition and classification fields, deep learning methods for feature abstraction have overpowered the traditional approach and have provided deep influence and development in numerous fields (Hortal et al., 2017).

In this chapter, we have deliberately chosen one deep learning algorithm and one machine learning algorithm in order to compare and contrast the two. While learning both methods, we will discuss some concepts and then move on to the practical approach. For deep learning, we have chosen CNN for emotion (feature) extraction from input images, whereas for the machine learning approach, we have chosen "DeepFace", a lightweight machine learning method for emotion recognition. This method is well known for accuracy.

In general, emotion recognition algorithms have three major stages: (i) detecting faces, (ii) extracting features, and (iii) classifying emotions, as shown in Figure 5.2.

The preprocessing stage is the first stage of emotion recognition. In the second stage, facial components are searched (Azizan & Fatimaah, 2020) The facial components are the action units, consisting of eyebrows, nose, mouth, and eyes. The second stage contains extraction of informative features from the face. The third phase is classifier training and testing. Many advanced emotion recognition procedures have been observed in recent years.

5.2.1 Types of Emotion Recognition

Facial expressions, speech signals, bodily movements, blood pressure, pulse rate, and textual information are among the many ways used to identify emotions in humans. Figure 5.3 shows the various types of emotion recognition techniques.

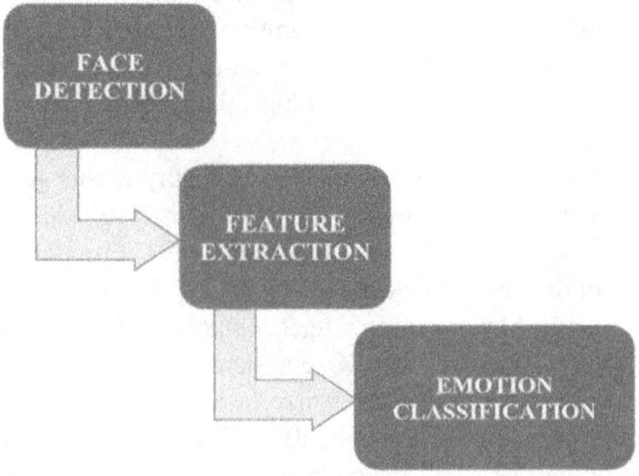

FIGURE 5.2 Stages of emotion recognition.

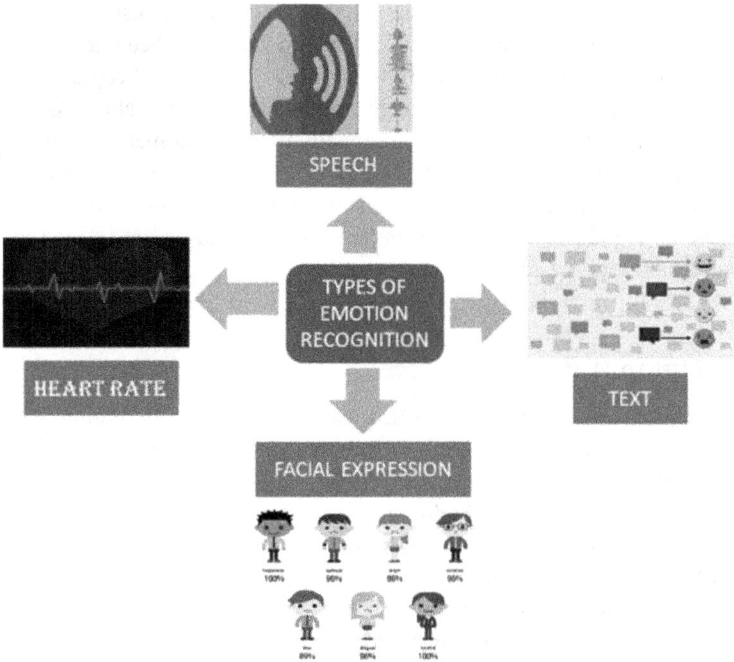

FIGURE 5.3 Types of emotion recognition.

Following are some of the major types of emotion recognition techniques:

a) Emotion recognition using speech: In this process, interpretation of emotion and states from speech are considered (Best et al., 2020). Speech emotion detection can be performed using a variety of methods, including the deep learning idea of CNNs or libraries such as librosa, sound file, and sklearn to build a model using a multilayer perceptron (MLP) classifier, among others.
b) Emotion recognition with text: The natural Language Toolkit software is utilized in the Word-based approach because it is the best for human language data. To begin, all of the emotion labels are applied to the text data, including joy, fear, rage, sadness, and happiness. Finally, textual input is analyzed, and output is generated using the labels that have been assigned. There are also learning-based approaches that include text training and testing.
c) Emotion recognition using heart rate: Different signs of cardiac activity include heart rate and ECG signals. Both can accurately depict the heart's pounding and variations in emotion (Chen & Yu, 2015). An emotion database can be created based on the features of heart sound signals or ECG signals. Ekman biofeedback sensors are also commonly employed in heart rate field emotion recognition.

d) Emotion recognition using facial expressions: Humans use facial expressions to express their emotions. Powerful models have been developed and trained using a variety of machine learning algorithms. People can easily detect the current emotional state of a person under observation. As a result, the facial expression method is frequently used in automatic emotion identification systems (Majkowski et al., 2017). To get the highest level of accuracy, large datasets are used. Various machine learning–based frameworks have emerged as a result of recent advances, including pre-trained models such as DeepFace for emotion recognition.

5.2.2 Literature Review

Real-time, lightweight and high-performance "DeepFace" emotion recognition was proposed by Hortal et al. (2017), which signifies that CNNs are known for their accurate generalization properties. Deep learning approaches have now conquered the traditional methodologies used for emotion recognition. Emotion recognition using CNN was proposed by Sugur & Prasad (2019), which shows that CNN-based models can be employed for emotion detection. These models of CNN are mainly used for facial feature extraction and face detection. Also, human–computer interaction plays a vital role in emotion recognition. A brief review of facial emotion recognition was initiated by Azizan & Fatimaah, 2020), where a brief overview of the stages, techniques, and datasets used for facial emotion recognition was illustrated, and the stages of emotion recognition were proposed and implemented accordingly. Emotion recognition using facial features was put forward by Majkowski et al. (2017), which provides the results of emotional states. Facial features are extracted for a three-dimensional (3-D) face model. Classification of features was implemented using the k-NN classifier and MLP neural network. DeepFace reduces the gap between machine and human identification levels, which shows that DeepFace has achieved accuracy on a par with the human level. It is trained using a large dataset of faces that have been collected from various populations and is able to work better than all the traditional systems with their minimally adapted approach. A place for emotions in behavior research systems was proposed by Burghardt (2019), signifying the importance of emotion detection or human behavior prediction and how emotion recognition can be integrated with deep learning and proposed CNN models in order to achieve human behavior prediction.

5.3 TECHNOLOGIES USED IN EMOTION RECOGNITION:

A number of technologies are currently being developed, out of which image processing, OpenCV, and Python play crucial roles in computer vision applications. Here, we discuss how these technologies together helped us achieve our desired results. Figure 5.4 lists a few of the technologies we used in the whole process.

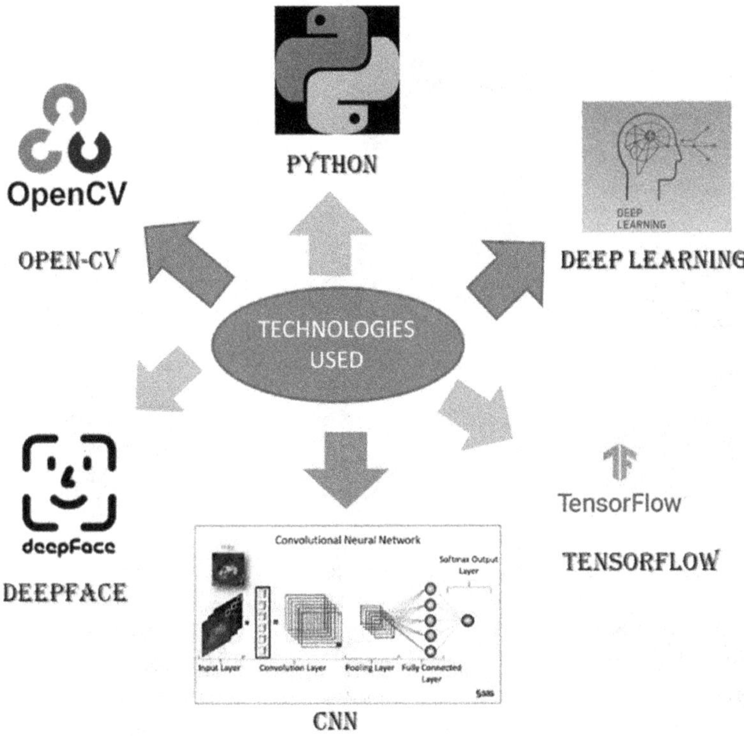

FIGURE 5.4 Technologies used for emotion recognition.

5.3.1 Image Processing

Image processing seeks to convert an image into ones and zeros. The purpose of processing is to acquire a better image and to get convenient information from it. The process of transformation of image into information is shown in Figure 5.5.

The method of image processing is utilized to obtain a better image and get some important information from it, which is integrated to predict human behavior by processing facial attributes in order to perform emotion detection. All signal processing algorithms are applicable to images because an image is considered as a two-dimensional (2-D) signal. Real-time image processing applications are expected to increase fast as they become more prevalent in our daily lives. Image processing involves modifying or processing an existing image to achieve a desired result as well as converting the image to a readable format (Sumitra & Buvana, 2015). Hence, for training the large dataset in order to perform face recognition and detection, image processing is widely used. It produces accurate results. The basic emotions shown in Figure 5.1 (happy, sad, disgust, surprise, etc.) are classified using a neural network–based method integrated with visual processing. The prototype system receives colored frontal face photographs as input. Following the detection of the face, a feature point extraction approach based on image processing is employed, and

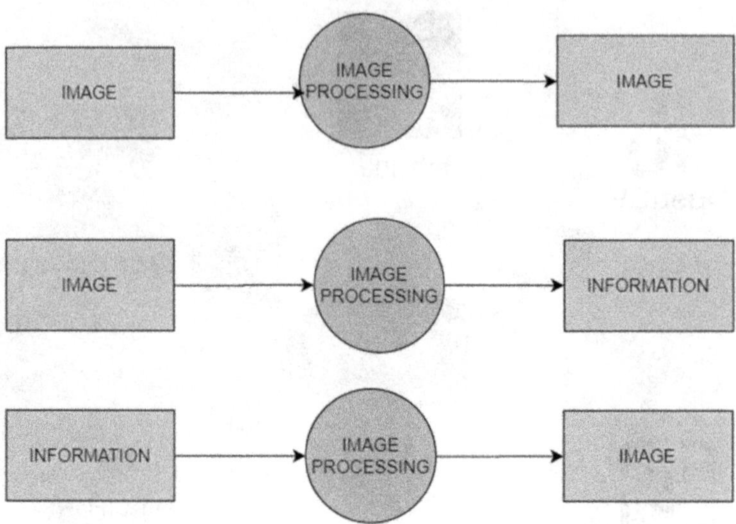

FIGURE 5.5 Transformation in image processing.

essential feature points are collected. After processing the retrieved feature points, a series of values is provided as input to the neural network to recognize the emotion. The process of image recognition includes analyzing and recognizing features from a video frame or an image, such as objects (Sumitra & Buvana, 2015). During the final emotion recognition mechanism, database images and the current image are compared, and if there is a match, the process is continued in real time. It aids in the process of authentication and authorization. So, in this way, image processing is used in human behavior detection, which analyzes features in videos, images, or objects. Emotion detection incorporates image processing in order to process features in an image. Applications of image processing include security surveillance, biometric identification, tool plaza monitoring, and automation in industry. It plays a major role in authorization processes, which involve the study of human behavior by analyzing facial expressions to promote security purposes.

5.3.1.1 Benefits of Image Processing

- Visualization tools can be used for the identification of objects that are not appropriate.
- Image processing is more efficient and less expensive.
- There is no noise. Image restoration and sharpening affect the efficiency. Images can be quickly retrieved from the database.

5.3.2 OpenCV

Computer vision is an interesting visualization concept in artificial intelligence. The concept is applicable to almost all fields of life, providing a useful visual conception of everything around. This conceptualization to comprehension journey is useful for automating routine but skillful tasks in various fields. Thus, computer vision comprehends the surrounding visuals and chooses the best possible action as per the chosen algorithm. OpenCV (Open Computer Vision Library) plays a major role in the image processing domain (Naveenkumar & Vadivel, 2015). It is the most important part of computer vision and machine learning projects, as it is responsible for processing the visual information. OpenCV is a real-time computer vision programming library. It is a large, open-source library containing over 2500 algorithms that are specially developed for computer vision and machine learning projects. Face recognition, object identification, camera movement tracking, scene recognition, and other tasks can all be accomplished with these algorithms. It is a vast community, with an estimated 47,000 people contributing to the library. It is used by a variety of private and public companies. In OpenCV, other sorts of classes can be used for emotion recognition, but we'll be focusing on the DeepFace class. We can install OpenCV using the command pip install Open-CV. Face identification with Haar cascades is a machine learning strategy that involves training a cascade function with a collection of input data (Naveenkumar & Vadivel, 2015). Many pre-trained classifiers for the face, eyes, grins, and other facial features are already included in OpenCV. We need to get the trained classifier XML file (haarcascade frontalface default.xml) from OpenCv's GitHub Repository. Only grayscale photos are used for detection. As a result, converting a color image to grayscale is essential. Faces are a collection of coordinates for rectangular regions where faces were discovered. The rectangles in our image are drawn using these coordinates. We developed a CNN to detect emotions using an open-source data set from Kaggle called Face Emotion Recognition (FER). There are seven types of emotions: happy, sad, fear, disgust, anger, neutral, and surprise.

5.3.3 Python

For coding purposes, we have used the Python programming language. Python is a high-level programming language. It has syntax that appears similar to understandable English language. The detection of emotion is done using machine learning concepts and algorithms. Python is a widely used programming language for machine learning, and processing all the algorithms becomes quite easy using the Python language. A Python program can be used to detect the real-time emotion of a human being using the camera. Selection of required datasets, installing related OpenCV frameworks and libraries, and developing codes for deploying and maintaining Python code are the processes in the implementation of our project. All these steps are also well documented online at various sites but are scattered here and there. The very important point to highlight is strong community support for flexible and platform-independent machine learning and artificial intelligence emotion detection–related projects. The simplicity of Python allows coders to focus on solutions and not on the technical glitches of the language.

5.3.4 Deep Learning and Convolutional Neural Networks

The developments in deep learning field are bio inspired. This means that in artificial neural networks, the human brain, its neurons, and their interrelation are mimicked.

This is a totally revolutionary advancement in the traditional approach of coding and machine learning. Deep learning is hierarchical feature learning. The features are learned automatically by extracting the features from raw data. Deep learning methods extract knowledge from a hierarchical feature structure composed of layers. These layer hierarchies extract meaningful data from lower levels (Hortal et al., 2017). Figure 5.6 describes the process of a deep learning algorithm.

Thus, it is an approach of decomposition of features, automatically learning complex concepts from small, simple levels, and implementing the hierarchy of perceptions. So, deep learning is simply a very large neural network made up of numerous processing layers that learns data representations at various levels of abstraction. Deep learning is applied to a wide range of activities that demand human-like performance, including speech and image recognition in smartphone applications, self-driving vehicles, and home automation systems (Jutus & Esam, 2017). The goal of the computer vision area is to empower and construct mimicking human-like machines that visualize the world in the same way that people do, and to use that information for a variety of tasks and processes, like face recognition, emotion detection, picture analysis, and classification. To make this vision a reality, data enthusiasts from all around the world have worked on breakthroughs in computer vision using deep learning, which had great success, especially with CNN. As a result, it became the popular technique in the field of computer vision. Figure 5.7 shows the implementation of deep learning for emotion recognition.

CNN is an excellent visual processing output. We have employed CNNs due to their excellent generalization properties (Hortal et al., 2017). CNN is a feed-forward neural network that detects and categorizes objects in images. It's a kind of deep neural network that concentrates on extracting and interpreting image features. There has been a revolution in the computer vision world due to the advent of work on CNNs. The most important contribution of computer vision is the diversity and quality of applications (Parkhi & Vedaldi, 2015). A neural network has layers of artificial neurons called nodes. This neural learning framework is made up of multiple neurons and user defined layers. Input data travels with predefined weights via the nodes. An activation function works on this data and returns the outcome based on the related function. Because it is made up of numerous layers that are stacked on top of each other, the term "deep" is employed. Most vision algorithms have a convolution mathematical function. The linear operation of MAC (multiply and accumulate) is done in co-evolution. The MAC operation has two functions, which are multiplied

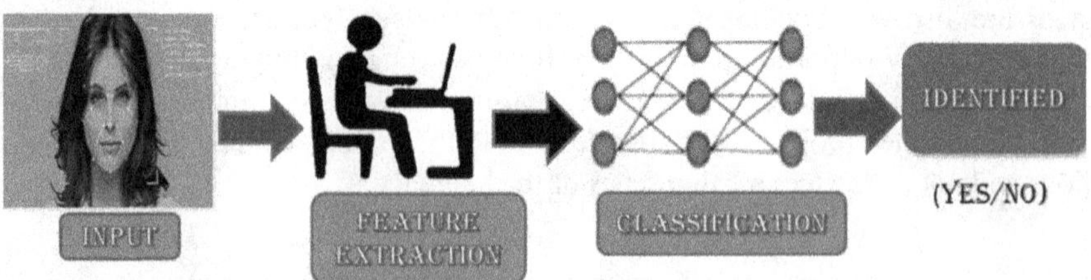

FIGURE 5.6 Deep learning in computer vision applications.

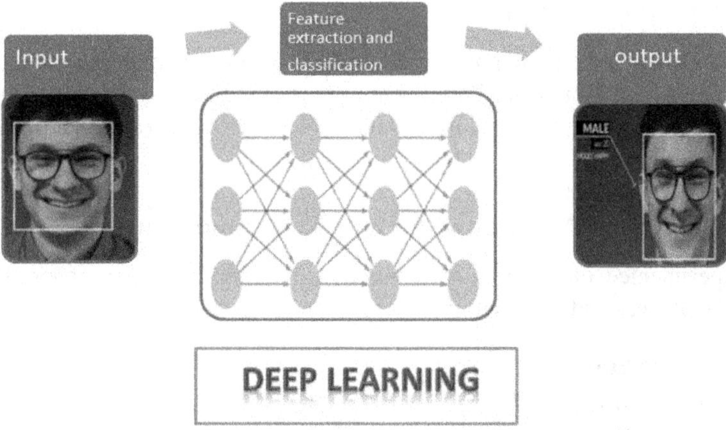

FIGURE 5.7 Deep learning in emotion recognition.

to form a third function that expresses how the shape of one function is transformed by the other. This is denoted "convolution" in CNN. Simply put, two matrices are multiplied to provide an output, and this is used to extract features from the image.

The input data layer, an unknown number of hidden layers, and eventually, the output layer are the main parts of a CNN model. The image has multiple shades of gray level, but the information stored is in terms of arrays. Hence, the input layer of the neural network gets data in the form of arrays. So, the dimensions of the input layer are solely dependent on the array.

By conducting various calculations and manipulations, the hidden layers extract features. This enables neural networks to understand complex data connections. There are *n* hidden layers, such as the convolution layer, the ReLU layer and the pooling layer, which extract features from an image. An output layer such as a class prediction will provide the ultimate result. That is, the object in the image is finally identified by a fully connected output layer. The size of the output layer is determined by the sort of required output, such as the number of classes we want to predict.

Automatic facial expression recognition could be an important factor in human–machine interactions. It can be employed in clinical practice and behavioral science. Humans can understand facial emotions almost instantly and without effort, but machine recognition of facial expressions remains a challenge. There have been several advancements in the field of face detection, feature extraction mechanisms, and expression classification techniques in recent years. Model training becomes computationally heavy as the number of data increases. Hence, developing an automatic framework in these fields is a tough task. However, CNNs are very good at reducing the number of parameters without sacrificing model quality, which effectively reduces training complexity (Hortal et al., 2017). CNNs' architecture is another reason for their enormous popularity. The nicest part is that feature extraction is not required. The system learns to extract features on its own, saving time and effort. Thus, in the realm of image processing, CNNs are the go-to models.

5.4 METHODOLOGY

We have performed emotion recognition using a couple of different techniques: one is emotion recognition using CNN, where we built and trained the model ourselves, and the second is the DeepFace framework, which uses a pre-trained model for emotion recognition.

5.4.1 Hands on Approach of Emotion Recognition with CNN

We experimented on emotion recognition with CNN. The following subsections show the process and essential procedure for implementation.

5.4.1.1 Data Source

The dataset used is provided by "Jonathan oheix" in a Kaggle competition named "Challenges in Representation Learning: Facial Expression Recognition Challenge" (Jonathan oheix, 2019). The Kaggle dataset is open source and free to use. The dataset consists of 48 × 48 pixel grayscale images of faces. The face images are automatically registered. This automatic registration has the benefit that approximately the same space is occupied in each image. Each image belongs to a facial expression in one of seven categories (0 showing angry, 1 showing disgust, 2 showing fear, 3 showing happy, 4 showing neutral, 5 showing sad, 6 showing surprise). The dataset has approximately 36,000 images.

In our dataset, raw images related to each emotion of the seven categories are separated into different folders; 80% of images are in the training folder, and the remaining 20% are in the validation folder, as shown in Figure 5.8.

5.4.1.2 Preprocessing

The preprocessing steps involve quickly visualizing the data from the dataset and setting up the generators to feed the dataset in order to train the model.

1) Data visualization: It is the very first phase, where the images from our dataset are plotted and visualized using powerful libraries such as matplotlib, keras, numpy, seaborn, etc. The number of images in each expression set is counted, as diversity of images and balance between the different expressions will contribute to making a more generalized model. Among these, we have displayed images in happy and disgust folders, as shown in Figure 5.9.
2) Data generator setup: In the next stage, two data generators are built: one for training and another for validation of images. Keras has a very useful class called ImageDataGenerator, which is used to automatically feed the data from the directory. After feeding the data, it shows the classification of images in various classes as output, which is shown in Figure 5.10. It can also be used for data augmentation.

5.4.1.3 Convolutional Neural Network (CNN) Setup

Being a deep learning algorithm, CNN specializes in processing visual imagery. It is widely useful for applications of image processing and practical cases of pattern

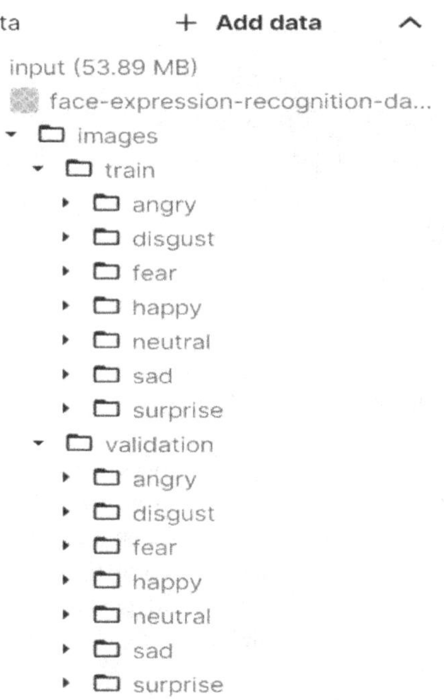

FIGURE 5.8 Dataset used for model training.

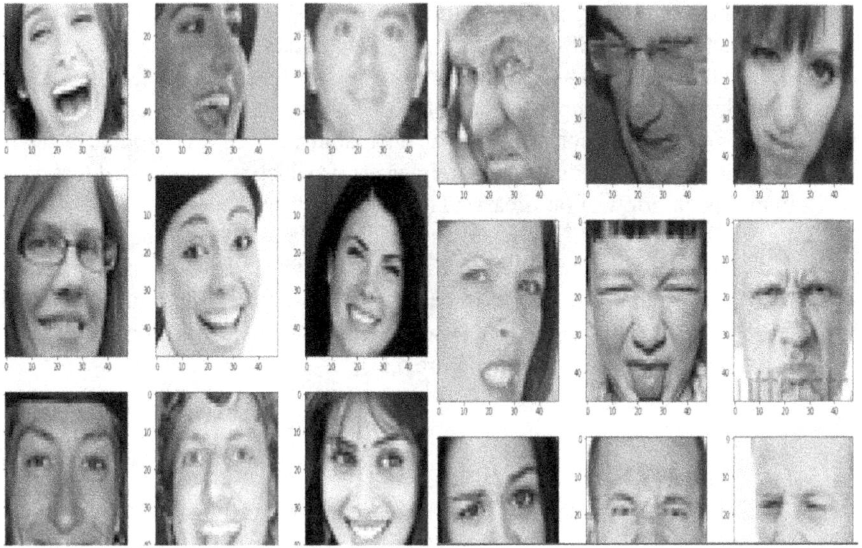

FIGURE 5.9 Data visualization output.

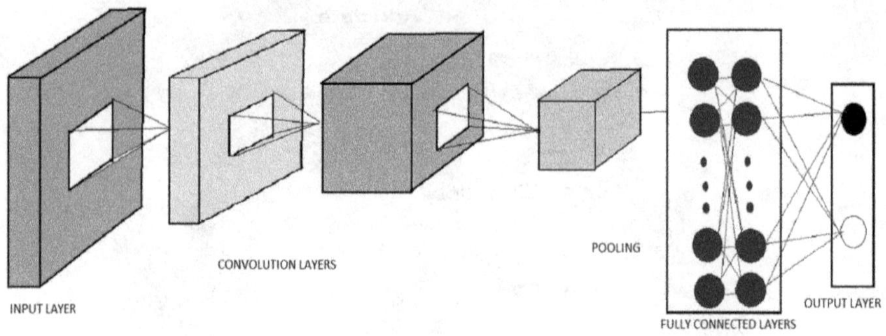

FIGURE 5.10 Convolutional neural network setup.

recognition. It is a class of deep neural networks that focuses on extracting and analyzing features of images. This architecture was inspired by the work on the subject: https://github.com/jrishabh96/Facial-Expresssion-Recognition. The term "deep" is used because it consists of numerous layers of artificial nodes. Each node individually does computations, assigns biases and weights to various aspects of the image, and finally, outputs the results. We defined our CNN with a global architecture consisting of six convolutional layers.

- Convolutional layer: This layer can be called the most fundamental element of the neural network. The first layer of CNN is used to extract the relevant features of the images. When we pass an image as input, it applies a filter to create a feature map. The feature map reviews images for the existence of detected features (Sugur & Prasad, 2019). Filters are of the same dimensions but have length and width smaller than the input volume. Each filter is convolved with the input to compute the map. In other words, each filter is multiplied with the portion of the image that fits, and the dot product is calculated. This dot product is stored in an array, which is the required feature map. The multiplication is carried out until the whole image is processed.
- ReLU Function: ReLU is a activation function having non-linearity. This activation function is applied to each element to increase the property of non-linearity in our network. In our work, ReLU or rectified linear unit is used to replace all negative values with zeros. The ReLU function is defined using the following equation (Figure 5.11).

$$f(\mathrm{u}) = \max(0, \mathrm{u})$$

In simple terms, ReLU returns the actual value if the input provided is greater than 0.0; else, it will return 0.0 if the value provided is less than 0.0, which is visible in the graph of ReLU function as shown in Figure 5.11. Due to its computational simplicity and representational sparsity, it is preferred over tanh and sigmoid functions.

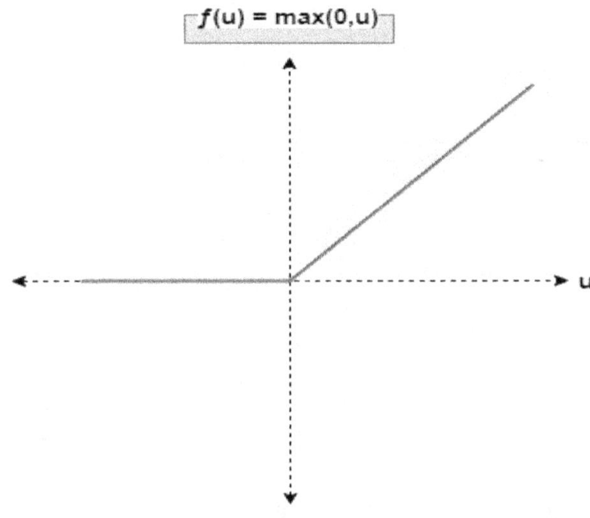

FIGURE 5.11 ReLU function.

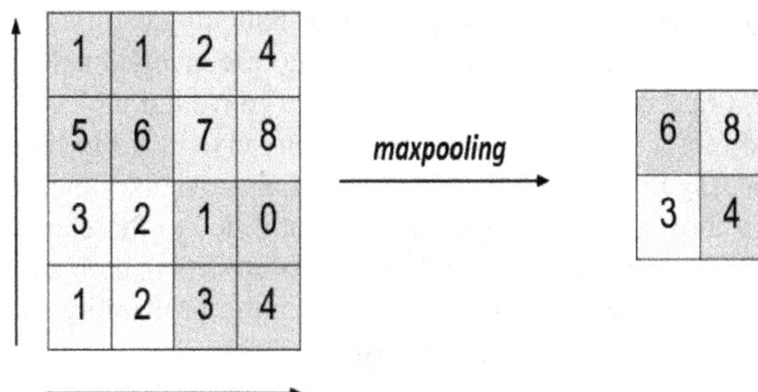

FIGURE 5.12 Operation of pooling layer.

- Pooling layer: After every convolutional layer, a pooling layer is added. For computation-intensive processes like image processing, reduction in the dimensions of the image while keeping redundant information intact is important. Image features' dimensions are reduced, and most important information is kept intact, with a pooling layer. This significantly reduces the power consumed in processing the data. For each convolutional step, a sliding function is applied on our data. There are many functions, such as min, max, sum, etc. We use a MaxPool2D layer of size (2,2) for the pooling, as max function usually performs better. The operation of the pooling layer is visualized in Figure 5.12. Max Pooling selects only the highest value (dominant features) from the portion the pool covers on the feature map.
- Fully connected layer: These are output layers situated at the last few layers of CNNs. In these layers, actual classification of our images is performed

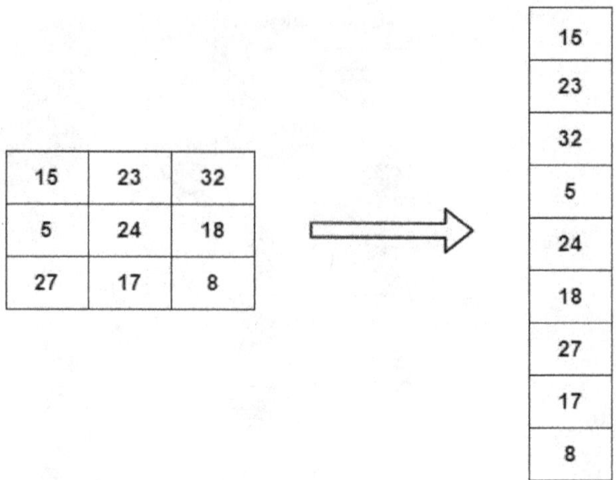

FIGURE 5.13 Operation of fully connected layer.

using features of images that were extracted in the convolutional layer. Flattened output from the pooling layer or convolutional layer is input to the fully connected layer. The process of flattening can be explained with an example. Say the output of the pooling or convolutional layer is in the form of a matrix (of size 3 × 3). It is then converted into a single column flattened vector. The flattened vector is then fed to the fully connected layer. After passing through the layer, the SoftMax activation function is applied, which in turn, assigns weights and probabilities to the class as shown in Figure 5.13, so now we have probabilities of a particular object in the image belonging to different classes. The higher the probability the higher the chance of the image belonging to that particular class.

Some common techniques that are used for each layer include:

- Batch normalization: This provides flexibility to every layer of the CNN to work individualistically without interrupting the learning of other layers. It basically normalizes the output of previous layers. Hence, it significantly increases the learning rate of the model. Thus, batch normalization has application for neural networks mainly in the case of advancing performance and offering stability.
- Dropout: It is a regularization technique. Overfitting in a model must be avoided. Hence, to lessen the possibility of overfitting and improving the model, dropout functions are used. Dropout does not permit to update the weights of some of the nodes selected on random ground.

5.4.1.4 Model Training

After CNN is defined, a few more parameters such as adam optimizer are compiled with it. This is a computationally effective algorithm used for training deep learning models.

```
Epoch 1/48
225/225 [==============================] - 170s 733ms/step - loss: 1.9331 - accuracy: 0.2634 - val_loss: 1.7059 - val_accuracy: 0.3564
Epoch 2/48
225/225 [==============================] - 26s 114ms/step - loss: 1.4733 - accuracy: 0.4343 - val_loss: 1.3616 - val_accuracy: 0.4821
Epoch 3/48
225/225 [==============================] - 26s 115ms/step - loss: 1.2878 - accuracy: 0.5115 - val_loss: 1.2539 - val_accuracy: 0.5226
Epoch 4/48
225/225 [==============================] - 25s 113ms/step - loss: 1.1919 - accuracy: 0.5432 - val_loss: 1.2051 - val_accuracy: 0.5403
Epoch 5/48
225/225 [==============================] - 26s 114ms/step - loss: 1.1292 - accuracy: 0.5732 - val_loss: 1.1390 - val_accuracy: 0.5615
Epoch 6/48
225/225 [==============================] - 26s 115ms/step - loss: 1.0754 - accuracy: 0.5927 - val_loss: 1.3052 - val_accuracy: 0.4935
Epoch 7/48
225/225 [==============================] - 25s 112ms/step - loss: 1.0216 - accuracy: 0.6158 - val_loss: 1.3517 - val_accuracy: 0.5030
Epoch 8/48
225/225 [==============================] - 25s 112ms/step - loss: 0.9637 - accuracy: 0.6342 - val_loss: 1.1310 - val_accuracy: 0.5750
Epoch 9/48
225/225 [==============================] - 26s 114ms/step - loss: 0.9306 - accuracy: 0.6506 - val_loss: 1.0508 - val_accuracy: 0.6036
Epoch 10/48
225/225 [==============================] - 25s 112ms/step - loss: 0.8947 - accuracy: 0.6596 - val_loss: 1.1392 - val_accuracy: 0.5729
Epoch 11/48
225/225 [==============================] - 26s 114ms/step - loss: 0.8427 - accuracy: 0.6840 - val_loss: 1.3011 - val_accuracy: 0.5268
Epoch 12/48
225/225 [==============================] - 26s 114ms/step - loss: 0.8068 - accuracy: 0.6957 - val_loss: 1.0834 - val_accuracy: 0.5973
Restoring model weights from the end of the best epoch.

Epoch 00012: ReduceLROnPlateau reducing learning rate to 0.00020000000949949026.
Epoch 00012: early stopping
```

FIGURE 5.14 CNN model training.

An epoch is a machine learning term that indicates one complete pass of the training dataset through the algorithm. There are 48 epochs involved in our training process. At each epoch, keras is used to check whether our model performed better than the previous epoch. New best models are saved in a file. We obtained validation accuracy of 72% for classes of seven different emotion categories. The process of model training is shown in Figure 5.14.

The output of emotion recognition with CNN for different images is provided at the end of the chapter. The aim of keeping all output of input images at the end of the chapter is to visually compare the results of both methods.

5.4.2 Emotion Recognition Using DeepFace Framework

In the preceding section, we learned about CNN-based emotion recognition. In this section, let us learn about one of the popular machine learning emotion recognition algorithms. DeepFace is a Python hybrid framework for face recognition and facial feature analysis (age, gender, emotion, and race). Models like VGG-Face, Google FaceNet, OpenFace, Facebook DeepFace, DeepID, ArcFace, Dlib, VGG-Keras and TensorFlow are the major components of the package. DeepFace is a system that has narrowed most of the gap in facial recognition and is approaching human-level accuracy DeepFace is a Facebook research group's deep learning facial recognition technology. It is trained on a big dataset of faces, which outperformed all previous systems with relatively minor improvements. The whole process of emotion recognition happens in the background when we import DeepFace and call the required functions. We have tried to demonstrate the basic methodology in Figure 5.15.

According to the Facebook research team, "the DeepFace technique has an accuracy of 97.35 % ± 0.25% on Labeled Faces in the Wild (LFW) data set where human beings have 97.53%" Modern CNNs have convolutional/pooling layers. These layers provide more meaningful description than pixel values can provide (Hortal et al., 2017). This suggests that DeepFace can outperform humans in some situations. DeepFace's applications have stayed relatively unchanged since its introduction in

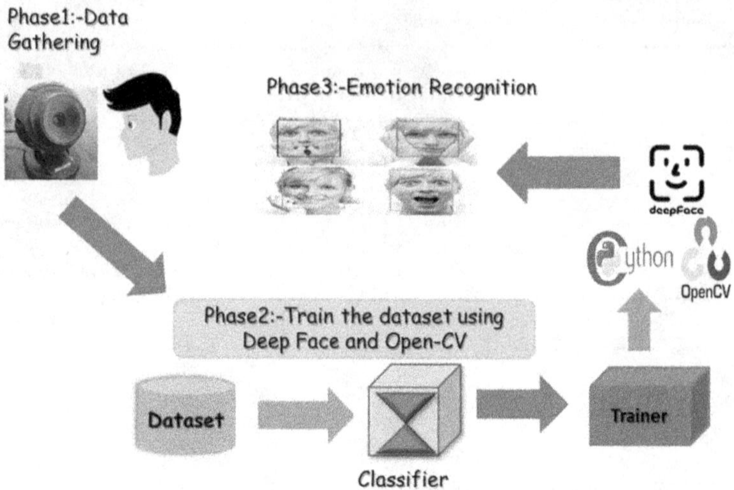

FIGURE 5.15 Procedure of emotion recognition using DeepFace.

2015. The algorithm has become more accurate as more people have posted photographs to Facebook. DeepFace, a facial recognition dataset created by Facebook, is the world's largest facial recognition dataset.

5.4.2.1 Hands on for Installation of DeepFace

We can easily install DeepFace using the command pip install deepface. One can install DeepFace by downloading DeepFace from PyPI.

Command used: – **"pip install deepface"**

By using this command, the necessary dependencies in the DeepFace framework will be installed in the system. In order to implement the DeepFace framework functions, we can use Anaconda IDE, which incorporates DeepFace functions and provides Jupyter Notebook. Python programming is used for implementation purposes.

5.4.2.2 Functions Used in DeepFace

DeepFace provides many in-built functions for face recognition and emotion recognition purposes, a few of which are listed in Figure 5.16.

5.4.2.2.1 Analyze Function

The Analyze function in DeepFace can be used to identify the age, gender, race, and dominant emotion of a person. Basically, it is a function through which we can predict the dominant race, gender, age, and emotion using the real-time webcam demo. It is used widely in order to analyze the facial attributes of a person more effectively. Hence, we can make use of this function to predict these facial attributes of a person.

5.4.2.2.2 Streaming and Real-Time Analysis

DeepFace can also be used to search for real-time videos. The DeepFace interface's stream function accesses the webcam of the computer. DeepFace performs face

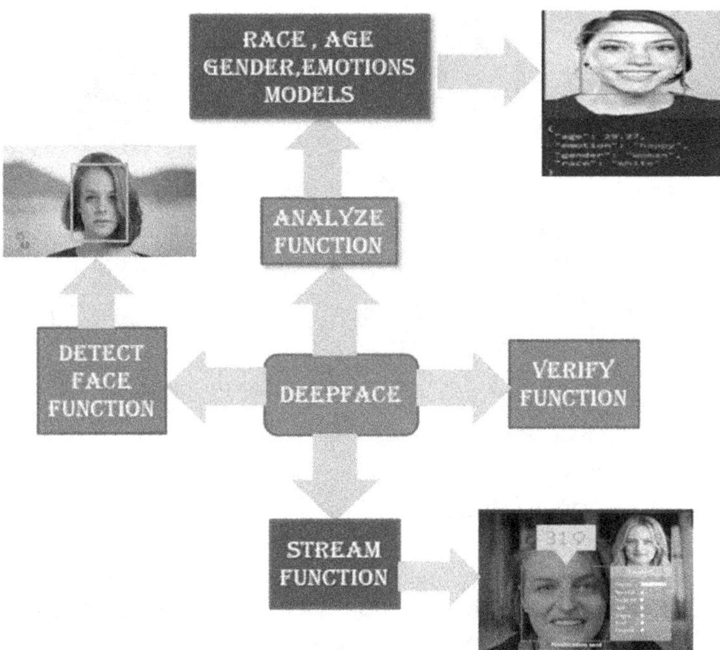

FIGURE 5.16 Functions of DeepFace framework.

recognition. Further, DeepFace performs facial attribute analysis. A face photos database folder is taken as input by the stream function. The default facial recognition model used by DeepFace is VGG-Face, and the default distance metric is cosine similarity, which is identical to the verify function. Figure 5.17 demonstrates the proper architecture of the DeepFace framework.

5.4.2.2.3 Architecture of DeepFace Framework

DeepFace worked with different current databases to advance the algorithms and produce a standardized output. The current models did not prove to be effective for facial recognition in all circumstances. However, DeepFace uses point detectors based on existing databases and aligns faces (Hortal et al., 2017). DeepFace performs the very first alignment as a 2-D alignment, then 3-D alignment, and then frontalization. Deep Face modifies the angles of an image so that the face in the photo appears to look forward. The process of fast 3-D alignment of faces is factor of achievement of any pretrained networks having performance on robust condition (Taigman & Yan, 2014). To accomplish this, it uses a 3-D model of a face. These 3-D model values are converted to surface representative numerical values. If DeepFace comes up with a similar enough description for two images, it assumes that these two images share a face.

5.4.2.2.4 2-D Alignment

DeepFace starts by detecting six fiducial points on the detected face: the center of the eyes, corners of eyes, corners of mouth, the tip of the nose, position of eyebrow,

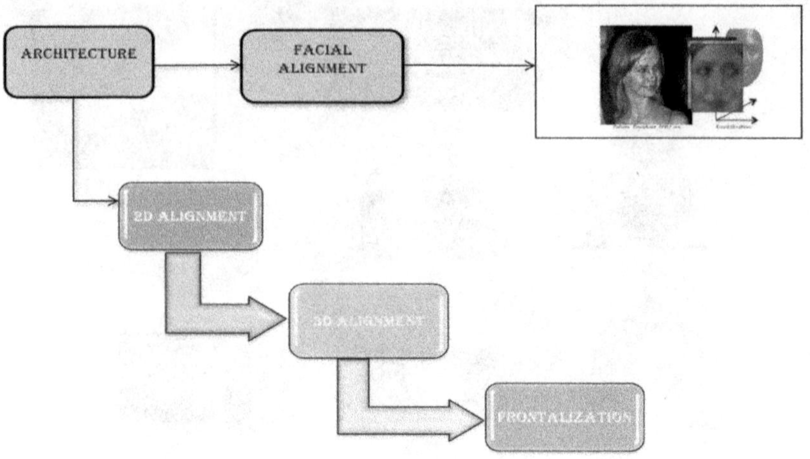

FIGURE 5.17 DeepFace architecture.

nose and chin and the position of the lips. To aid in the detection of the face, these points are transferred onto a warped image. 2-D transformation, on the other hand, is unable to correct for out-of-place rotations.

5.4.2.2.5 3-D Alignment

DeepFace employs a general 3-D model to align faces, in which 2-D photos are cropped into 3-D copies. Sixty-seven facial action unit points are manually placed on the image after it has been distorted to match the 67 fiducial points. After that, a 3-D-to-2-D conversion camera is installed to minimize losses. This step is necessary because 3-D identified points may be wrong on the face. The next step is frontalization.

4.4.2.3 Current Uses

In the current age of information exchange, it is observed that the information shared may or may not be for good reasons. Social media, like Facebook, are using DeepFace to notify users about shared content. Individuals can have control over their facial emotion recognition feature.

5.5 APPLICATIONS

The opportunities brought by this technology are infinite, and they go beyond market research and digital advertising. Figure 5.18 visually demonstrates a few of the applications of emotion recognition for human behavior analysis in the real world.

- Business firms can use photo scans or real-time videos to monitor video feeds for hunting possible business customers and opportunities using facial emotion recognition.

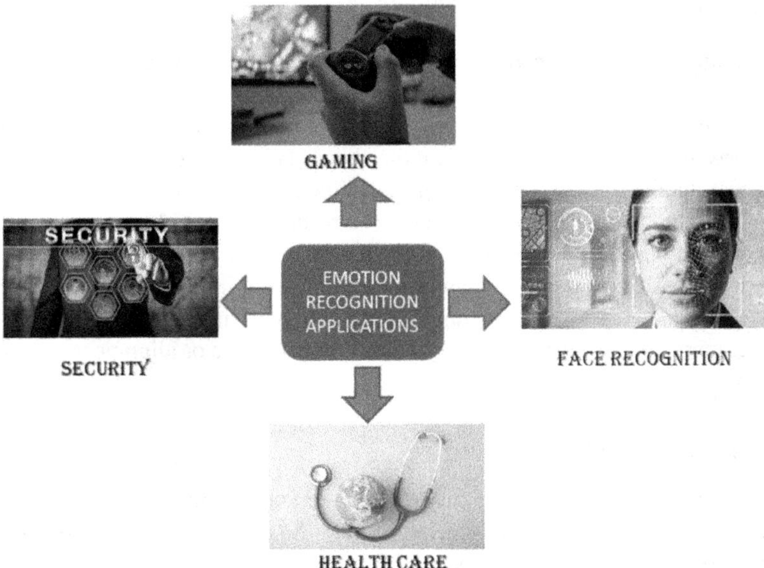

FIGURE 5.18 Applications of emotion recognition for human behavior analysis.

- Through virtual assistant gadgets, the technology assists businesses in establishing strong emotional ties with their customers. Furthermore, an emotion-sensing wearable aids in the monitoring of users' mental and other health characteristics.
- It plays an essential role in security. It can recognize people in a crowd and track residents' identities, age, gender, and present emotional state to look for suspicious conduct, and it can be used to prevent criminals and possible terrorists from entering the country.
- Automated customer service employees can understand callers' emotional states thanks to real-time voice-based emotion analysis. This enables them to adjust to changing circumstances.
- Emotion recognition can be employed in healthcare, with artificial intelligence–powered recognition software assisting in determining whether patients require medicine or assisting physicians in choosing who to examine first.
- Emotion recognition technology is also being used in the automotive sector. Car manufacturers all over the world are providing safe cars with person-specific needs. The car manufacturers have started research and development in this sector.
- Video games are created with a certain target audience in mind, with the goal of inducing a specific behavior and set of feelings in the players. Users are requested to play the game for a set amount of time during the testing process, and their feedback is used to improve the final product.

5.5.1 Drawbacks

As with any technology, there are potential drawbacks to using emotion recognition.

- Face and emotion recognition technology poses a huge risk to individual privacy. People don't appreciate their faces being photographed and stored in a database for unknown purposes in the future.
- Technology isn't always accurate. As a result, the algorithms develop inadvertent biases. False positives come with their own set of risks.
- It is possible to deceive technology. People with the wrong intentions can take advantage of bad lighting illumination, poor image or video quality, or camera conditions like a blurry picture or bad angle or alignment.

5.6 TEST RESULTS

We have discussed all aspects of one CNN-based and one machine learning emotion detection algorithm. In this section, we discuss the test results of emotion detection techniques implemented using CNN and DeepFace.

5.6.1 Emotion Recognition Using DeepFace Result

- We performed emotion recognition using the DeepFace framework. This is one of the best models of all time. It has an accuracy of 97%, which is the same as a human.
- The output of the analyze function consists of the age, gender, race, and emotion of the human, as shown in Figure 5.19. It is one of the advantages of DeepFace over self-trained CNN models.
- From all these values, we have displayed the dominant emotion in a real-time situation which gives us the desired result of emotion recognition. Various emotions were recognized, as shown in Figures 5.20 and 5.21.

Another approach used is the "stream" function, which gave us the result in three lines of code. Following Figures 5.20 and 5.21 are the results obtained using the stream function (Figure 5.22).

5.6.2 Emotion Recognition Using Convolutional Neural Network

Thus, we have successfully performed emotion recognition using CNN. The model was trained with seven emotions and had accuracy around 70%. Our results obtained using CNN are shown in Figure 5.23.

- Evaluation of loss and accuracy after each epoch is plotted using the matplotlib library. We get output at each step of the training phase. After each epoch, loss decreases and accuracy increases, as shown in Figure 5.24.

```
ıpyter   Untitled64 Last Checkpoint: Yesterday at 8:54 PM  (unsaved changes)

 Edit   View   Insert   Cell   Kernel   Widgets   Help

         ▶ Run  ■  C  ▶▶  Code

        nat relaxes argument shapes that can avoid unnecessary retracing. For (3
        ction#controlling_retracing and https://www.tensorflow.org/api_docs/pyth
        WARNING:tensorflow:7 out of the last 33 calls to <function Model.make_pr
        43EE9D8> triggered tf.function retracing. Tracing is expensive and the e
        g @tf.function repeatedly in a loop, (2) passing tensors with different
        For (1), please define your @tf.function outside of the loop. For (2), @
        hat relaxes argument shapes that can avoid unnecessary retracing. For (3
        ction#controlling_retracing and https://www.tensorflow.org/api_docs/pyth

        Action: race: 100%|

In [22]: predictions

Out[22]: {'emotion': {'angry': 2.4659577933962017e-09,
           'disgust': 6.494392999624314e-13,
           'fear': 5.499762084933112e-08,
           'happy': 99.93420838917544,
           'sad': 2.0703596516642303e-05,
           'surprise': 5.528023701887889e-07,
           'neutral': 0.06577089471680789},
          'dominant_emotion': 'happy',
          'age': 27,
          'gender': 'Man',
          'race': {'asian': 5.462395027279854,
           'indian': 17.69561916589737,
           'black': 7.04357847571373,
           'white': 17.72809624671936,
           'middle eastern': 12.223904579877853,
           'latino hispanic': 39.846405386924744},
          'dominant_race': 'latino hispanic'}
```

FIGURE 5.19 Analyze function output.

Neutral Emotion Detected By Deep Face

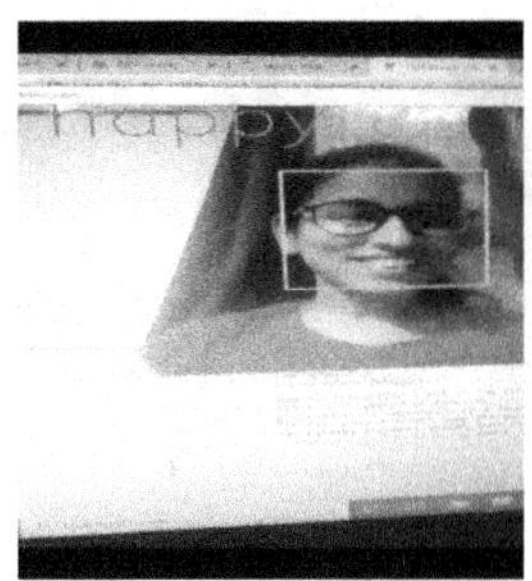

Happy emotion detected by deepface

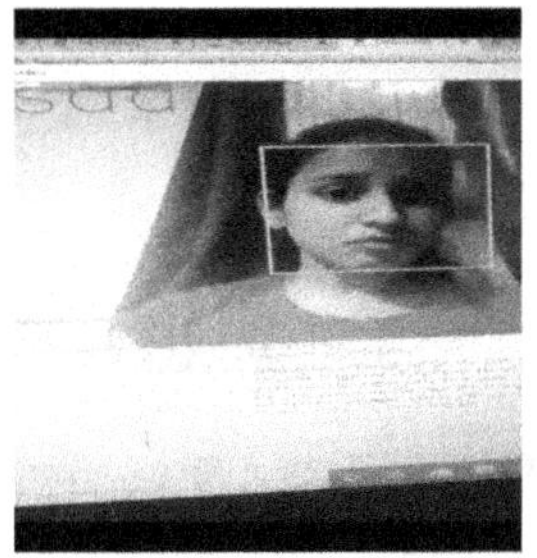

Sad emotion detected by deepface

Fear Emotion detected by deepface

Angry emotion detected by deepface

FIGURE 5.20 Emotion recognition using DeepFace output.

FIGURE 5.21 Stream function output 1.

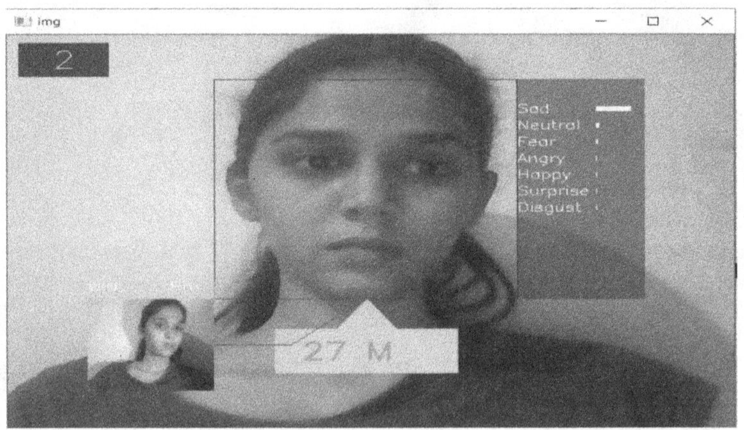

FIGURE 5.22 Stream function output 2.

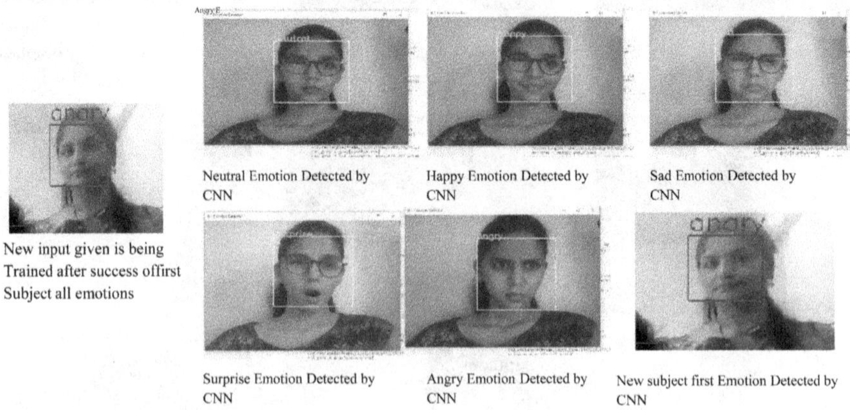

FIGURE 5.23 Emotion recognition using CNN output.

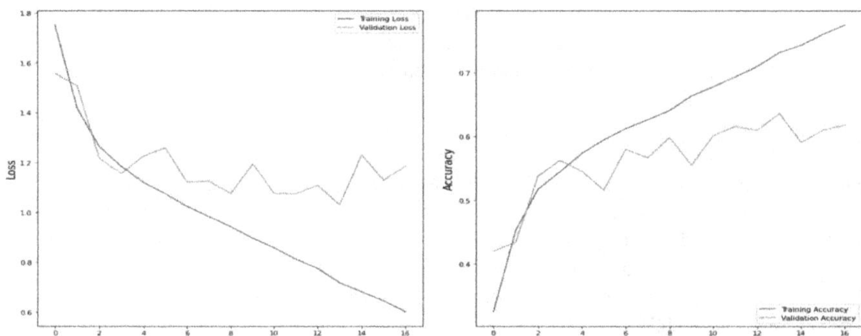

FIGURE 5.24 Evaluation of loss and accuracy with each training epoch.

BIBLIOGRAPHY

A. Basnet, S. Shakya and S. Sharma, 2016, "Human behavior prediction using facial expression analysis", 2016 International Conference on Computing, Communication and Automation (ICCCA), pp. 399–404, doi: 10.1109/CCAA.2016.

Abeer Alsadoon, D. Yang, and Prasad P.W.C., January 2018, "An emotion recognition model based on facial recognition in virtual learning environment", *Procedia Computer Science* 125:2–10 DOI:10.1016/j.procs.2017.12.003.

Agata Kolakowaska, Agnieszka Landowaska, Mariusz Szwoch, et al., July 2014, "Emotion recognition and its applications", *Advances in Intelligent Systems and Computing* 300:51–62. DOI:10.1007/978-3-319-08491-6_5.

Ahmad Fakhri Ab. Nasir et al., 2020, "Text-based emotion prediction system using machine learning approach", *IOP Conf. Ser.: Mater. Sci. Eng.* 769:012022.

Akash Sarvanan, Gurudatta Perichetala, and Dr. K. S. Gayathri, n.d., "Facial emotion recognition using convolutional neural network".

Andrzej Majkowski, Marcin Kołodziej, Paweł Tarnowski, and Remigiusz J. Rak, 2017, "Emotion recognition using facial expressions", International Conference on Computational Science, ICCS 2017, 12–14 June 2017, Zurich, Switzerland.

Awais Mahmood, Khalid Iqbal, Shariq Hussain, and Wail S. Elkilani, 2019, "Recognition of facial expressions under varying conditions using dual-feature fusion", *Mathematical Problems in Engineering*, 2019: 9185481, 12 pages. https://doi.org/10.1155/2019/9185481

Baris Akis and Tanner Gilligan, n.d., "Emotion AI, real-time emotion detection using CNN", tanner12@stanford.edu. Baris Akis. B.S. Computer Science. Stanford University bakis@stanford.edu.

Christopher Best, Margaret Lech, Melissa Stolar et al., May 26, 2020, "Real-time speech emotion recognition using a pre-trained image classification network: effects of bandwidth reduction and companding"*Frontiers of Computer Science.*

Claudio Loconsole, Domenico Chiaradia, and Vitoantonio Bevilacqua, 2014, "Real-time emotion recognition: an improved hybrid approach for classification performance intelligent computing theory", 8588. ISBN: 978-3-319-09332.

Dongwei Jiang, Kun Han, Miao Cao et al., n.d., "Speech SIMCLR: combining contrastive and reconstruction objective for self-supervised speech representation learning", arXiv:2010.13991 [cs.CL] (or arXiv:2010.13991v1 [cs.CL] for this version),Tue, 27 Oct 2020 02:09:06 UTC.

Enrique Hortal, Essam Ghaleb, Justus Schwan, et al., July 2017, "High-performance and lightweight real-time deep face emotion recognition", in 12th International Workshop on Semantic and Social Media Adaptation and Personalization (SMAP). DOI:10.1109/SMAP.2017.8022671.

Gordon M. Burghardt, Sep 2019, "A place for emotions in behavior systems research", 166: 103881. doi: 10.1016/j.beproc.2019.06.004. Epub 2019 Jun 5.

Heekyung Yang, Jongdae Han, and Kyungha Min, Oct 31, 2019, "A multi-column CNN model for emotion recognition from EEG signals", 19(21): 4736. doi: 10.3390/s19214736.

Hongli Niu, Kunliang Xu, Weiqing Wang, Hongli Niu, and Xiangrong Miao, 2020, "Emotion recognition of students based on facial expressions in online education based on the perspective of computer simulation", *Complexity*, 2020: 4065207, 9 pages. https://doi.org/10.1155/2020/4065207

IIiana Azizan and K. Fatimaah, August 2020, "Facial emotion recognition: a brief review", in Conference: International Conference on Sustainable Engineering, Technology and Management 2018 (ICSETM-2018), Mantin, Negeri Sembilan.

J. Deepthy, P. Suja, and Shikha Tripathi, 2014, "Emotion recognition from facial expressions using frequency domain techniques advances in signal processing and intelligent recognition systems", 264, ISBN: 978-3-319-04959-5

Jason Brownlee, August 16, 2019, "What is deep learning?".

Jonathan oheix, 7 Jan 2019, "From raw images to real-time predictions with Deep Learning-Face expression recognition using Keras, Flask and OpenCV".

K. Sumithra, R. Somasundaram, and S. Buvana, March 2015, "A survey on various types of image processing technique", *International Journal of Engineering Research & Technology (IJERT)*, 4(3). http://dx.doi.org/10.17577/IJERTV4IS030552

M.A.H. Akhand, Nazmul Siddiquie, Shuvendu Roy et al., 2021, "Facial emotion recognition using transfer learning in the deep CNN", Academic Editors: Nikolaos Mitianoudis, Georgios Tzimiropoulos and Juan M. Corchado, *Electronics*, 10(9): 1036; https://doi.org/10.3390/electronics10091036 Received: 3 March 2021 / Revised: 12 April 2021 / Accepted: 24 April 2021 / Published: 27 April 2021.

MinSeop Lee, Yun Kyu Lee, Myo-Taeg Lim, and Tae-Koo Kang, 2020, "Emotion recognition using convolutional neural network with selected statistical photoplethysmogram features", Received: 21 April 2020; Accepted: 14 May 2020; Published: 19 May 2020.

Navinkumar Mahamkali, and Vadivel Ayyasamya, March 20, 2015, "Open-CV for computer vision applications", in Conference: Proceedings of National Conference on Big Data and Cloud Computing (NCBDC'15), Trichy, Project: Human Action Recognition.

P. A. Riyantoko et al., 2021, "Facial emotion detection using haar-cascade classifier and convolutional neural networks", *Journal of Physics: Conference Series*, 1844: 012004.

Parkhi et al., n.d., "Deep face recognition", Visual Geometry Group Department of Engineering Science University of Oxford, Cited by 4208 – Omkar M. Parkhi omkar @robots.ox.ac.uk

Rishabh Jain, n.d., *Facial-Expresssion-Recognition*.

Rui Xia, and Zixiang Ding, n.d., "Emotion-cause pair extraction: A new task to emotion analysis in texts", School of Computer Science and Engineering, Nanjing University of Science and Technology, China {rxia, dingzixiang}@njust.edu.cn

Seok-Pil Lee, and Sung-Woo Byun, 2020, "Human emotion recognition based on the weighted integration method using image sequences and acoustic features", Received: 4 June 2020 / Revised: 27 July 2020 / Accepted: 9 September 2020 # The Author(s) 2020.

Shu-Feng Chen, and Sung-Nien Yu, Aug 2015, "Emotion state identification based on heart rate variability and genetic algorithm", in Annual International Conference of the IEEE Engineering in Medicine and Biology Society 2015, 538–41. doi: 10.1109/EMBC.2015.7318418.

Suwicha Jirayucharoensak, Setha Pan-Ngum, and Pasin Israsena, 2014, "EEG-based emotion recognition using deep learning network with principal component based covariate shift adaptation", *The Scientific World Journal*, 2014: 627892, 10 pages. https://doi.org/10.1155/2014/627892.

Y Taigman, n.d., "DeepFace: Closing the gap to human-level performance in face verification", https://www.cs.toronto.edu › taigman_cvpr14. Y Taigman. Cited by 5577—wolf @cs.tau.ac.il ... Faces in the Wild (LFW) dataset, reducing the error.

Zhang H., Jolfaei A., and Alazab M. 2019, "A face emotion recognition method using convolutional neural network and image edge computing", *IEEE Access*, 7: 8884205, pp. 159081–159089.

6 A Comparative Study of Convolutional Neural Networks for Plant Phenology Recognition

Shivani Gaba, Shally Nagpal, and Alankrita Aggarwal

CONTENTS

6.1 INTRODUCTION

Cultivation is necessary for the sustained survival of humanity, as humans' lives are directly dependent on agriculture for the fabrication of nutrients. As soon as there is an increase in population, that results in a rapid increase in food requirements. Using precision agriculture, the required production rate can be achieved. The main focus

DOI: 10.1201/9781003221333-6

is to determine the optimal time to plant and harvest the crops so that agriculture can be rid of crop diseases, which lead to a significant threat to food security [1].

As we all know, the risk to food security is due to diseases occurring in crops; however, the people of many areas are not aware of these diseases, and they have no idea about the necessary arrangements. The combination of growing worldwide smartphone use and recent research in computer preparedness has paved the way for analysis of these diseases. Overall, the deep learning models present a vibrant way towards smartphone-supported disease analysis of crops on a worldwide measure [2].

Deep learning has provided an up-to-date method for image processing at a significant level. Deep learning has enhanced many research areas, and nowadays, it has also arrived in the agriculture domain. This chapteranalyzesvarious research efforts employing convolutional neural networks (CNNs),a specific domain of deep learning, which are applied to numerous challenges in agriculture and food production. CNNs are associated with other current methods, and various pros and cons of using CNNs in agriculture are examined, listed, and compared [3]. Furthermore, the future capability of this strategy is talked about, together with the authors'own experience of utilizing CNN to solve an issue of determining the optimal time to plant and harvest. The general discoveries show that CNNs comprise a promising strategy, superiorin terms of precision to existing regularly utilized picture handling methods. Notwithstanding, the accomplishment of each CNN model is dependent on the superiority of the dataset utilized [4].

A database of images was gathered utilizing isolated detecting strategies, and the pictures were examined to build up a model to choose appropriate treatment plans for various yield types and districts. Features of images from vegetation should be separated, ordered, and portioned. Various techniques and strategies have been used for this, but the outcomes of the deep learning approach of the CNN system has been received for the classification of various images. CNN is utilized to give responses on agriculture images as well as on the formation on the plantation[5].

Onthe whole, we realize that CNNhas accomplished substantial results in the field of classifying images and in recognizing different plant diseases. We are now concerned with one of the methodologies utilizing CNNto decide the ideal time for planting and harvesting. In this chapter, crop models are refined by CNN to estimate the capacity of deep learning to figure out how to recognize the plant physiology information behind such crop models basically by learning. Because of the difficulty in gathering the enormous amount of information on crop development, a harvest model was utilizedto create datasets. The produced datasets were incorporated into CNN to refine the harvest model. The outcomes also showed that CNN could be used to develop new plant physiology hypotheses basically by learning [6].

CNN also recognized the significant individual natural elements influencing grain yield. Such deep learning approaches can speed up the comprehension of yield models and make the models increasingly convenient. Additionally, the outcomes demonstrated that CNN could be utilized to develop new speculations about plant physiology basically by learning [7].

The demonstration of CNN in object recognition and classification of images has massivelyincreased in the last few years. Previously, the traditional methodology

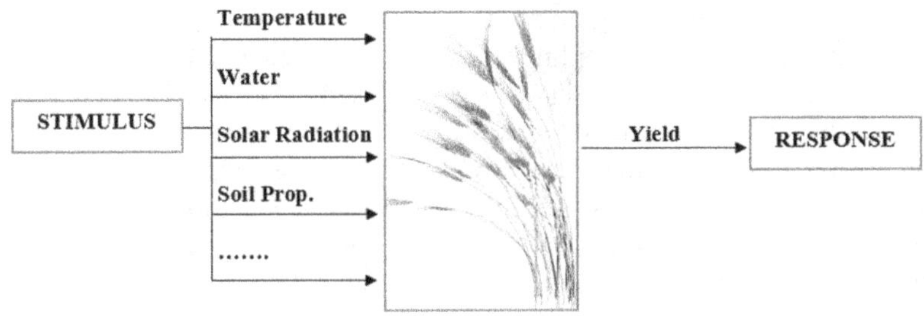

FIGURE 6.1 System of input and output: simplified.

for image classification has been initiated for planned highlights, such as SIFT, HoG, SURF, and to utilize some learning design in these constituent places. For this reason,each of these methodologies is reliant upon predefined hidden features. Feature extraction is itself a complicated and repetitive procedure that needs to be revisited each time the current issue or the related dataset changes significantly. The basic structure of input and output for any crop or plant is presented in Figure 6.1[8].

This issue has happened in every conventional endeavor to recognize plant infections using computer vision.networks, AlexNet started a few years ago to utilize design using a deep convolutional neural [9, 10].Utilizing CNN engineering, we prepared a model on pictures of plant leaves with the objective of grouping both harvest crops and disease characteristicsin the pictures. This presents a vibrant route towards the diagnosis of crops on enormous, comprehensive scales. So, by determining the optimal time to plant and harvest using CNNs, we can enhance our crops.

6.2 RELATED WORKS

A. Mohapatra et al. [11] suggested progressing designs using CNN, and deep learning plans in provincial applications are discussed. Before the development of deep learning, pictures dealing with artificial intelligence (AI) methodologies were used to describe unmistakable plant diseases.Overall, most of these systems have followed technological progress in their development. First efficient pictures picked up using a boosted camera. Then, picture preparation methodologies, such as picture update, division, and isolating, are applied to make the photos sensible for the accompanying stages. After that, notable features are expelled after the image and used as a commitment to the classifier. CNNs are multi-layer directed frameworks that can pick up regularly incorporated datasets. Traditionally, CNNs have achieved top tier execution in all intentions and commitments for gathering deeds. It can perform both part extractions and requests under a comparable structure according to S. Yang et al.[12]. CNN is an extraordinary kind of neural network commonly applied to a grouping of model assertion issues, such as computer vision. CNN relies upon the human visual structure, first charged by (Hubel and Wiesel 1962) and continually executed by various investigators. CNN merges three compositional designs to promise some degree of the move, scale, and turning invariance: close by open fields, shared burdens, and spatial or characteristic subsampling according to J. Chen et al.[13].

Distinctive CNN structures have been proposed for object identification. The primer CNN is LeNet presented byJ. Li et al.[1414] able to see physically composed digits.It contains two convolutional layers and two subsampling layers followed by a related multi-layer perceptron. Rarely have authors planned to use CNNs for plant declaration and illness. The sequence of action for plants arranges CNN to distinguish plants and pictures of leaves.

Y. Li et al. [15] used a deep learning approach in which the entire system was discovered. The organized structure has fiveconvolutional layers followed by two related layers.

Z. Ma et al. [16] used the current deep CNN structuresAlex andGoogLeNetto portray plant diseases. Using an open dataset of nearly 54 pictures of leaves from undesirable and healthy plants assembled under controlled conditions, the CNN was set up to identify 14 yield species and 26 diseases. The model achieved 99.35% precision. However, when shown pictures taken at anunexpected area, in contrast to the photos used for the arrangement, the model's precision dropped to 31.4%. Overall, the result shows the feasibility of deep CNN for plant disease queries.

A shading-based technique, where the readiness characterization framework accomplishes high precision, has been proposed byY. Sun et al. [17].

H. Huang et al. [18] utilized AI innovation for programmed location and arrangement of red tomatoes and structured a classifying framework for improving precision and saving time.

G. Gui et al. [19] received a picture-prepared approach for red banana's aging. Even though these techniques gave excellent execution intrials in specific situations, they are hard to compare Currently, the characterization framework dependent on CNN has accomplished excellent outcomes in numerous areas. Analysts have proposed various techniques to increase information in this way.

Y.Li et al.[20] suggested a learning based on word reference for cross-mark concealment in facial acknowledgment to safeguard the property.

Z.-H. Tan et al. [21] recommended a new methodology which relates to deep convolutional neural systems techniques for distinguishing the imperfection of latches.

X. Sun et al. [22] proposed a network called FingerNet, which consists of distinctive de-convolution portions to improve a unique mark to successfully stifle the exceptions and precisely reproduce the picture from compressive estimated information.

M. Brahimi et al. [23] introduced a new calculation based on a multiplier arrangementto accomplish better execution in picture remaking.

A. Fuentes et al. [24] introduced a methodology that uses a vibrational Bayesian learning strategy and execution for hopeful segments on the Dirichlet technique with changed Dirichlet scatterings, and also, CNN-based examinations for face affirmation, remote correspondences, modified speaker checks dictated by the striking accomplishment of deep learning. Importantly for a convincing strategy to overcome overfitting issues, various deep learning–based assessments misuse data extension methods. For example, another technique subject to CNN for alcohol fixation area assignments with data development systems uses 0 pictures.

S. Bargoti and J. Underwood [25] proposed another novel system for constrained assistance data used in various very large applications.

The same authors [26] suggested a practical answer for difficulty in picture division and picture acknowledgment of feeding horse a leaf. This technique was first used to remove an aggregate of 129 highlights, and then, a support vector machine (SVM) model was prepared with the most significant highlights. The outcomes showed that picture acknowledgment of the four-horses feeding a leaf could be achieved with an average precision of 94.74%.

S.H. Wang et al. [27]introduced an example acknowledgment framework for distinguishing and arranging three cotton leaf diseases. Utilizing the dataset of regular pictures, a functioning shape model was utilized for picture division, and Hu's minutes were separated as highlights for the preparation of a versatile neuro-fluffy induction framework. The example acknowledgment framework accomplished a reasonable precision of 85%.

J. Muñoz Bulnes et al. [28]introduced a methodology that incorporated picture handling and AI to permit the identification of diseases from leaf pictures. This robotized technique arranges infections of potato plants from "Plant Village", an openly accessible plant picture database. The division approach and use of an SVM showed disease arrangement in more than 300 pictures and achieved a standard precision of 95%.

S. Hussein et al.[29]proposed a self-governing altered Cuckoo search-based SVM (SVM-CS) model to recognize healthy and diseased areas ID and characterization approaches of these investigations are self-loader and complex and manages progression of picture. Simultaneously, it is hard to precisely recognize particularly noisy pictures with the suitable order highlights relying intensely upon master understanding. Recently, a few analysts have considered that recognizable proof of plant diseases dependent on deep learning is becoming a reality.

Z. Ma et al.[30]planned a new recognizable proof methodology for diseases of rice dependent on deep convolutional neural systems. Utilizing a dataset of 500 regular pictures of diseased and healthy rice leaves and stems, CNNs were prepared to recognize 10 basic rice diseases. The trial results indicated that the proposed model accomplished a standard precision of 95.48%.

L. van der Maaten and G. Hinton[31]introduced a methodology dependent on CNNs to perceive apple disease pictures and utilized a self-supervised multimodal versatile force rule to refresh the CNNs' parameters. The output showed that the acknowledgment precision of the proposition was up to 96.08%, with a genuinely quick assembly.

M. A. Hearst et al. [32]presented a new cucumber leaf disease discovery framework dependent on CNN. Under the fourfold cross-approval technique, the suggested CNN-based framework accomplished a standard precision of 94.9% in arranging cucumbers into two common disease classes and a healthy class. The exploratory outcomes shows that CNNs-based models can consequently separate the imperious classification and obtain ideal execution.

C. Tomasi et al. [33]introduced a new methodology dependent on Deep Convolutional Neural Network (DCNN) to distinguish infections between plants. By separating the leaves of plants from environmental factors, 13common plant diseases were perceived by the suggested CNN-based model. The trial outcomes indicated that the proposed model based on CNNs was capable of a decent acknowledgment execution and achieved an average 96.3% precision.

6.3 BACKGROUND

In this section, we present the foundation information required to handle the basic ideas utilized in this part. Specifically, we talk about deep learning and convolutional neural systems.

6.3.1 Deep Learning

This is a period of actions in AI advancements, consisting of different preparation layers that permit the rendering of learning for various levels of information. The significance of deep learning is its ability to make and extrapolate new highlights from crude portrayals of information without being advised unequivocally which highlights to utilize and how to remove them. In the plant ID space, various investigations have concentrated on methodology or calculations that expand the utilization of databases of leaves, and this prompts a standard that highlights the modification with information about leaf and extraction procedures. Up until now, we have been locked in with the uncertainty encompassing the subsection of highlights that best speak to the leaf information. Consequently, in a current examination, rather than digging into the formation of highlight portrayal as in past methodologies, we figure out the procedure by requesting that deep learning decipher and inspire the specific highlights that best speak to the leaf information. By understanding outcomes the psychological complexities of vision for leaf disease identification and mirroring the insignificant information scientists naturally convey heir innovative vision [34].

6.3.1.1 Deep Learning Usage in Crop Production

Deep learning take a period of field data: bits of information on how crops have acted in various airs and procured specific properties. It then uses this data to develop a probability model. In emerging area, they are used to anticipate the crop and nature of harvests.

So Temperature and Soil in irrigation is a mix for recognizing remote picture data as sensor data is given to CNNs and various fields related to agriculture lead to, for instance, processing information related to agriculture, production system of agriculture perfect control, Smart Agriculture Machinery gear, cultivating monetary structure the board and significant learning can apply to plants, creatures, lands, and mechanization, as shown in Figures 6.2 and 6.3[35].

6.3.1.2 Various Methods in Plant Subject Area

When the sensor data related to agriculture passes through artificial methods such as CNN, then it leads to Agronomy. In Agronomy as output, there are some fields such as Plant, Animal, Land, and Mechanization, as described in Figure 6.3. As we are discussing Agriculture, there are many categories in plants, such as Crop classification, Phenology recognition, Disease detection, Weed detection, Counting of fruits,and Crop prediction, as shown in Figure 6.4.

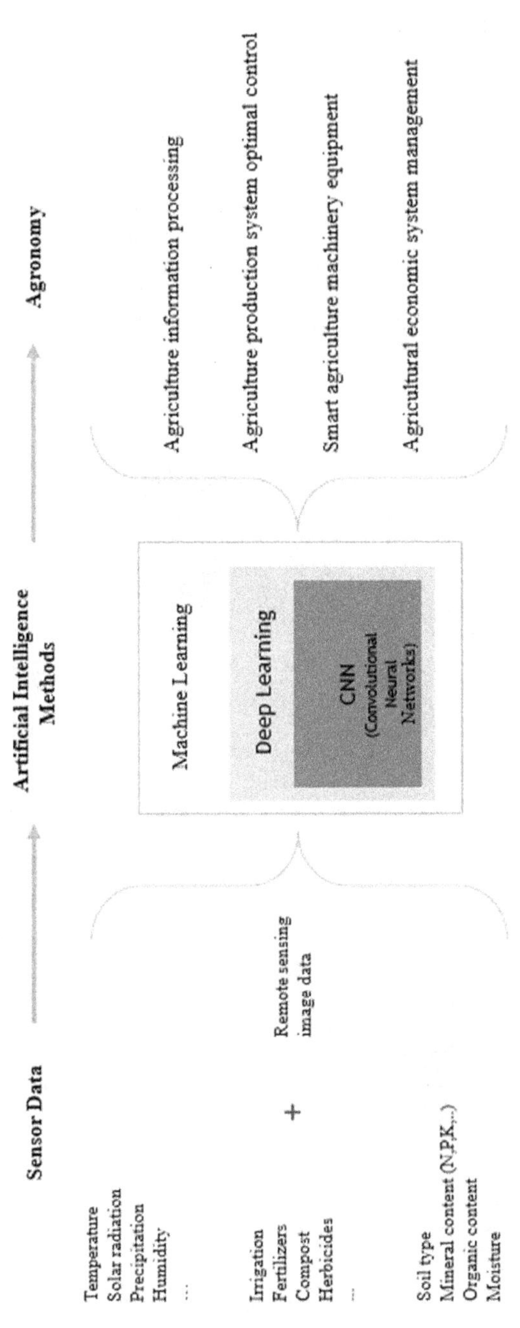

FIGURE 6.2 Mechanism of input and output: using CNN.

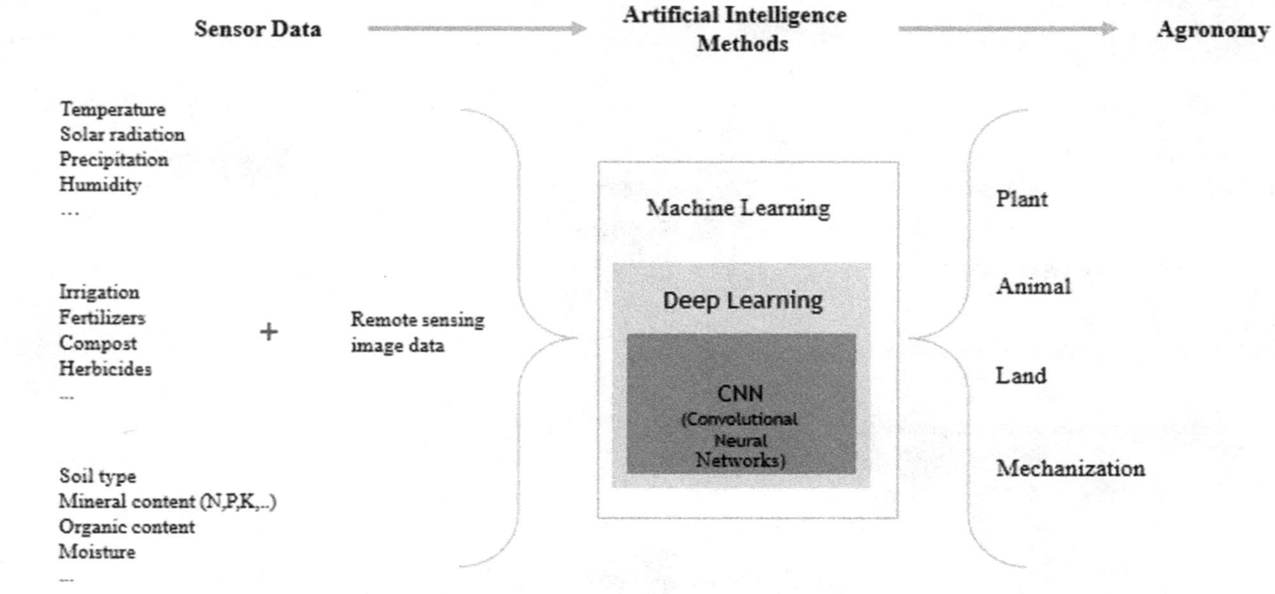

FIGURE 6.3 Mechanism of input and output.

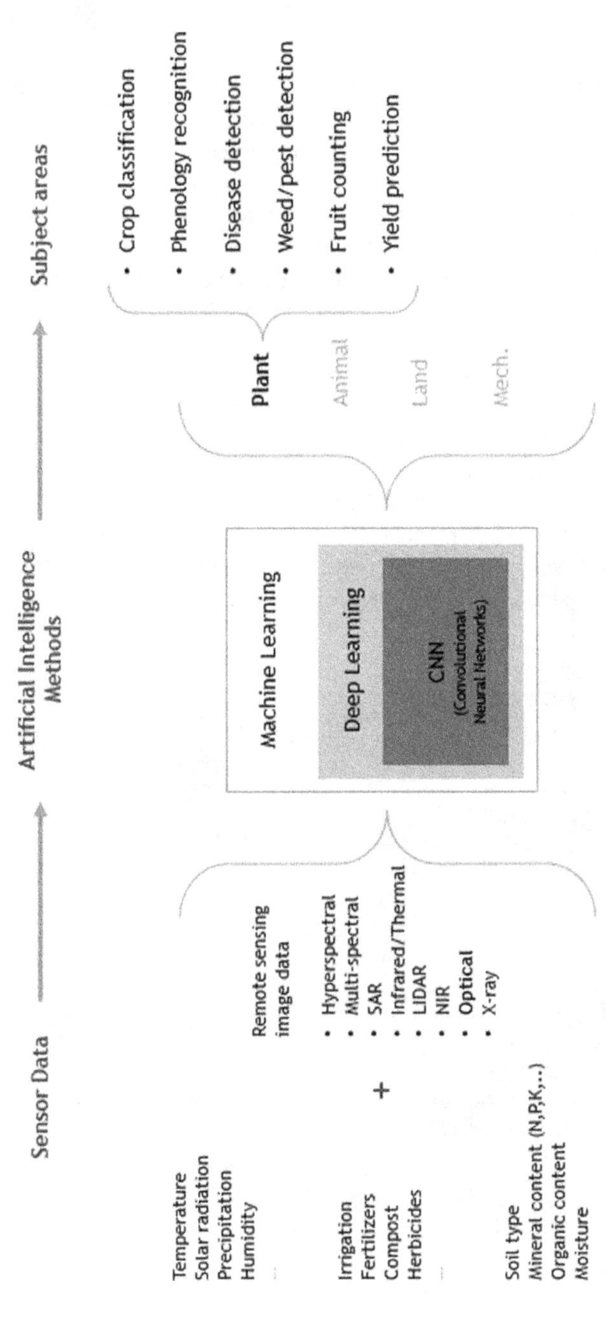

FIGURE 6.4 Different mechanisms in plant subject area.

6.3.1.2.1 Crop Classification

CNN and visual geometry gathering (VGG) are utilized to arrange crops dependent on the various quantities of information groups made by optical and synthetic aperture radar (SAR) information. Various mechanisms come under Plant subject areas, as already shown in Figure 6.4. As we know, CNNsare divided into two-dimensional (2-D) CNNs and three-dimensional (3-D) CNNs. Figure 6.5 shows that there are many crops, such as season wheat, midwinter rapeseed, spring crops, maize, water, and many more,as described bySalakhutdinov and Hinton [36]. So, various crops are classified in Figure 6.5. Yield items are characterized for the most part as indicated by the sort of harvest. The harvest grouping alludes to which yields are developed, while the item arrangement alludes to the product(s) created from that crop. In this manner, "mustard" is an oilseed crop, while "mustard seed" is the oilseed item [37].

6.3.1.2.2 Phenology Recognition

The already prepared CNN engineering was utilized to remove the highlights of pictures naturally. Test results propose that CNN design outflanks the AI calculations dependent on available made highlights for the segregation of phenological stages. So,for the crops of cotton, pepper, and corn shown in Figure 6.6, in the case of cotton, the precision is 87.32%, recall is 86.14%, Fi-score is 86.58%, and accuracy is 86.54%; however, the figures are different in the case of pepper and corn. In the case of pepper, the precision is 88.12%, recall is 87.24%, Fi-Score is 87.28%, and accuracy

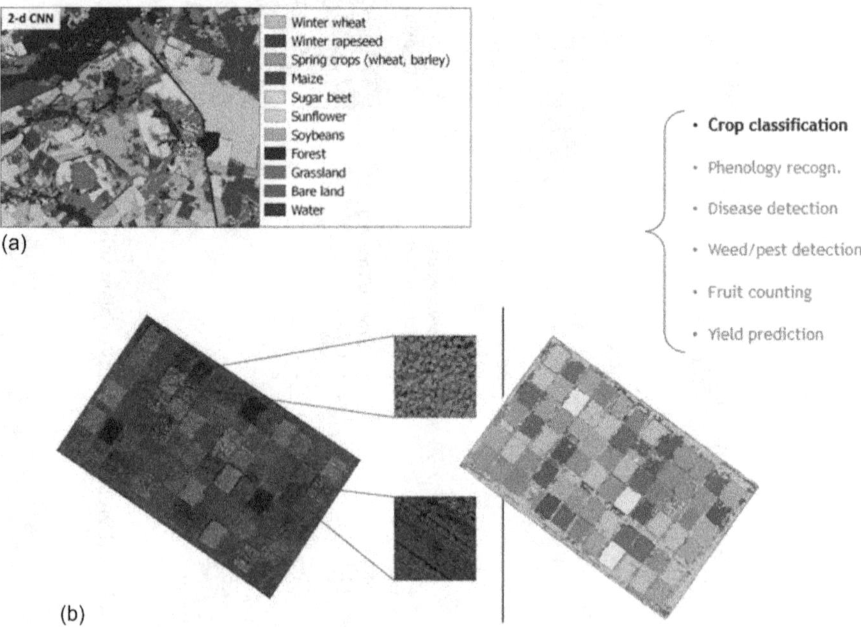

FIGURE 6.5 Crop classification in 2-D CNN. (Panel (a) from Kussul et al., 2017; panel (b) from Rebetez et al., 2017.)

Dataset	Method	Precision (%)	Recall (%)	FI-Score (%)	Accuracy (%)
Cotton	CNN-BA	87.32	86.14	86.58	86.54
Pepper	CNN-BA	88.12	87.24	87.28	87.14
Corn	CNN-BA	87.32	86.14	86.58	86.54

- Crop classification
- **Phenology recogn.**
- Disease detection
- Weed/pest detection
- Fruit counting
- Yield prediction

FIGURE 6.6 Phenology recognition. (From Geol and Seghal, 2015.)

is 87.14%, whereas in the case of corn, the precision is 87.32%, recall is 86.14%, FI-Score is 86.58%, and accuracy is 86.54% [38].The observation of phenology of plants is a basic comprehension in accuracy farming. Fundamental improvements can be accomplished with exact recognition of phenological changes of plants, which would from this time forward improve the planning for harvest, bug control, yield expectation, ranch observations, warning of disasters, and so on.

6.3.1.2.3 Disease Detection

Farmers experience an extraordinary variety of yields. Different factors, for example, climatic conditions, soil conditions, different diseases, and so on, influence the harvest. The technique we are proposing to recognize a plant disease is picture preparation utilizing CNN. So, different types of diseases are shown in Figure 6.7. The authorshave utilized an open dataset of thousands of pictures of infected and healthy plant leaves gathered under controlled conditions; we train DCNNs to distinguish 14 yield species and 26 diseases (or nonappearance thereof).

6.3.1.2.4 Weed/Pest Detection

Weeddetection via CNN performs a useful role in the field of agriculture because CNNsare frequently useful for examining an optical image. There are so many techniques for detecting weeds/pests in agriculture. Weed detection utilizes picture handling procedures, which are used in an agricultural setting. Utilizing procedures like division, highlight extraction, and bunching can help to investigate pictures of the harvests. Picture preparation methods have been utilized over a considerable range of rural production settings. In agriculture, research into programmed discovery of leaf attributes is fundamental to identifying crop names, weeds, pests, diseases, and

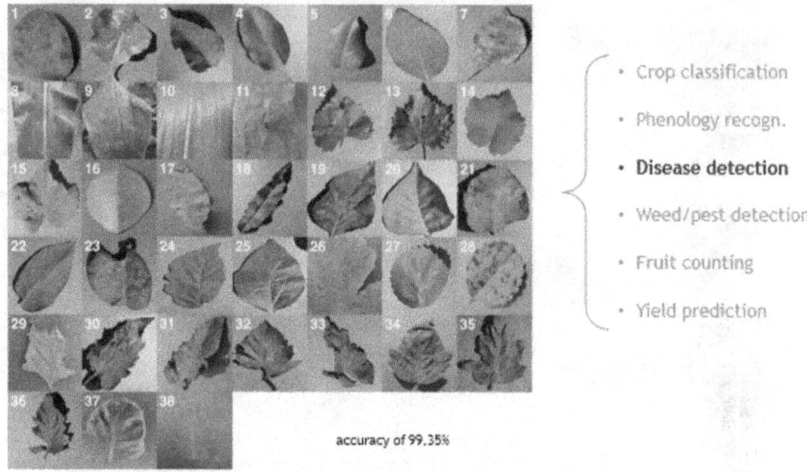

FIGURE 6.7 Disease detection in different leaves. (From Pavrithra et al., 2015.)

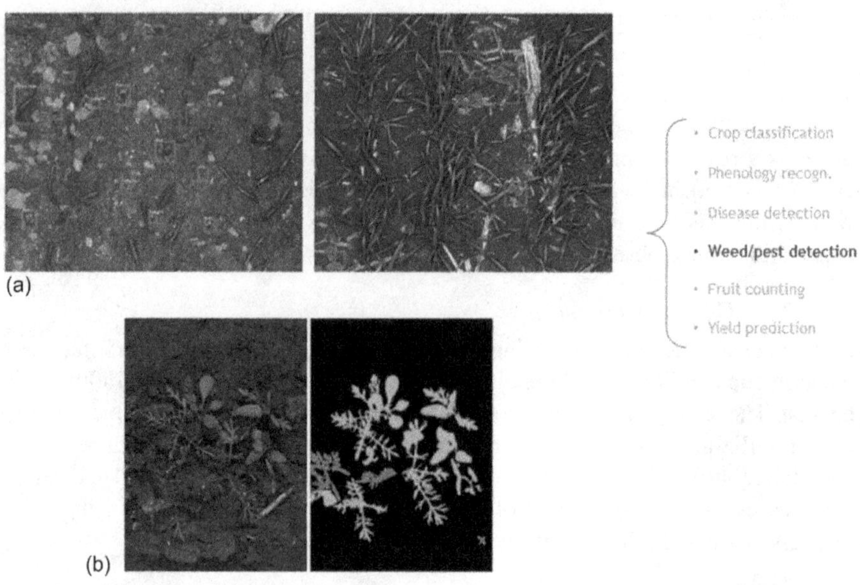

FIGURE 6.8 Weed/pest detection in different crops. (Panel (a) from Dyrmann et al., 2017; panel (b) from McCool et al., 2017.)

supplement deficiencies.Theprecision of directives varies, relying on the calculations of pictures and security of picture, as shown in Figure 6.8.

6.3.1.2.5 Fruit Counting/Yield Prediction

Optical methods to deal with robotized natural product tallying have been usedto assess yield with a demonstrably factual approach. CNN is a kind of feedforward imitation neural system that utilizes neurons; the classifier predicts the proximity of organic products. All these are different types of techniques that come under the subject area of plants. Even within CNNs, there are so many techniques that follow the concept of fruit counting or yield prediction – the technique followed by Li and Tam [39], as shown in Figure 6.9. In agriculture, the programmed strategies for tallying the quantity of organic products assume a basic job in cropping the executives. Tree shade mapping with a mechanized ultrasonic framework is cheap, genuinely direct, and could be utilized to gauge natural product yield inside a forest to design site-specific administration practices.

6.3.2 Convolutional Neural Networks

CNN is a neural network methoddesigned for handling2-Dorthree-dimensional (3-D) created information, for example, pictures and records. In the usual neural system (NN), entire information divisions are identified with all the yield units through a fullyconnected (FC) layer (Figure 6.10a). Regardless, in a CNN, every yield part is identified with just a subdivision of the whole information parts through a convolutional layer. These subsections are recognized as responsive areas, as shown in Figure 6.10b [40].

In this subsection, firstly, 2-D and 3-D CNNs are presented. After that, there follow a few sections which cover the convenient aspects of CNNs, such as dropout and bunch normalization.

6.3.2.1 2-D CNNs

2-DCNNs are individually used for the extraction of pieces for pictures. A CNN is ordinarily prepared out of convolutional layers, non-linear institutions.

6.3.2.1.1 Convolutional Layer

A convolutional layer utilizes channels to accomplish convolutions on the information. Every levelgains from the previous layer, and convolutional channels perceive various sorts of highlights at various layer depths in the system. In the fundamental

FIGURE 6.9 Fruit counting/yield prediction using CNN. (Panel (a) from Chen et al., 2017; panel (b) from Bargoti and Underwood, 2016.)

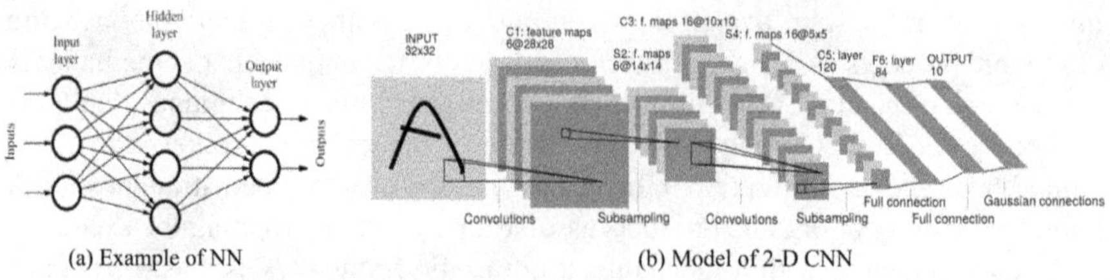

FIGURE 6.10 (a)Architectural instance of neural networks and convolutional neural networks. (b) Preparation of neural network from convolutional layers.

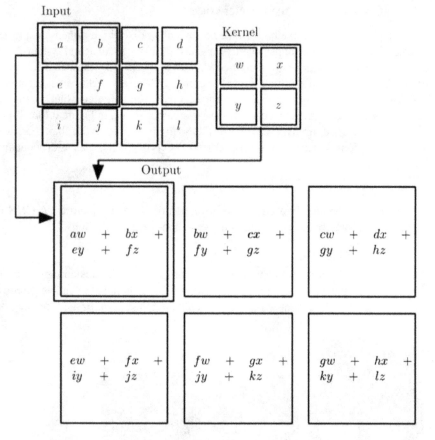

FIGURE 6.11 Effective convolution with stride 1 on an input of size 4 × 3, kernel of size 2 × 2.

layer, for instance, the channels perceive edges and tints; in the subsequent layer, channels perceive a mix of edges, for example, corners; and in the third layer, channels perceive a mix of corners, for example,shapes that are unclearlike circles or squares [41] (Figure 6.11).

6.3.2.1.2 *Non-Linear Activation*

This is applied afterwards to convolutional layers so as not to present linearity in the structure. The sustained non-straight enactment work is the Rectified Linear Unit (ReLU) [42]. ReLU channels incitation's with negative qualities to zero. Showing up diversely corresponding to sigmoid nonlinearity, ReLU has non-doused yields and it permits a grade to stream improved while arranging huge systems.

6.3.2.1.3 *Max-Pooling Layer*

A pooling layer is from time to time included after a ReLU. When performing convolutions with somewhat advance, the responsive fields spread and the data can get dull. Pooling layers are utilized to extract the utmost appropriate data and down sampled the three-dimensions. Also, pooling enables us to pass on interpretation-invariant portrayals. The most typically utilized pooling strategy is max-pooling. A case of max-pooling appears in Figure 6.12.

6.3.2.1.4 *Batch Normalization*

These layers are utilized to invigorate the improvement of, particularly critical CNNs. Social affair standardization lessens the covariate move of the hidden estimations of each convolutional or completely related layer. The standardization of individual beginning is finished by deducting the mean and separating with the change that is enlisted from each less assembling.

6.3.2.2 3-DCNNs

3-DCNNis a tremendous improvement on 2-D CNN. Figure 6.13 looks at 2-D and 3-D convolutions. The same applies to 3-D pooling.

6.3.2.3 Methods of Regularization

NNs are inclined to overfit. The twomost popular regularization approaches are dropout and early closure.

6.3.2.3.1 *Dropout*

It is planned as a productive strategy to standardize huge neural networks by presenting stochasticity for establishments of neurons. In particular, dropout erratically

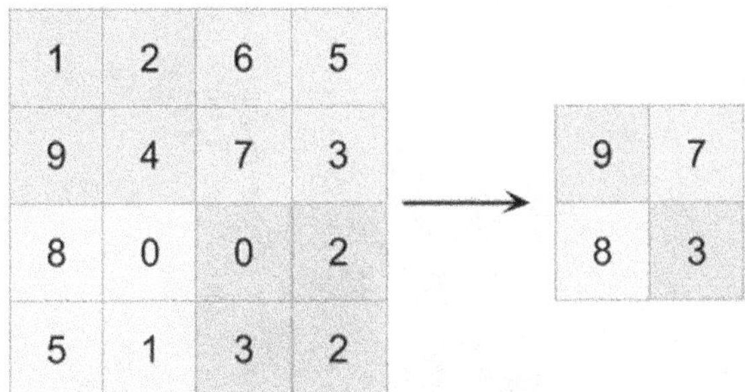

FIGURE 6.12 Max-pooling process with kernel and stride of size 2 × 2.

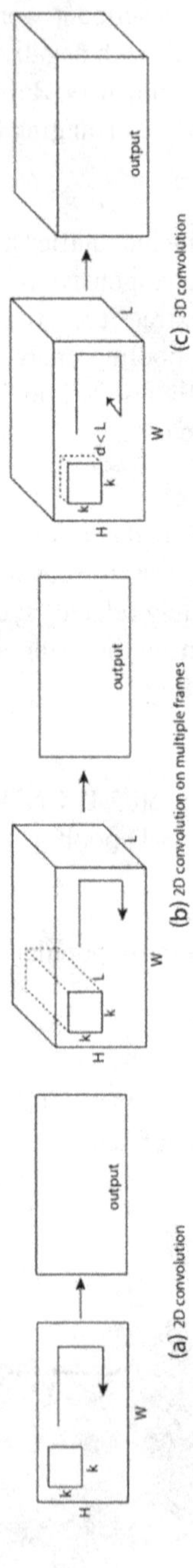

FIGURE 6.13 Comparisons of 2-D and 3-D convolutions. (a) 2-D convolutions applied on 2-D input,results in a 2-D output. (b) 2-D convolution on a3-D input, fallouts in 2-D output. (c) 3-D convolution on 3-D input, results in 3-D output.

beads the estimation of neurons through a specific ratio and sets the qualities to zero through system streamlining. As per necessities may be, the system can sum up improved and is least arranged for overfitting.

6.3.2.3.2 Early Closure

As clarified beforehand, an overfitting model executes fit on preparing information, yet insufficiently on unnoticeable information. During set up, the introduction of the structure on inconspicuous information is assessed with an underwriting set. An overfitting structure accomplishes improvement in the underwriting mess up while the course of action spoils dependably decreases. By watching the most immaterial underwriting bungle later every age, one can stop the game plan when the model execution on the support set hasn't upgraded for various ages[43].

6.4 CNN PERFORMANCE

In this subsection, we will discuss the performance of CNN and also compare it with other approaches.

6.4.1 Comparing CNN with Other Methods

In picture characterization, CNNs beat conventional picture handling techniques in a few applications. This general pattern is likewise seen in the programmed distinguishing proof of yield illnesses. Some of the chosen examinations contrasted CNNs with different strategies. In these examinations, the results of CNNs are superior to the other approaches. Figure 6.14 shows the finest outcomes of CNNs. The distinction in precision ran from 3% to 28.89% [44].

6.4.2 Generalized Productivity

To overview the generalized performance of a prototype, it must essentially be estimated on a dataset of pictures that it has never seen. Only 2 of the 19 assessments picked relied upon an unequivocally self-sufficient dataset to coordinate this appraisal. This can be explained by the fact that the test pictures were picked up close to the completion of the period – where the signs are commonly obvious – yet also by virtue of legitimate planning practices[45].

6.5 MATERIALS AND METHODS

6.5.1 Convolutional Neural Network Models

Artificial neural networks (ANNs) are scientific representations that imitate the generalized way in which the brain works, with its neurons and neurotransmitters interrelated among themselves. Their fundamental trademark is their capacity prepared by the procedure of ordered learning. For the duration of that procedure, NNs are "prepared" by showing a few frameworks with the utilization of existing information

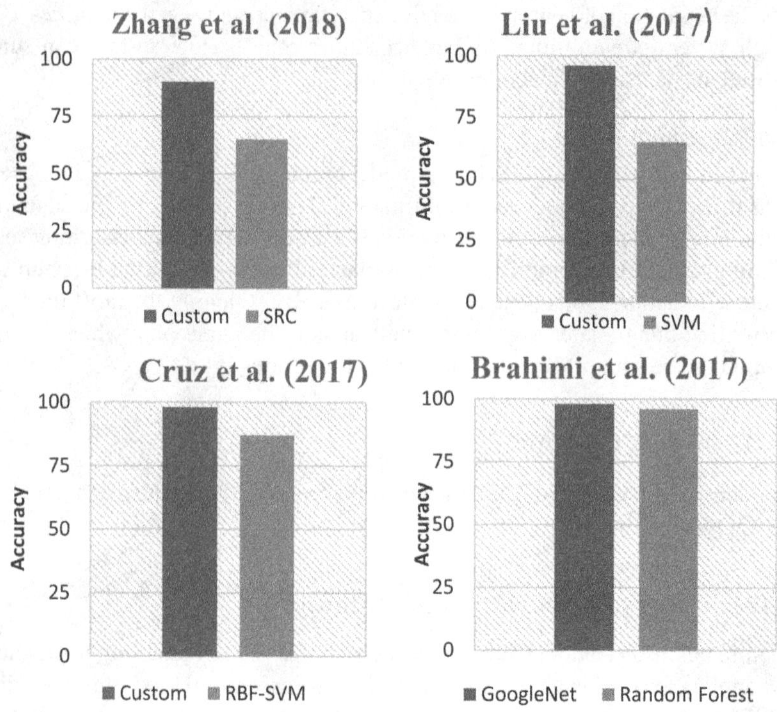

FIGURE 6.14 Performance of CNN.

that contains explicit matchings of data sources and the framework's yield. CNNs are an advancement of conventional artificial neural systems, concentrating for the most part on applications rehashing designs in various regions of the demonstrating space, particularly picture acknowledgment. Their fundamental trademark is that with the philosophy utilized in their layering, they decrease the necessary number of auxiliary components (number of artificial neurons) compared with conventional feedforward neural systems. For picture acknowledgment applications, a few CNNs have created pattern models, which have been effectively applied to entangled undertakings of visual symbolism. The five essential CNN designs resulting from this were investigated in this work regarding the recognizable proof of plant infections from pictures of the respective plants: AlexNet, AlexNet OWTBn, GoogLeNet, Over feat and VGG.

6.5.2 Datasets of Training and Testing

As discussed, a database of nearly 87,848 photos of solid leaves and plants is available for the analysis of CNN models. The modified database contains less pictures. The database that is utilized here comprises 58 distinct classes, each of which is depicted as a few plants and a differentiating infection, whereas some of the classes contain healthy plants. Table 6.1 presents data on the 58 classes, including some quantifiable information, for example, the number of accessible pictures per class, and the images taken for investigation and for plantation. These 58 classes incorporate 25 distinctive healthy or diseased plants. As shown in Table 6.1, one-third of the explicit pictures (37.3%) have caught certain improvements conditions in the field. The preliminary of an inconsistent class, containing two agent pictures to explore center conditions and two in favorable conditions. The all-inclusive flightiness of the last pictures is evident, with two or three focuses adding to it, for example, various leaves and different bits of the plants, irrelevant articles (e.g., shoes), indisputable ground surfaces, covering impacts [46–48].

A Python content made for the course of action of the two datasets, passing on dependably streamed pseudorandom numbers for the emotional choice of the photographs, like this the paces of "investigate center conditions" pictures and "genuine conditions" pictures for both datasets (arranging and testing) held like those introduced in Table 6.1.

Likewise, another way to deal with the movement of the preparation and testing information was, besides, thought of, by pre-treatment of the photographs. The option of utilizing grayscale sorts of the photographs for arranging not thought of as past works displayed that this framework doesn't improve the last strategy execution of huge learning models in provisional applications. The same holds for the division of leaves from the foundation of the photograph [49–51].

This further improvement, all the while, was additionally not thought of this. It holds since noteworthy learning frameworks like CNNs can see the key and non-huge highlights of a considerable amount of pictures, and somehow or another, overlook the last referenced. In this manner, the further improvement of the division of the objects of intrigue, which can also wind up being incredibly hazardous in complex pictures like the field-pictures in the present application, can maintain a decent critical path from it [52–54].

At long last, a third strategy to deal with the bundle of the database into arranging and testing datasets was finished, concentrating on investigating the vitality of the sort and spot of the catch of the leaves pictures, i.e., regardless of whether they have caught in lab conditions or authentic conditions in the progression field. In that capacity, from the 58 open classes of the structure [plant, disease], which contained photographs of the two sorts were chosen (the remainder of the classes exclusively comprise both of research facility conditions pictures or field conditions pictures, as appears in Table 6.2) [55, 56].

6.6 RESULTS AND DISCUSSION

Most applications for recognizable proof of crop disease are developed in uncontrolled conditions. Endeavors, accordingly, must be centered on shaping datasets,for example, what diseases have originated in the area, and ideally concerned with the instrument that procures information on the external situation,regardless of whether it is a cell phone, an automaton, a robot, or a tractor. When new information is procured, it ordinarily includes comments. This process is cumbersome; however, it is

TABLE 6.1
Statistics and Computable Data of Images Database

Class	Plant common name	Plant scientific name	Disease common name	Disease scientific name	Images (number)	Laboratory conditions (%)	Field conditions (%)
c_0	Apple	*Malusdomestica*	–	–	1835	89.7	10.3
c_1	Apple	*Malusdomestica*	Apple scab	*Venturia inaequalis*	630	100	0
c_2	Apple	*Malusdomestica*	Cedar apple rust	*Gymnosporangi umjuniperivirginianae*	276	100	0
c_3	Apple	*Malusdomestica*	Black rot	*Botryosphaeria obtuse*	712	87.2	12.8
c_4	Banana	*Musaparadisiaca*			1643	0	100
c_5	Banana	*Musaparadisiaca*	Black sigatoka	*Mycosphaerellafijensis*	240	0	100
c_6	Banana	*Musaparadisiaca*	Banana speckle	*Mycosphaerellamusac*	3284	0	100
c_7	Blueberry	*Vaccinium m* spp.	–	–	1735	86.7	13.3
c_8	Cabbage	*Brassica oleracea*	–	–	420	0	100
c_9	Cabbage	*Brassica oleracea*	Black rot	*Xanthomonas campestris*	64	0	100
c_10	Cantaloupe	*Cucumis melo*	–	-	1055	0	100
c_ll	Cassava (manioc)	*Manihot esculenta*	Brown leaf spot	*Cercosporidiumhenningsii*	43	100	0
c_12	Cassava (manioc)	*Manihot esculenta*	Cassava green spider mite	*Monony chellustanajoa* and *progresivus*	892	100	0
c_13	Celery	*Apiumgraveolens*	Early blight, Cercospora	*Cercosporaapii*	1204	0	100
c_14	Cherry (and sour)	*Prunus* spp.	–	–	854	100	0
c_15	Cherry (and sour)	*Prunus* spp.	Powdery mildew	*Podosphaera* spp.	1052	100	0
c_16	Corn (maize)	*Zea mays*	–		4450	26.1	73.9
c_17	Corn (maize)	*Zea mays*	Cercospora leaf spot	*Cercosporazeae-maydis*	1457	35.2	64.8
c_18	Corn (maize)	*Zea mays*	Common rust	*Puccinia sorghi*	1614	73.9	26.1
c_19	Corn (maize)	*Zea mays*	Northern leaf blight	*Exserohilumturcicum*	985	100	0
c_20	Cucumber	*Cucumis sativus*	–	–	267	0	100
c_21	Cucumber	*Cucumis sativus*	Downy mildew	*Pseudoperonosporacubensis*	1318	0	100
c_22	Eggplant	*Solanum melongena*			515	0	100
c_23	Gourd	*Langenaria spp.*	*Downy mildew*	*Pseudoperonosporacubensis*	114	0	100
c_24	Grape	*Vitis vinifera*	–	–	613	69	31
c_25	Grape	*Vitis vinifera*	Black rot	*Guignardiabidwellii*	1180	100	0
c_26	Grape	*Vitis vinifera*	Esca (Black measles)	*Phaeomoniellachlamydospora*	1384	100	0
c_27	Grape	*Vitis vinifera*	Leaf blight	*Pseudocercosporavitis*	1076	100	0
c_28	Onion	*Allium cepa*	–	–	154	0	100
c_29	Orange	*Citrus sinensis*	Huanglongbing	CandidatusLiberibacter	5507	100	0

(Continued)

TABLE 6.1 (CONTINUED)
Statistics and Computable Data of Images Database

Class	Plant common name	Plant scientific name	Disease common name	Disease scientific name	Images (number)	Laboratory conditions (%)	Field conditions (%)
c_30	Peach	*Prunus persica*	–	–	360	100	0
c_31	Peach	*Prunus persica*	Bacterial spot	*Xanthomonas campestris*	2297	100	0
c_32	Pepper, bell	*Capsicumannuum*			2029	72.8	27.2
c_33	Pepper, bell	*Capsicumannuum*	Bacterial spot	*Xanthomonas campestris*	997	100	0
c_34	Potato	*Solanum tuberosum*			152	100	0
c_35	Potato	*Solanum tuberosum*	Late blight	*Phytophthora infestans*	1000	100	0
c_36	Potato	*Solanum tuberosum*	Early blight	*Alternaria solani*	3167	31.6	68.4
c_37	Pumpkin	*Cucurbita* spp.	Cucumber mosaic	Cucumber mosaic virus (CMV)	2387	0	100
c_38	Raspberry	*Rubus* spp.	–	–	371	100	0
c_39	Soybean	*Glycine max*	–	–	6235	81.6	18.4
c_40	Soybean	*Glycine max*	Downy mildew	*Peronospora manshurica*	851	0	100
c_41	Soybean	*Glycine max*	Frogeyeleaf spot	*Cercosporasojina*	2023	0	100
c_42	Soybean	*Glycine max*	Septoria leafblight	*Septoria glycines*	3565	0	100
c_43	Squash	*Cucurbita*spp.		—	264	0	100
c_44	Squash	*Cucurbita*spp.	Powdery mildew	*Erysiphe cichoracearum, Sphaerothecafuliginea*	1835	100	0
c_45	Strawberry	*Fragaria* spp.	–	–	456	100	0
c_46	Strawberry	*Fragaria* spp.	Leafscorch	*Diplocarponcarlianum*	3396	29.7	70.3
c_47	Tomato	*Lycopersicum esculent um*			1592	100	0
c_48	Tomato	*Lycopersicum esculent um*	Bacterial spot	*Xanthomonas campestris*pv. Vesicatoria	2127	100	0
c_49	Tomato	*Lycopersicum esculent um*	Earlyblight	*Alternariasolani*	2579	38.8	61.2
c_50	Tomato	*Lycopersicum esculent um*	Late blight	*Phytophthora infestans*	1910	100	0
c_51	Tomato	*Lycopersicum esculent um*	Septoria leaf spot	*Septoria lycopersici*	1771	100	0
c_52	Tomato	*Lycopersicum esculent um*	Spider mites	*Tetranychusurticae*	1653	100	0
c_53	Tomato	*Lycopersicum esculent um*	Tomatomosaicvirus	Tomato mosaic virus (ToMV)	373	100	0
c_54	Tomato	*Lycopersicum esculent um*	Leaf mold	*Fulvia fulva*	952	100	0
c_55	Tomato	*Lycopersicumesculent um*	Target spot	*Corynesporacassiicola*	1404	100	0
c_56	Tomato	*Lycopersicum esculent um*	TYLCV	Begomovirus (Fam.Geminiviridae)	5357	100	0
c_57	Watermel on	*Citrulluslanatus*	–	–	172	0	100
TOTAL:					87.848	62.7	37.3

TABLE 6.2
Different Convolutional Neural Network Model Architectures: Performance for Identifying PlantDisease

From Model	Success rate (%)	Average error	Epoch	Time (s/epoch)	Success rate (%)	Average error	Epoch	Time (s/epoch)
AlexNet	99.06	0.0354	47	7034	98.64	0.0658	50	1022
AlexNetOWTBn	99.44	0.0192	46	7520	99.07	0.0332	45	1125
GoogLeNet	97.27	0.0957	45	7845	97.06	0.0984	40	2670
Overfeat	98.96	0.0412	45	6204	98.26	0.0848	49	1570
VGG	99.48	0.0223	48	7294	98.87	0.0542	49	4208

TABLE 6.3 (A)
Different Datasets and Crops

References	Culture	Training Dataset	Complexity	Number of Classes	Numberof Images	Min-Max Ni Samples/ C1
Atabay, 2017	Tomato	PlantVillage subset	*	10	19,742	373–5,357
Barbedo, 2018b	12 crop plants	Barbedo. 2016 (15% controlled, 85% in field)	*	56	1,383	5–77
Brahimi et al., 2017	Tomato	PlantVillage subset	*	9	14,82.8	325–4,032
Brahimi et al., 2018	14 crop species	PlantVillage	*	39	54,323	152–5,507
Cruzet al., 2017	Olive	Own dataset (controlled)	*	3	299	99–100
DeChant et al., 2017	Maize	Own dataset (field)	***	2	1,796	768–1,028
Ferentinos, 2018	25 crop species	PlantVillage dataset	*	58	87,848	43–6,235
Fuentes et al., 2017	Tomato	Own dataset (field)	***	10	5,000	338–18,899
Fuentes et al., 2018	Tomato	Own dataset (field)	***	12	8,927	338–18,899
Liu B. et al., 2017	Apple	Own dataset (Controlled and field)	*	4	1,053	2,366–4,147
Mohanty et al., 2016	14 crop species	PlantVillage	*	38	54,306	152–5,507
Oppenbeim and Shani, 2017	Potato	Own dataset (controlled)	*	5	400	265–738
Picon et al., 2018	Wheat	Johannes et al., 2017 extended (field)	**	4	8,178	1,116–3,338
Ramcharan et al., 2017	Cassava	Own dataset (field)	**	6	2,756	309–415
Sladojevic et al., 2016	Apple, Pear, Cherry, Peach, Grapevine	Own dataset (internet)	NA	15	4,483	108–1,235
Too et al., 2018	14 crop plants	PlantVillage	*	38	54,306	152–5,507
Wang et al., 2017	Apple	PlantVillage subset	*	4	2,086	145–1,644
Zhang S. et al., 2018	A/Tomato, B/ Cucumber	A/ PlantVillagesubset, B/ Own dataset (in field)	A/*B/**	A/8B/5	A/15,817B/500	A/366–5,350B/100
Zhang K. et al., 2018	Tomato	PlantVillage subset	*	9	5,550	405–814

TABLE 6.3 (B)
Different Datasets and Crops

References	Classification or detection[a]	Deep CNN architecture	Training strategy [	Best accuracy (%)	Evaluation quality[4]
Atabay, 2017	C	VGG 16,19, custom architecture	FS-TL	97.53	**
Barbedo, 2018b	C	GoogLeNet	TL	87 “	*
Brahimi et al., 2017	C	AlexNet, GoogLeNet	FS-TL	99.18	*
Brahimi et al., 2018	C	AlcxNet, DenseNetl69, Inception v3, ResNet34, SqueezeNetl-1.1, VGG 13	FS-TL	99.76	*
Cruz et al., 2017	C	LeNet	TL	98.6	**
DeChant et al., 2017	D	Custom three stages architecture	FS	96.7	**
Ferratinos, 2018	C	AlexNetAlexNetOWTBn, GoogLeNet, Overfeat, VGG	Unspecified	99.53	*
Fuentes et al., 2017	D	AlexNet, ZFNet, GoogLeNet, VGG16,ResNet 50, 101’	TL	85.98	**
Fuentes et al., 2018	D	Custom architecture with Refinement Filter Bank	TL	96.25	**
Liu B. et al.,2017	C	AlexNet, GoogLeNet, ResNet 20, VGG 16 and custom architecture	FS-TL	97.62	*
Mobanty et al., 2016	C	AlcxNet, GoogLeNet	FS-TL	31	**
Oppenbeim and Shani, 2017	C	VGG	Unspecified	96	*
Picon et al., 2018	C	Custom ResNetSO, Resnet50	TL	97	***
Ramcbaran et al., 2017	C	Inception V3	TL	93	**
Sladojevic et al.,2016	C	CaffeNet	TL	96.3	*
Too et al., 2018	C	Inception V4, VGG 16, RcsNet 50, 101 and 152, DenscNet	TL	99.75	**
Wang et al., 2017	c	VGG 16,19, Inception-V3, ResNet50	TL	90.4	*
Zhang S. et al., 2018	C	Custom Three Channels CNN, DNN. LeNet-5, GoogLeNet	FS	A/ 87.15 B/ 91.16	A/* B/*
Zhang K. et al., 2018	c	AlexNet, GoogLeNet, ResNet	TL	97.28	*

conceivable to improve it with dynamic learning. Dynamic learning is an iterative strategy intended to discover and clarify the most educational examples. The idea of driving this strategy is that clarifying genuine models can prompt comparative or surprisingly better exactness than commenting on all models,with lower cost and in less time. A model should initially be prepared from a small subset of commented-on models (Table 6.3a and b).

6.7 CONCLUSION

In this chapter, we recognized some of the critical issues and weaknesses of works that utilized CNNs to distinguish crop infections. We likewise gave rules and methods to follow to boost the capability of CNNs for certifiable applications. Some effectively distributed arrangements dependent on CNNs are not yet operational for field use, for the most part because of an absence of adjustment to a few significant AI ideas. This absence of similarity may cause poor speculation abilities for new information tests and imaging conditions, which reduces the usefulness of the prepared models. The works studied also show the potential for deep learning strategies for crop disease identification. Their discoveries are promising for the improvement of new agrarian devices that could aidincreasingly practical and secure food production.

REFERENCES

1. A. Sembiring, A. Budiman, and Y. D. Lestari, "Design and control of agricultural robot for tomato plants treatment and harvesting," *J. Phys. Conf. Ser.*, vol. 930, no. 1, p. 012019, 2017.
2. L. Wang et al., "Development of a tomato harvesting robot used in greenhouse," *Int. J. Agricult. Biol. Eng.*, vol. 10, no. 4, pp. 140–149, 2017.
3. W. M. Syahrir, A. Suryanti, and C. Connsynn, "Color grading in tomato maturity estimator using image processing technique," in Proceedings of the IEEE International Conference on Computer Science and Information Technology (ICCSIT), Aug. 2009, pp. 276–280.
4. O. O. Arjenaki, P. A. Moghaddam, and A. M. Motlagh, "Online tomato sorting based on shape, maturity, size, and surface defects using machine vision," *Turkish J. Agricult. Forestry*, vol. 37, no. 1, pp. 62–68, 2013.
5. Z. Ma, J.-H. Xue, A. Leijon, Z.-H. Tan, Z. Yang, and J. Guo, "Decorrelation of neutral vector variables: Theory and applications," *IEEE Trans. Neural Netw. Learn. Syst.*, vol. 29, no. 1, pp. 129–143, Jan. 2018.
6. P. M. Pieczywek, J. Cybulska, A. Zdunek, and A. Kurenda, "Exponentially smoothed Fujii index for online imaging of biospeckle spatial activity," *Comput. Electron. Agricult.*, vol. 142, pp. 70–78, Nov. 2017.
7. K. Jarrett, K. Kavukcuoglu, M. Ranzato, and Y. LeCun, "What is the best multi-stage architecture for object recognition?" in Proceedings of the IEEE International Conference on Computer Vision, Sep. 2009, pp. 2146–2153.
8. N. Goel and P. Sehgal, "Fuzzy classification of pre-harvest tomatoes for ripeness estimation: An approach based on automatic rule learning using decision tree," *Appl. Soft Comput. J.*, vol. 36, pp. 45–56, Nov. 2015.

9. V. Pavithra, R. Pounroja, and B. S. Bama, "Machine vision based automatic sorting of cherry tomatoes," in Proceedings of the 2nd International Conference on Electronics Communication Systems (ICECS), Feb. 2015, pp. 271–275.
10. H. Lu, F. Wang, X. Liu, and Y. Wu, "Rapid assessment of tomato ripeness using visible/near-infrared spectroscopy and machine vision," *Food Anal. Methods*, vol. 10, no. 6, pp. 1721–1726, 2017.
11. A. Mohapatra, S. Shanmugasundaram, and R. Malmathanraj, "Grading of ripening stages of red banana using dielectric properties changes and image processing approach," *Comput. Electron. Agricult.*, vol. 143, no. 382, pp. 100–110, 2017.
12. T. Zhou, S. Yang, L. Wang, J. Yao, and G. Gui, "Improved cross-label suppression dictionary learning for face recognition," *IEEE Access*, vol. 6, pp. 48716–48725, 2018.
13. J. Chen, Z. Liu, H. Wang, A. Núñez, and Z. Han, "Automatic defect detection of fasteners on the catenary support device using deep convolutional neural network," *IEEE Trans. Instrum. Meas.*, vol. 67, no. 2, pp. 257–269, Feb. 2018.
14. J. Li, J. Feng, and C.-C. J. Kuo, "Deep convolutional neural network for latent fingerprint enhancement," *Signal Process. Image Commun.*, vol. 60, pp. 52–63, Feb. 2018.
15. Y. Li, X. Cheng, and G. Gui, "Co-robust-ADMM-net: Joint ADMM framework and DNN for robust sparse composite regularization," *IEEE Access*, vol. 6, pp. 47943–47952, 2018.
16. Z. Ma, Y. Lai, W. B. Kleijn, Y.-Z. Song, L. Wang, and J. Guo, "Variational Bayesian learning for Dirichlet process mixture of inverted Dirichlet distributions in non-Gaussian image feature modeling," *IEEE Trans. Neural Netw. Learn. Syst*, vol. 30, no. 2, pp 449–463.
17. Y. Sun, X. Wang, and X. Tang, "Deep learning face representation from predicting 10,000 classes," in Proceedings of the CVPR, Jun. 2014, pp. 1891–1898.
18. H. Huang, J. Yang, Y. Song, H. Huang, and G. Gui, "Deep learning for super-resolution channel estimation and DOA estimation based massive MIMO system," *IEEE Trans. Veh. Technol.*, vol. 67, no. 9, pp. 8549–8560, Sep. 2018.
19. G. Gui, H. Huang, Y. Song, and H. Sari, "Deep learning for an effective nonorthogonal multiple access scheme," *IEEE Trans. Veh. Technol.*, vol. 67, no. 9, pp. 8440–8450, Sep. 2018.
20. Y. Li, J. Zhang, Z. Ma, and Y. Zhang, "Clustering analysis in the wireless propagation channel with a variational Gaussian mixture model," *IEEE Trans. Big Data*, vol. 6, no. 2, pp. 223–232.
21. H. Yu, Z.-H. Tan, Z. Ma, R. Martin, and J. Guo, "Spoofing detection in automatic speaker verification systems using DNN classifiers and dynamic acoustic features," *IEEE Trans. Neural Netw. Learn. Syst.*, vol. 29, no. 10, pp. 4633–4644, Oct. 2018.
22. X. Sun, G. Gui, Y. Li, R. P. Liu, and Y. An, "ResInNet: A novel deep neural network with feature re-use for Internet of Things," *IEEE Internet Things J.*, vol. 6, no. 1, pp. 679–691.
23. M. Brahimi, K. Boukhalfa, and A. Moussaoui, "Deep learning for tomato diseases: Classification and symptoms visualization," *Appl. Artif. Intell.*, vol. 31, no. 4, pp. 299–315, 2017.
24. A. Fuentes, S. Yoon, S. C. Kim, and D. S. Park, "A robust deep-learningbased detector for real-time tomato plant diseases and pests recognition," *Sensors*, vol. 17, no. 9, p. 2022, 2017.
25. S. Bargoti and J. Underwood, "Deep fruit detection in orchards," in Proceeding of the IEEE International Conference on Robotics and Automation, May 2017, pp. 3626–3633.
26. S. Bargoti and J. P. Underwood, "Image segmentation for fruit detection and yield estimation in Apple orchards," *J. Field Robot.*, vol. 34, no. 6, pp. 1039–1060, 2017.

27. S.-H. Wang, Y.-D. Lv, Y. Sui, S. Liu, S.-J. Wang, and Y.-D. Zhang, "Alcoholism detection by data augmentation and convolutional neural network with stochastic pooling," *J. Med. Syst.*, vol. 42, no. 1, p. 2, 2018.
28. J. Muñoz-Bulnes, C. Fernández, I. Parra, D. Fernández-Llorca, and M. A. Sotelo, "Deep fully convolutional networks with random data augmentation for enhanced generalization in road detection," in Proceedings of the IEEE 20th International Conference on Intelligent Transport System, Oct. 2017, pp. 366–371.
29. S. Hussein, R. Gillies, K. Cao, Q. Song, and U. Bagci, "TumorNet: Lung nodule characterization using multi-view convolutional neural network with Gaussian process," in Proceedings of the International Symposium on Biomedical Imaging, Apr. 2017, pp. 1007–1010.
30. Z. Ma, A. E. Teschendorff, A. Leijon, Y. Qiao, H. Zhang, and J. Guo, "Variational Bayesian matrix factorization for bounded support data," *IEEE Trans. Pattern Anal. Mach. Intell.*, vol. 37, no. 4, pp. 876–889, Apr. 2015.
31. L. van der Maaten and G. Hinton, "Visualizing high-dimensional data using t-SNE," *J. Mach. Learn. Res.*, vol. 9, pp. 2579–2605, Nov. 2008.
32. M. A. Hearst, S. T. Dumais, E. Osuna, J. Platt, and B. Schölkopf, "Support vector machines," *IEEE Intell. Syst. Appl.*, vol. 13, no. 4, pp. 18–28, Jul. 1998.
33. C. Tomasi, "Histograms of oriented gradients," *Comput. Vis. Sampler*, vol. 6, pp. 1–6, 2012.
34. T. Lindeberg, "Scale invariant feature transform," *Scholarpedia*, vol. 7, no. 5, p. 10491, May 2012.
35. V. Nair and G. E. Hinton, "Rectified linear units improve restricted Boltzmann machines," in Proceedings of the 27th International Conference on Machine Learning, vol. 3, 2010, pp. 807–814.
36. R. Salakhutdinov and G. Hinton, "Replicated softmax: An undirected topic model," in Proceedings of the Advanced Neural Information Processing Systems, 2009, pp. 1607–1614.
37. X. Li et al., "Supervised latent Dirichlet allocation with a mixture of sparse softmax," *Neurocomputing*, vol. 312, pp. 324–335, May 2018.
38. P.-T. de Boer, D. P. Kroese, S. Mannor, and R. Y. Rubinstein, "A tutorial on the cross-entropy method," *Ann. Oper. Res.*, vol. 134, no. 1, pp. 19–67, 2005.
39. C. H. Li and P. K. S. Tam, "An iterative algorithm for minimum cross entropy thresholding," *Pattern Recognit. Lett.*, vol. 19, no. 8, pp. 771–776, 1998.
40. Dutot, M.L., Nelson, M., and Tyson, R.C., "Predicting the spread of postharvest disease in stored fruit, with application to apples," *Postharvest Biol. Technol.*, vol. 85, pp. 45–56, 2013. CrossRef
41. Zhao, P., Liu, G., and Li, M.Z., "Management information system for apple diseases and insect pests based on GIS," *Trans. Chin. Soc. Agric. Eng.*, vol. 22, pp. 150–154, 2006.
42. Es-Saady, Y., Massi, I.E., Yassa, M.E., Mammass, D., and Benazoun, A., "Automatic recognition of plant leaves diseases based on serial combination of two SVM classifiers," in Proceedings of the 2nd International Conference on Electrical and Information Technologies, Tangiers, Morocco, 4–7 May 2016, pp. 561–566.
43. Padol, P.B., and Yadav, A.A., "SVM classifier based grape leaf disease detection," in Proceedings of the 2016 Advances in Signal Processing, Pune, India, 9–11 June 2016, pp. 175–179.
44. Sannakki, S.S., Rajpurohit, V.S., Nargund, V.B., Kumar, A., and Yallur, P.S., Diagnosis and classification of grape leaf diseases using neural networks," in Proceedings of the 4th International Conference on Computing, Tiruchengode, India, 4–6 July 2013, pp. 1–5.

45. Qin, F., Liu, D.X., Sun, B.D., Ruan, L., Ma, Z., and Wang, H.Identification of alfalfa leaf diseases using image recognition technology. *PLoS ONE*, vol. 11, p. e0168274, 2016. CrossRefPubMed
46. Rothe, P.R. and Kshirsagar, R.V., "Cotton leaf disease identification using pattern recognition techniques," in Proceedings of the 2015 International Conference on Pervasive Computing, Pune, India, 8–10 January 2015, pp. 1–6.
47. Islam, M., Dinh, A., Wahid, K., and Bhowmik, P., "Detection of potato diseases using image segmentation and multiclass support vector machine," in Proceedings of the 30th IEEE Canadian Conference on Electrical and Computer Engineering, Windsor, ON, Canada, 30 April–3 May 2017, pp. 1–4. Symmetry 2018, 10, 11 16 of 16
48. Gupta, T.Plant leaf disease analysis using image processing technique with modified SVM-CS classifier. *Int. J. Eng. Manag. Technol.*, vol. 5, pp. 11–17, 2017.
49. Dhakate, M. and Ingole, A.B., "Diagnosis of pomegranate plant diseases using neural network," in Proceedings of the 5th National Conference on Computer Vision, Pattern Recognition, Image Processing and Graphics, Patna, India, 16–19 December 2015, pp. 1–4.
50. Gavhale, M.K.R. and Gawande, U., "An overview of the research on plant leaves disease detection using image processing techniques," *J. Comput. Eng.*, vol. 16, pp. 10–16, 2014.
51. Gaba, S., Aggarwal, A., and Nagpal, S., "Role of machine learning for ad hoc networks," Cloud and IoT Based Vehicular Ad-Hoc Networks, 269, 2021.
52. Aggarwal, A., Gaba, S., Nagpal, S., and Vig, B., "Bio-inspired routing in VANET," in Cloud and IoT Based Vehicular Ad-Hoc Networks, 199, 2021.
53. AggarwalA., GabaS., and MittalM., "A comparative investigation of consensus algorithms in collaboration with IoT and blockchain," in AgrawalR. and GuptaN. (eds) *Transforming Cybersecurity Solutions using Blockchain. Blockchain Technologies.* Springer, Singapore, 2021. https://doi.org/10.1007/978-981-33-6858-3_7
54. Aggarwal, A., Dhindsa, K. S., and Suri, P. K., "Performance-aware approach for software risk management using random forest algorithm," *Int. J. Software Innovation*, vol. 9, no. 1, pp. 12–19, 2021.
55. Aggarwal, A., Dhindsa, K. S., and Suri, P. K., "Design for software risk management using soft computing and simulated biological approach," *Int. J. Secur. Privacy Pervasive Comput.*, vol. 12, no. 2, pp. 44–54, 2020.
56. Gaba,S., Singla, S., and KumarD, "A genetic improved quantum cryptography model to optimize network communication," *Special Issue*, vol. 8, no. 9S, pp. 256–259, 2019. doi:10.35940/ijitee.i1040.0789s19

7 IoT and Wearable Sensors for Health Monitoring

Radhika G. Deshmukh, Akanksha Pinjarkar, and Arun Kumar Rana

CONTENTS

7.1 INTRODUCTION

Our way of living has altered dramatically in modern times. The advancement of modern technology has made our lives easier, and our average longevity has increased. A person's health is an important aspect of their life. Due to growing development, natural landscapes and ecosystems are deteriorating, resulting in pollution. These events also have an impact on human health in numerous nations around the world [1–5]. A decent and clean atmosphere is necessary for a healthy physique. The healthcare industry consumes/uses a significant number of chemicals globally [6]. Chemical processing has the potential to damage the environment indirectly.

DOI: 10.1201/9781003221333-7

The environment is determined to be responsible for one-quarter of all health disorders [7]. The use of toxic by-products of medication, infectious waste products, radioactive materials, and water waste in solid form in modern healthcare systems harms our environment and ecology. The by-products of these are released into the environment, causing further harm to human health [8]. These can have an impact on environmental factors, including pollution levels, temperature, drinking water, humidity, and so on [9–16]. Recently, pandemics such as Covid-19 have made things more difficult for everyone, particularly in the medical area. It is observed that there have been several patients in society who were infected with COVID-19 during their stay in hospital. Some hospitals have undergone complete transformations to become Covid-19 institutions [17–25]. As a result, many typical activities at diagnostic centres have been halted. People from rural areas are unable to access doctors or hospitals, because the majority of hospitals and diagnostic centres are located in urban areas. They are largely neglected and afflicted with ailments. Similarly, elderly folk require more precautions regarding health and proper treatment than others due to their age. However, for some people, travelling from their residence to hospital is an unpleasant memory. They must also be monitored at all times, whether in the hospital or at home. These pandemic conditions have halted practically all doctor and patient activities in the hospital [26–30]. Micro- and nanotechnology have advanced dramatically in recent decades, and micro-/nanosensors and electronics can now be found in virtually every instrument. The size of these micro-/nanosensors is small, and they use less electricity [31]. We can employ electronic health (e-health) technologies to overcome all of these challenges in healthcare. Smartphones are now carried by the majority of people. Patients can get information about their health status by using a health app. Advanced smartphones can serve to connect doctors and patients in this way. Patients in remote areas will use wearable microsensors and devices to capture data in real time. If any problems are discovered in the data, it will be transferred to patients and clinicians via Wi-Fi modules as soon as possible. After that, doctors will then analyse these data sets and give reports to patients via their mobile device. Patients do not need to attend the hospital to evaluate their health; hence, wearable sensors and e-healthcare systems cut energy consumption in hospitals. In this chapter, we'll look at the future of nanosensors and wearable sensors, as well as how technologies such as artificial intelligence, machine learning, and the Internet of Things (IoT) will help with patient treatment in the future. These IoT-enabled sensors/devices can help to minimise pollution and create a more sustainable environment.

7.2 COVID-19: IMPORTANCE OF WEARABLE SENSING TECHNOLOGY

In the current circumstances, the spread of the Covid-19 pandemic has had a significant influence on the world. To lessen its impact, the only possible solution is to slow the disease's spread and contain it. The best strategy to limit and manage the disease is to keep an eye on the potentially infected patient (PIP) in quarantine by enforcing the predetermined location for the duration of the quarantine. This is only possible with the support of IoT technology, such as remote collection, monitoring, management, and analysis of disease symptoms. The accelerometer sensor, electrocardiography (ECG) data, temperature sensor, and other sensors operate together as a network to monitor the patient's health status, with a local server acting as a node to collect data and inform the clinician. The e-sensor platform is being utilised as a wearable sensor to monitor the patient's health in both static and dynamic modes. Simply put, it's a patient monitor that uses edge, fog, and cloud levels. As a result, this wearable technology is extremely useful in monitoring a Covid-19-infected patient. Wearable sensors could be used as Covid-19 viral indicators [32]. Heart rate, respiration rate, motion activity, body temperature, oxygen saturation level, cough symptoms, and stress are all measured using various sensors. Sensors transfer physiological data to the cloud, where it is processed and analysed to anticipate the Covid-19 viral scenario among individuals. By incorporating artificial intelligence technology into traditional healthcare, Covid-19 has accelerated the creation of health and fitness apps. Covid-19 has raised personal hygiene awareness, and wearable gadgets are assisting users in infection prevention. For example, Apple recently updated the Cardiogram app, which analyses users' heart rates, with a new sleeping beats per minute feature for users with Covid-19 to monitor heart rate changes. Another example is a German smartwatch app that tracks Covid-19's distribution. The Robert Koch Institute has teamed with a to introduce Corona – App, the Corona Datenspende, which collects vital signs like pulse, temperature, and sleep length from users and analyses whether they are Covid-19 symptomatic. Smartwatch manufacturers like Apple, Samsung, and Fitbit, as well as newcomers like Huami and Oppo, are emphasising their devices' capabilities to consumers by adding functions like heart rate tracking and accelerometry to make them more holistic health monitors. Other possible healthcare-related wearable tech use cases are currently being developed. Omron's Heart Guide smartwatch, for example, includes a complete oscillometric blood pressure test that measures systolic and diastolic pressures in the same way as medical hospitals do. The latest Apple smartwatch can detect blood oxygen saturation (SpO_2) and deliver ECG readings. However, because healthcare is a highly regulated business that is sluggish to adopt new technologies, convincing healthcare practitioners to use these devices will be difficult. The majority of doctors and medical professionals doubt the functionality and utility of wearable technology [33–39].

There are questions about the accuracy of the data they collect as well as the lack of a sufficient medical record system. Other obstacles, such as data privacy restrictions, could stymie the adoption of wearable tech devices like smartwatches in mainstream healthcare. Physicians are frequently unsure what to do with the data provided by wearables and are hesitant to make judgments based on data that has yet to be shown to be helpful. In order for the wearable tech market to remain relevant and expand in the healthcare industry, it must meet certain criteria. Companies that make wearable technology devices must publish high-quality research papers and undertake controlled trials to document and demonstrate the devices' accuracy across populations. Healthcare providers and people with health and physical limitations must have easy access to wearable tech devices. The percentage of wearable health gadgets has become critical during the Covid-19 pandemic. Covid-19 pre-symptomatic patients can be detected using heart rate, daily steps, and sleep time. The researchers examined the data (from Fitbits and Apple watches) of 32 afflicted people, who were chosen from a group of over 5,000 people, and looked for correlations to discover abnormal physiology. They focused on raised resting heart rates and increased heart rates in relation to the number of steps taken. The researchers also developed an online detection method that uses real-time heart rate monitoring to detect early stages of infection. One of the most difficult aspects of stopping SARS-CoV-2 transmission is the ability to promptly identify, detect, and isolate cases before they disseminate the virus to others. According to the researchers, "As communities around the country begin to reopen businesses, schools, and other activities, many rely on current COVID-19 screening protocols, which typically involve a combination of symptom and travel-related survey questions, as well as temperature measurements." They found that a high temperature isn't as prevalent as previously thought, and that only roughly a third of patients tested positive for Covid-19 during their hospitalisation. The Covid-19 pandemic has highlighted the importance of leveraging and harnessing our digital infrastructure for remote patient monitoring [40–42].

7.3 SENSORS AND TYPES OF SENSORS

In this section, we will discuss the wearable sensors and devices that are employed in healthcare monitoring systems. The materials utilised in sensor fabrication and various types of devices will be discussed. Sensors are created based on the therapeutic application in question. Wearable sensing technology has quickly evolved from a science fiction concept to a wide range of well-established consumer and medical devices. A sensor is a device that detects changes in the environment and reacts to some other system's output.

A sensor turns a physical phenomenon into a digital signal, which is then displayed or sent for reading or further processing. Light, temperature, motion, and pressure are all examples of possible inputs. Sensors produce useful data, which they can exchange with other connected devices and management systems if they are connected to a network. Sensors are vital to the success of many modern organisations. They can alert you to possible issues before they turn into major issues, allowing firms to do preventative maintenance and avoid costly downtime. Sensors output valuable information, and if they are connected to a network, they can share data with other connected devices and management systems. Sensors come in many shapes and sizes. Some are purpose-built, containing many built-in individual sensors, allowing you to monitor and measure many sources of data. Wearable sensors are shown in Figure 7.1.

Wearable sensors are used to collect physiological and mobility data, allowing continuous monitoring of the patient's condition. Almost any analyte that a clinician could want to assess from a patient can now be measured with diagnostic tools. Unfortunately, such devices are not wearable, and blood draws and traditional benchtop analysis techniques are still required. As a result, the central question in many people's thoughts is: how might wearable sensor technology begin to bridge the gap

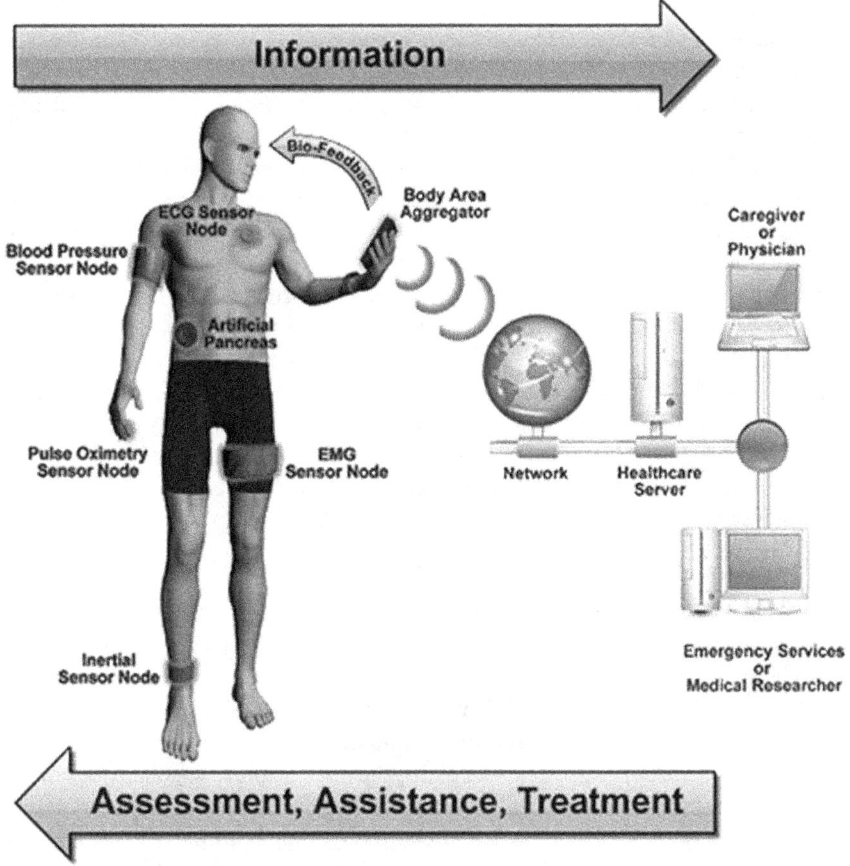

FIGURE 7.1 Wearable sensors.

into modalities that record more detailed physiological events? Wearable technology relies heavily on sensors. Wearables are pointless without sensors. Consumers want monitoring systems that provide particular information. Sensor data is collected and processed for the intended user. Body-wearable sensor networks are now being used for remote health and activity monitoring thanks to wearable embedded system technologies. This provides for a more tailored approach to health and wellness. These networks improve people's quality of life and lifestyle, and could also save the lives of people who are at risk of cardiac failure.

7.3.1 Types of Sensors Used in Wearable Technology

7.3.1.1 Accelerometer

It is a device that measures how fast something is moving. Proper acceleration refers to a body's acceleration (rate of change of velocity) in its own instantaneous rest frame, as opposed to coordinate acceleration, which refers to acceleration in a fixed coordinate system. In wearables, accelerometers are employed as sensors. Their detecting abilities are demonstrated by their choice of acceleration, such as gravity and linear. Meanwhile, their ability to measure allows the programming of measured data for many applications. For example, a runner can see his or her speed output and acceleration. Accelerometers can also monitor sleep patterns. They offer a number of applications in industry. Inertial navigation systems for aircraft and missiles use highly sensitive accelerometers. An accelerometer is a type of electrical sensor that controls the acceleration forces acting on it and determines its position to track its movement. The rate of change of an object's velocity, which is a vector quantity, is called acceleration. Static and dynamic acceleration forces are the two forms of acceleration forces. Frictional or gravity forces constantly applied to an item are static. Dynamic forces are "moving" forces that are applied to an object at different speeds. This is why, for example, accelerometers are employed in automotive collision safety systems. When a car is hit by a strong dynamic force, the accelerometer (which detects rapid deceleration) sends an electronic signal to an embedded computer, which activates the airbags. Low power consumption and a cheap price are combined to produce good results [43–46]. Accelerometer sensors are shown in Figure 7.2.

7.3.1.2 Gyroscopes

A gyroscope sensor is a device that can measure and keep track of an object's orientation and angular velocity. Accelerometers are less advanced than these sensors, which can detect the object's tilt and lateral orientation, whereas an accelerometer can only detect linear motion. Gyroscopes are another popular type of wearable sensor. They are distinct from accelerometers in that they only capture angular accelerations. The accelerometer is employed to detect rotational acceleration in some implementations, whereas some systems would prefer to combine both for filtering mistakes. Gyroscopes improve data tracking precision and come in a variety of kinds, including gas bearing, mechanical, and optical. A gyroscope sensor is shown in Figure 7.3.

7.3.1.3 Magnetometers

A magnetometer is a device that measures magnetic dipole moment or magnetic field. Magnetometers measure the direction, strength, or change in the strength of a

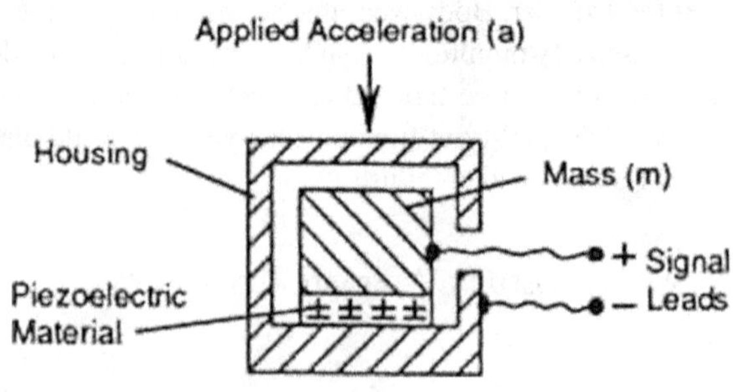

FIGURE 7.2 Accelerometer.

FIGURE 7.3 Gyroscope sensor.

magnetic field at a specific area. For inertial measurement unit (IMU), they can be used with accelerometers and gyroscopes. All these sensors have three axes, work similarly to a compass, and can help with balance. Magnetometers match them by filtering the motion orientation, but gyroscopes and accelerometers are commonly employed along with them. A magnetometer sensor is shown in Figure 7.4.

7.3.1.4 Global Positioning System (GPS)

Many products, such as smartphones and smart watches, employ GPS as a sensor. It is used to scan users and tell them of their current position. To quantify the exact position and time, data is transferred to a satellite.

7.3.1.5 Heart Rate Sensors

Photoplethysmography (PPG) is the method used by most wearables with heart rate monitors nowadays to measure heart rate. The term PPG refers to shining light into the skin and measuring the amount of light dispersed by blood flow. PPG sensors work on the principle that light entering the body scatters in a predictable manner as blood flow dynamics change, such as with variations in blood pulse rates (heart rate) or blood volume (cardiac output). Heart rate can be measured using a variety of techniques and sensors. The electrode (sensor) and the skin are idealised as two portions of a standard capacitor in one method, which uses capacitive sensing. PPG is a method of measuring blood flow volume variations using light. Fitbit and other fitness trackers use a photodiode to achieve this. A constant green light is supplied to the wearer's skin, which measures the photodiode's light absorption. This data is passed on so that the pulse can be calculated. The more blood that flows through

FIGURE 7.4 Magnetometer sensor.

the user's veins, the more light the diodes absorb. A heart rate sensor is shown in Figure 7.5.

To monitor heart rate, PPG sensors rely on four main technical components:

1. Optical emitter – a device that sends light waves into the skin and is usually made up of at least two light-emitting diodes (LEDs). Because customers have such a wide range of skin tone, thickness, and morphology, optical heart rate monitors (OHRM) employ different wavelengths of light from these optical emitters that interact differently with different thicknesses of skin and tissue.
2. Digital Signal Processor (DSP) – the DSP takes the light refracted by the device's user and converts it into ones and zeros that may be used to generate relevant heart rate data.
3. Accelerometer – the accelerometer measures motion and is utilised as an input to PPG algorithms together with the DSP signal.
4. Algorithms – The algorithms convert the DSP and accelerometer information into motion-tolerant heart rate data, as well as other biometrics like calories burnt, R-R interval, heart rate variability, blood metabolite concentrations, blood oxygen levels, and even blood pressure. The operation of wearable sensors is shown in Figure 7.6.

7.3.1.6 Pedometers

Pedometers (shown in Figure 7.7), which can count the user's steps while jogging or walking, are generally seen in wearables focusing on physical health. Pedometers come in two types: electrical and mechanical. The former is the most common

FIGURE 7.5 Heart rate sensor.

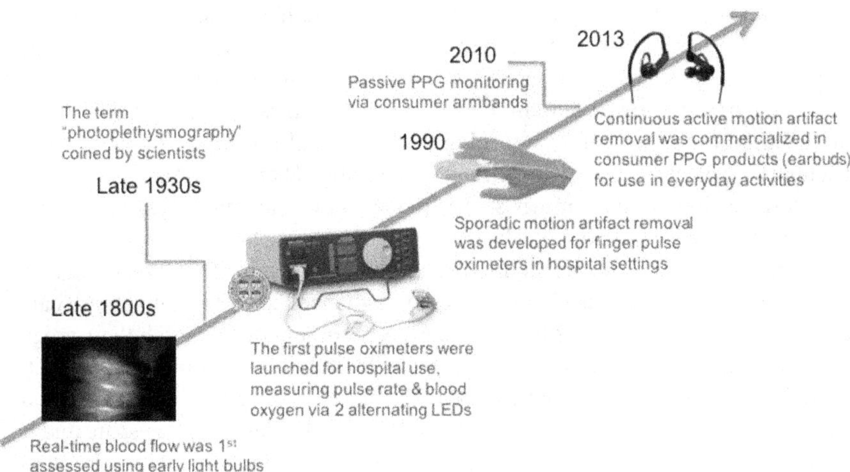

FIGURE 7.6 Operation of wearable sensors.

today, and it is based on mechanical pedometer principles but uses micro-electro-mechanical systems (MEMS) technology for efficiency. A pedometer is a gadget that counts each step a person takes by detecting motion of the user's hands or hips. It is usually portable and electronic or electromechanical. Because every person's steps vary in length, an informal calibration by the user is required if the distance covered

FIGURE 7.7 Pedometer.

is to be displayed in a unit of length, though there are now pedometers that use electronics and software to automatically determine how a person's step varies. A GPS receiver can immediately measure the distance travelled. The pedometer user's steps are measured using the pendulum function. Two-ended pedometers, one with a screw, use a tiny metal pendulum. Every time a user takes a step, the hammer swings and smacks the other before returning to its original position. The device is connected to an electronic counting circuit by a spring. There is no current at the start; therefore, each time the hammer strikes the other side, an open circuit is closed. As a result, current begins to flow. The circuit closes once the pendulum returns to its original position, and the pendulum revolution begins again. This enables the circuit to recognise each step. People who desire to enhance their physical activity can use pedometers as a motivator. Many websites exist to help people track their progress; however, many people will find that recording their daily step count and heart rate on a calendar is extremely motivating. Clinical studies have demonstrated that wearing a pedometer increases physical activity while also lowering blood pressure and body mass index. One complaint about the pedometer is that it does not track intensity; however, this can be remedied by setting time limits on step objectives (for example, 1,000 steps in 10 minutes is considered moderate exercise).

7.3.1.7 Pressure Sensors

Strain gauges are commonly used to power pressure sensors. The circuit changes resistance as pressure is applied to the sensors. Mechanical qualities like force are measured using a variety of methods before being converted into resistance-dependent electrical measurements. The building of a Wheatstone Bridge, which can track static or dynamic resistance changes, is used to create this way of sensing pressure. In the Wheatstone Bridge design, the sensing device will have one, two, or four arms with variation as to how the items are used.

7.3.1.8 Integration of Sensors into Wearables (Microcontroller)

A microcontroller is a crucial component in the operation of wearable technology. It's usually thought of as a little computer (the chip system) that allows the IoT to be integrated with the required application. Most importantly, it reduces the need for a large number of electronic components to perform multiple operations on a single chip. It is best employed in wearable technology due to its ease of programming, reprogramming, cost, size, connectivity with various sensors, and capacity to handle sophisticated operations, including graphic displays. Microcontrollers can be customised to meet the needs of customers because of their adaptability.

7.4 INTERNET OF THINGS

The phrase "Internet of Things" was coined by Kevin Ashton in 1999; hence, it is still relatively new. It has only recently gained popularity (since 2010) [47]. Sensor sensing, data storage, communication, power, computing, and protocols are all part of the IoT. Applications include smart cities, remote monitoring, agriculture, and healthcare, to name a few. IoT operations are improved by advances in information technology combined with industrial skills. This chapter focuses on wearable sensors for IoT-based healthcare monitoring. Until today, most countries' healthcare systems relied on paper records kept by doctors or clinics. In third-world countries, data exchange in the healthcare area is extremely rare. In the healthcare area, on the other hand, IoT has a lot of promise. IoTs combine sensing, storage, computation, and communication to sense and respond to physical systems in real time. These systems also link nodes to cloud or fog computing, necessitating a new IoT chip architecture. To put it another way, in the IoT era, huge central processing units aren't always the best choice for fog and edge devices. Because of their low cost, IoT devices are being employed in a variety of industries, including healthcare, smart transportation, environmental monitoring, and so on. Low power consumption and machine learning execution at the device, as well as security and safety, are becoming increasingly problematic as the number of devices grows rapidly. The following are some of the most often used terminologies, despite the fact that technical phrases have yet to be standardised:

- Edge devices that provide communication and sensing are known as nodes.
- Devices that connect one or more nodes are known as hubs and gateways.
- Fog refers to nodes that exist between edge devices and the cloud.
- Data is stored on the cloud, and distant computations are performed there.

Traditional sensor systems on chip (SoC) designs employ large chipsets, whereas IoT device designs consume less power and have a smaller footprint. System-in-package designs must use both traditional manufacturing and innovative technologies to deliver a mix of low power consumption, computation, sensing, and connectivity. The IoT-based healthcare system is made up of several components. Sensors that can read data and deliver it to a machine will be required first, followed by wearable sensors. The data will be processed by machines, which will provide the results. IoT-enabled healthcare systems require a lot of communication, networking, and security. There are three communication protocols for IoT-based healthcare systems [48]: people-to-people, machine-to-person, and machine-to-machine are the three types of connections.

7.4.1 Network of the IoT

It is necessary to convey the sensing data to the system for further processing. Because of the variety of physical sensors, there are different communication techniques accessible, notably in healthcare systems. Device-to-device, device-to-cloud, and other applications use these communication technologies [49]. Healthcare, smart cities, and agriculture are just a few of the areas that employ IoT. As a result, the amount of traffic in IoT-based communication and networking systems is growing. As a result, on an IoT-based system, numerous network needs have developed, including factors like identifying individual products and their connections, object location, security and individuality with complete privacy, dependability, autonomous networking, and so on.

7.4.2 IoT-Based Wearable Healthcare System

Web technologies, sensor and power supply controllers, system components' proposed applications, security model, modelling exercise, implementation details, experimental details, and implementation details are all part of an IoT-based system's framework. There are two types of sensors that can be used in healthcare IoT. Wearable flexible sensors are one type, while wearable sensors and devices based on traditional MEMS are another. MEMS-based sensors and devices can easily be integrated into a circuit board despite their lack of flexibility. As a result, these devices can have their own microcontrollers and coin-type batteries to power the controller circuit and send data to smartphones through Zigbee or Bluetooth. Pragmatic recently built a flexible microcontroller circuit or random access memory on a flexible substrate. They can customise the design to their customers' needs and applications. In IoT-based applications, 32-bit microcontrollers are extensively used. A huge number of communication and network connections are required by every IoT system. System components are included in the IoT. The human body is detected by sensors, which transform it into digital data. Following that, the digital data will be sent using various network protocols and gateways. During data transmission from the patient to the doctor and from the doctor to the patient, a serious security-enabled procedure is required. For each person and device, a password should be set, and the Transport Layer Security protocol should be utilised for mobile and cloud connection. Smart homes, smarter healthcare, intelligent transportation, intelligent buildings, and smart cities with diverse capacities, as well as a wide range of products, are all examples of edge IoT devices. As the majority of the computing, storage, and networking resources of these power data centres come from application-service providers who work directly with web servers on a small number of dispersed larger data centres, it is common in today's cloud and application facilities for such large numbers of edge devices to be used. With sensing devices, a similar technique can be employed [50–52].

7.5 FUTURE PERSPECTIVE

Wearable electronic devices such as smartwatches currently provide data such as heart rate, sleep time, and activity patterns. This could be supplemented in the future with new classes of wearable devices that monitor, for example, cortisol concentrations for tracking stress (using electronic epidermal tattoos), biomarkers of inflammation and levels of blood O_2 (microneedle patches), skin temperature (electronic textiles), blood pressure (smart rings), ion concentration (wristbands), intraocular pressure (smart contacts), and skin temperature (electronic textiles). Emerging low-cost wearable sensing technologies could be utilised to identify symptomatic and pre-symptomatic cases in future pandemics by monitoring both physical characteristics and biochemical indicators. The gadgets could potentially be used to track the progress of people in treatment or who are self-isolating at home.

7.6 CONCLUSION

Wearable sensors and their various types have been explored in this chapter for continuous health monitoring systems based on the IoT. In the current circumstances, the spread of the Covid-19 pandemic has had a significant influence on the world. To lessen its impact, the only possible solution is to slow the disease's spread and contain it. The best strategy to limit and manage the disease is to keep an eye on the PIP in quarantine by enforcing the predetermined location for the duration of the quarantine. This is only possible with the support of IoT technology such as remote collection, monitoring, management, and analysis of disease symptoms. Our goal is to use IoT-based healthcare systems to slow the spread of the Covid 19 epidemic. Another issue in IoT-based healthcare systems is security. In the future, having an individual lab on the body with IoT for day-to-day healthcare monitoring systems may be achievable. These wearable healthcare devices based on the IoT will help to reduce pollution in the environment and create a more sustainable world for future generations.

REFERENCES

1. Kumar, A., Sharma, S., Goyal, N., Singh, A., Cheng, X., Singh, P., 2021. Secure and energy-efficient smart building architecture with emerging technology IoT. *Comput Commun.* 2021;176:207–217.
2. Kumar, A., Sharma, S., 2021. IoT with energy sector-challenges and development. In *Electrical and Electronic Devices, Circuits and Materials* (pp. 183–196). CRC Press.
3. Lalit, G., Emeka, C., Nasser, N., Chinmay, C., Garg, G., 2020. Anonymity preserving IoT-based COVID-19 and other infectious disease contact tracing model. *IEEE Access.* 2020;8:159402–159414

4. Kumar, K., Gupta, E.S., Rana, E.A.K., 2019. Wireless sensor networks: A review on "Challenges and opportunities for the future world-LTE. *Amity Journal of Computational Sciences (AJCS)* 2019;1(2). ISSN: 2456-6616 (Online).
5. Rana, A.K., Salau, A., Gupta, S., Arora, S., 2018. A survey of machine learning methods for IoT and their future applications. *Amity Journal of Computational Sciences*, Amity University, 2018;2(2):1–5. ffhal-01983429f
6. Rana, A.K., Sharma, S., 2020. Industry 4.0 manufacturing based on IoT, cloud computing, and big data: Manufacturing purpose scenario. In *Advances in Communication and Computational Technology* (pp. 1109–1119). Springer.
7. Kumar, A., Sharma, S., 2021. IFTTT rely based a semantic web approach to simplifying trigger-action programming for end-user application with IoT applications. In *Semantic IoT: Theory and Applications: Interoperability, Provenance and Beyond* (p. 385). Springer Nature.
8. Kumar, A., Salau, A.O., Gupta, S., Paliwal, K., 2019. Recent trends in IoT and its requisition with IoT built engineering: A review. In *Advances in Signal Processing and Communication* (pp. 15–25). Springer.
9. Kumar, A., Sharma, S., 2021. Demur and routing protocols with application in underwater wireless sensor networks for smart city. In *Energy-Efficient Underwater Wireless Communications and Networking* (pp. 262–278). IGI Global.
10. Gabriel, K., Jarvis, J., Trimmer, W., 1988. *Small Machines, Large Opportunities: A Report on the Emerging Field of Micro Dynamics: Report of the Workshop on Microelectromechanical Systems Research*. AT & T Bell Laboratories.
11. Kaltenbrunner, M., Sekitani, T., Reeder, J., et al., 2013. An ultra-lightweight design for imperceptible plastic electronics. *Nature*. 2013;499:458–463. doi: 10.1038/nature12314. [PubMed] [CrossRef] [Google Scholar].
12. Sekitani, T., Someya, T., 2010. Stretchable, large-area organic electronics. *Adv Mater.* 2010;22:2228–2246. doi: 10.1002/adma.200904054. [PubMed] [CrossRef] [Google Scholar].
13. Rana, A.K., Sharma, S., 2019. *Enhanced Energy-Efficient Heterogeneous Routing Protocols in WSNs for IoT Application*. IJEAT.
14. Rim, Y.S., Bae, S.H., Chen, H., De Marco, N., Yang, Y., 2016. Recent progress in materials and devices toward printable and flexible sensors. *Adv Mater.* 2016;28:4415. doi: 10.1002/adma.201505118. [PubMed] [CrossRef] [Google Scholar].
15. Zhu, B., Wang, H., Liu, Y., Qi, D., Liu, Z., Wang, H., Yu, J., Sherburne, M., Wang, Z., Chen, X., 2016. Skin-inspired haptic memory arrays with an electrically reconfigurable architecture. *Adv Mater.* 2016;28:1559. doi: 10.1002/adma.201504754. [PubMed] [CrossRef] [Google Scholar].
16. Kanao, K., Arie, T., Akita, S., Takei, K., 2016. An all-solution-processed tactile memory flexible device integrated with a NiOReRAM. *J Mater Chem C.* 2016;4:9261. doi: 10.1039/C6TC03321K. [CrossRef] [Google Scholar].
17. Casula, G., Cosseddu, P., Bonfiglio, A., 2015. Pressure-triggered memory: integration of an organic resistive memory with a pressure-sensitive element on a fully flexible substrate. *Adv Electron Mater.* 2015;1:1500234. doi: 10.1002/aelm.201500234. [CrossRef] [Google Scholar].
18. Rana, A.K., Krishna, R., Dhwan, S., Sharma, S., Gupta, R., 2019, October. Review on artificial intelligence with internet of things – problems, challenges and opportunities. In 2019 2nd International Conference on Power Energy, Environment and Intelligent Control (PEEIC) (pp. 383–387). IEEE.
19. Bariya, M., Nyein, H.Y.Y., Javey, A., 2018. Wearable sweat sensors. *Nat Electron.* 2018;1:160–171. doi: 10.1038/s41928-018-0043-y. [CrossRef] [Google Scholar].

20. Rana, A.K., Sharma, S., 2021. ContikiCooja security solution (CCSS) with IPv6 routing protocol for low-power and lossy networks (RPL) in internet of things applications. In *Mobile Radio Communications and 5G Networks* (pp. 251–259). Springer.
21. Yang, Y., Gao, W., 2019. Wearable and flexible electronics for continuous molecular monitoring. *ChemSoc Rev.* 2019;48:1465–1491. doi: 10.1039/C7CS00730B. [PubMed] [CrossRef] [Google Scholar].
22. Gaoetal, W., 2016. Fully integrated wearable sensor arrays for multiplexed in situ perspiration analysis. *Nature.* 2016;529:509–514. doi: 10.1038/nature16521. [PMC free article] [PubMed] [CrossRef] [Google Scholar].
23. Rana, A.K., Sharma, S., 2021. The fusion of blockchain and IoT technologies with industry 4.0. In *Intelligent Communication and Automation Systems* (pp. 275–290). CRC Press.
24. Emaminejad, S., et al., 2017. Autonomous sweat extraction and analysis applied to cystic fibrosis and glucose monitoring using a fully integrated wearable platform. *Proc Natl Acad Sci.* 2017;114:4625–4630. doi: 10.1073/pnas.1701740114. [PMC free article] [PubMed] [CrossRef] [Google Scholar].
25. Green, A.A., Dodds, P., Pennock, C., 1985. A study of sweat sodium and chloride; criteria for the diagnosis of cystic fibrosis. *Ann ClinBiochem.* 1985;22:71–176. doi: 10.1177/000456328502200212. [PubMed] [CrossRef] [Google Scholar].
26. Takei, K., Gao, W., Wang, C., Javey, A., 2019. Physical and chemical sensing with electronic skin. *Proc IEEE.* 2019;107:2155–2167. doi: 10.1109/JPROC.2019.2907317. [CrossRef] [Google Scholar].
27. Tai, L.-C., et al., 2018. Methylxanthine drug monitoring with wearable sweat sensors. *Adv Mater.* 2018;30:1707442. doi: 10.1002/adma.201707442. [PubMed] [CrossRef] [Google Scholar].
28. Chun, K.Y., Oh, Y., Rho, J., Ahn, J.H., Kim, Y.J., Choi, H.R., 2010. Highly conductive, printable and stretchable composite films of carbon nanotubes and silver. *Nat Nanotechnol.* 2010;5:853–857. doi: 10.1038/nnano.2010.232. [PubMed] [CrossRef] [Google Scholar].
29. Xu, F., Zhu, Y., 2012. Highly conductive and stretchable silver nanowire conductors. *Adv Mater.* 2012;24(37):5117–5122. doi: 10.1002/adma.201201886. [PubMed] [CrossRef] [Google Scholar].
30. Yao, S., Zhu, Y., 2015. Nanomaterial-enabled stretchable conductors: strategies, materials and devices. *Adv Mater.* 2015;27:1480–1511. doi: 10.1002/adma.201404446. [PubMed] [CrossRef] [Google Scholar].
31. Sekitani, T., Noguchi, Y., Hata, K., Fukushima, T., Aida, T., Someya, T., 2008. A rubberlike stretchable active matrix using elastic conductors. *Science.* 2008;321:1468–1472. doi: 10.1126/science.1160309. [PubMed] [CrossRef] [Google Scholar].
32. Fan, F.-R., Tian, Z.-Q., Wang, Z.L., 2012. Flexibletriboelectric generator. *Nano Energy.* 2012;1:328. doi: 10.1016/j.nanoen.2012.01.004. [CrossRef] [Google Scholar].
33. Lin, Z., Chen, J., Li, X., Zhou, Z., Meng, K., Wei, W., Yang, J., Wang, Z.L., 2017. Triboelectric nanogenerator enabled body sensor network for self-powered human heart-rate monitoring. *ACS Nano.* 2017;11:8830. doi: 10.1021/acsnano.7b02975. [PubMed] [CrossRef] [Google Scholar].
34. Park, D.Y., et al., 2017. Self-powered real-time arterial pulse monitoring using ultrathin epidermal piezoelectric sensors. *Adv Mater.* 2017;29:1702308. doi: 10.1002/adma.201702308. [PubMed] [CrossRef] [Google Scholar].
35. Park, S., et al., 2018. Self-powered ultra-flexible electronics via nano-grating-patterned organic photovoltaics. *Nature.* 2018;561:516–521. doi: 10.1038/s41586-018-0536-x. [PubMed] [CrossRef] [Google Scholar].

36. Yun, J., et al., 2018. Stretchable array of high-performance micro-supercapacitors charged with solar cells for wireless powering of an integrated strain sensor. *Nano Energy.* 2018;49:644.
37. Wang, R., Mu, L., Bao, Y., Lin, H., Ji, T., Shi, Y., Zhu, J., Wu, W., 2020. Holistically engineered polymer–polymer and polymer–ion interactions in biocompatible polyvinyl alcohol blends for high-performance triboelectric devices in self-powered wearable cardiovascular monitorings. *Adv Mater.* doi: 10.1002/adma.202002878. Accessed 28 June 2020 [PubMed].
38. Noh, S., Yoon, C., Hun, E., Yoon, H.N., Chung, T.J., Par, K.S., Kim, H.C., 2014. Ferroelectret film-based patch-type sensor for continuous blood pressure monitoring. *Electron Lett.* 2014;50:143–144. doi: 10.1049/el.2013.3715. [CrossRef] [Google Scholar].
39. Luo, N., Dai, W., Li, C., Zhou, Z., Lu, L., Poon, X.C.Y., Chen, S., Zhang, Y., Zhao, N., 2016. Flexible piezoresistive sensor patch enabling ultralow power cuffless blood pressure measurement. *Adv Funct Mater* 2016;26:1178–1187.
40. Wang, C., et al., 2018. Monitoring of the central blood pressure waveform via a conformal ultrasonic device. *Nat Biomed Eng.* 2018;2:687–695. doi: 10.1038/s41551-018-0287-x. [PMC free article] [PubMed] [CrossRef] [Google Scholar].
41. Huang, X., et al., 2018. Stretchable, wireless sensors and functional substrates for epidermal characterization of sweat. *Small.* 2014;10:3083–3090 [PubMed].
42. Kim, J., et al., 2018. Simultaneous monitoring of sweat and interstitial fluid using a single wearable biosensor platform. *Adv Sci.* 2018;5:1800880. doi: 10.1002/advs.201800880. [PMC free article] [PubMed] [CrossRef] [Google Scholar].
43. Alizadeh, A., et al., 2018. A wearable patch for continuous monitoring of sweat electrolytes during exertion. *Lab Chip.* 2018;18:2632–2641. doi: 10.1039/C8LC00510A. [PubMed] [CrossRef] [Google Scholar].
44. Anastasova, S., Crewther, B., Bembnowicz, P., Curto, V., Ip, H.M., Rosa, B., Yang, G.Z., 2017. A wearable multisensing patch for continuous sweat monitoring. *BiosensBioelectron.* 2017;93:139–145 [PubMed].
45. Oh, S.Y., et al., 2018. Skin-attachable, stretchable electrochemical sweat sensor for glucose and pH detection. *ACS Appl Mater Interfaces.* 2018;10:13729–13740. doi: 10.1021/acsami.8b03342. [PubMed] [CrossRef] [Google Scholar].
46. Cho, E., Mohammadifar, M., Choi, S., 2017. A single-use, self-powered, paper-based sensor patch for detection of exercise-induced hypoglycemia. *Micromachines.* 2017;8:265. doi: 10.3390/mi8090265. [PMC free article] [PubMed] [CrossRef] [Google Scholar].
47. Lee, S., Son, I., Choi, J., Nam, D., Hong, Y., Lee, W., 2011. Estimated blood pressure algorithm for a wrist-wearable pulsimeter using hall device. *J Korean Phys Soc.* 2011;58:349–352. doi: 10.3938/jkps.58.349. [CrossRef] [Google Scholar].
48. Hsu, Y., Young, D.J., 2013. Skin-surface-coupled personal health monitoring system. In Proceedings of the 2013 IEEE Sensors, Baltimore, MD, 4–6 Nov 2013, pp 1–4.
49. Liu, X., Lillehoj, P.B., 2016. Embroidered electrochemical sensors for biomolecular detection. *Lab Chip.* 2016;16:2093–2098. doi: 10.1039/C6LC00307A. [PubMed] [CrossRef] [Google Scholar].
50. Tokuda, T., Ishizu, T., Nattakarn, W., Haruta, M., Noda, T., Sasagawa, K., Sawan, M., Ohta, J., 2018. 1 mm3-sized optical neural stimulator based on CMOS integrated photovoltaic power receiver. *AIP Adv.* 2018;8:4. [Google Scholar].
51. Dhawan, S., Gupta, R., Rana, A., Sharma, S., 2021. Various swarm optimization algorithms: Review, challenges, and opportunities. *Soft Computing for Intelligent Systems*, pp.291–301.
52. Kumar, A., Sharma, S., n.d. Participation of 5G with wireless sensor networks in the internet of things (IoT) application. In *Wireless Sensor Networks and the Internet of Things* (pp. 229–244). Apple Academic Press.

8 Analysis of Interpolation-Based Image In-Painting Approaches

Mustafa Zor, Erkan Bostanci,
Mehmet Serdar Güzel and Erinç Karataş

CONTENTS

8.1 INTRODUCTION

When analogue cameras were widely used, the photographs we kept in print were at risk of aging, fading, wear, and thus, loss of information. With the advent of digital cameras, the development of computer storage, and even cloud storage, our habit of storing photographs in print has evolved accordingly. However, this has not eliminated the risk of information loss from our photos. Errors or loss of information have been observed in digital photographs during the acquisition and transmission of the photograph. Some of these may be related to the direct quality of the photograph,

DOI: 10.1201/9781003221333-8

such as blur and noise, and in some cases, loss of information in certain areas of the picture. In addition, there may be errors such as the unintentional incorporation of undesirable objects into the photo frame during the photo shoot.

Some methods have been developed due to requirements such as eliminating the lack of information in photographs or eliminating unwanted areas. Bertalmio et al. [1] developed in-painting terminology. In-painting is the art of modifying a picture or video in a way that cannot be easily detected by an ordinary observer and has become a major research area in image processing [2]. The missing or undesirable portion of the image is completed by using intensity levels on adjacent pixels. However, the image in-painting will not be able to restore the original form of the missing part in the picture; it will only fill the missing or undesirable parts closely to the original [3].

The methods developed for in-painting can be grouped under three main headings:

- **Texture synthesis:** The basis of this approach is the self-similarity principle. It is based on the assumption that similar structures in a picture are often repeated [2].
- **Exemplar-based approach:** The missing region is filled with information from the known region at the patch level.
- **Partial differential equations and variation-based diffusion techniques:** The method of partial differential equations fills the regions to be uniformly propagated along the isophot directions from neighbouring regions along the direction of isophot. The method, which gives good results for small areas, causes turbidity in larger areas. Variation methods address the problem in the form of finding the extremes of energy functions. However, these models only aim at dealing with non-textural in-painting. The difficulty of real in-painting problems is due to the rapid changes of the isophot and the roughness of the image functions [4].

In addition to this, methods have been developed that consider interiors as an interpolation problem and apply different interpolation techniques to complete the missing areas in the images.

In this study, we aimed to evaluate the interpolation methods to handle the image in-painting process as an interpolation problem, as the main contribution. State-of-the-art methods were tested on a generated dataset that was subject to both noise and corruption. These methods were then assessed based on their Structural SIMilarity (SSIM), peak signal-to-noise ratio and mean square error (MSE) in a quantitative evaluation. A qualitative evaluation was also performed on the results to ensure that the interpolation approach yielded visually appealing images.

The rest of the chapter is structured as follows. Section 8.2 presents the current literature and background on the interpolation approaches employed in the chapter. This section is followed by Section 8.3, where the dataset and the evaluation approach are elaborated. Section 8.4 presents both quantitative and qualitative results, and finally, the paper is concluded in Section 8.5.

8.2 LITERATURE REVIEW AND BACKGROUND

8.2.1 Cubic Interpolation

This interpolation is one of the common methods used to estimate unknown points. It generates results with smoother transitions and lower error rate than other interpolation techniques with polynomials. It is a commonly used method for filling lost pixels. When an interpolation of surface values is desired in a two-dimensional field, it can be formulated as follows.

On a unit square, if the surface values at the corners and the partial derivative values at these points are known, the convergence values of the points within the square can be obtained by the polynomial Equation (8.1):

$$p(x,y) = \sum_{i=0}^{3}\sum_{j=0}^{3} a_{ij}x^{i}y^{j} \tag{8.1}$$

The known corner values of the surface and the partial derivative values (f_x, f_y, f_z) in the x and y direction can be obtained from the corner points. The polynomial is obtained by replacing the required equations with unknown coefficients for a_{ij}, which can be solved using matrix format to express the polynomial exactly.

8.2.2 Kriging Interpolation

Kriging is a geostatistics interpolation method that takes into account the distance and degree of variation between known points when estimating values at unknown points [5]. This approach uses the values of the entire sample to calculate an unknown value. Kriging assumes that the distance or direction between sample points reflects a spatial correlation that can be used to explain variation on the surface. In Kriging interpolation, a mathematical function is applied to all points of a specified number or a specified radius to determine the estimation value of each unknown point. The closer the point, the higher the value of the weights. Kriging is the most appropriate approach when there is a spatially related distance or directional trend in the data (spatial autocorrelation) and calculated as follows:

$$\hat{P}^{*} = \sum_{i=1}^{N} \lambda_i P_i \tag{8.2}$$

where $N{:}k \times k$ (sample) is the total number of intact pixels, $\hat{P}^{*}$:the pixel to be interpolated, $P_i{:}k \times k$ the intact pixels in the block, $\lambda_i{:}k \times k$ the weights of the intact pixels in the block($\sum_{i=1}^{N} \lambda_i = 1$). Kriging interpolation chooses values of λ_i in order to minimize the interpolation variance($\sigma^2 = E\left[\left(P - \hat{P}^{*}\right)^2\right]$).

Two steps are required for Kriging interpolation:

1. Dependency rules: Constructing variograms and covariance functions to predict statistical dependence (spatial autocorrelation) based on the autocorrelation model (fitting a model).

   The Variogram is a function of distance and direction that separates the two positions used to measure dependence. It is defined as the variance of the difference between two variables at two different points and is calculated as follows:

$$2\gamma(h) = \frac{1}{n}\sum_{i=1}^{N}\left[P(x_i) - P(x_i + h)\right]^2; P(x_i), P(x_i + h); x_i, x_i + h \tag{8.3}$$

2. Estimation: Estimating unknown values.

The data is used twice to perform these two steps.

Jassim [5] creates damaged images by using 4 different masks on 10 different greyscale images and uses Kriging interpolation to remove them. Sapkal and Kadbe[6] also do in-painting using Kriging interpolation. In this study, the results are obtained by using the same masks on five different greyscale images.

Sapkal et al.[7] used Kriging interpolation to remove several types of masks. In the study, the same dataset was used with a different set of masks.

Awati et al. [8]conducted a study on troubleshooting colour images using modified Kriging interpolation. They separate colour images into red, green, blue (RGB) components and apply separate interpolation to each component. In practice, 3×3 matrices are used. The masks they use consist of vertical, horizontal and curved lines. They make separate trials for one-,two-,three-, and four-pixel thicknesses as line thicknesses in each mask and evaluate the results.

8.2.3 Radial Basis Functions

Chang and Chongxiu [4]define a mapping between the coordinates and colours of the image pixels, and implement an algorithm based on radial-based functions to generate the best approximation of this mapping in a given neighbourhood. Radial-based functions are a means of approximating a multivariable function as a linear combination of univariate functions. It is one of the methods that provides good results for the interpolation of scattered data. In order to increase the accuracy of their solutions and reduce the complexity of the algorithm, researchers created the pixel-by-pixel zoom function. For larger loss areas, interpolation of different overlapping coefficients is used.

Wang and Qin [9] propose an algorithm for image in-painting based on compactly supported radial basis functions (CSRBF). The algorithm transforms the two-dimensional (2-D) image in-painting problem from a three-dimensional point set into a surface reconstruction problem. First, a covered surface is constructed for approximation to the set of dots obtained from the damaged image using radial-based functions. The values of the lost pixels are then calculated using this surface.

8.2.4 High-Dimensional Model Representation and Lagrange Interpolation

Karaca and Tunga[10], who consider image in-painting as an interpolation problem, have designed this problem using the high-dimensional model representation (HDMR) method and Lagrange interpolation, which allows a multivariate function to be expressed as the sum of multiple functions with fewer variables.

Normally, greyscale images are represented as functions with two variables $f(x, y)$, where x: number of rows, y: number of columns. Similarly, coloured images are represented by a function of three variables: $f(x, y, z)$. In order to apply HDMR, the copy of the image itself is added as an additional dimension. In other words, a greyscale image, $f(x, y, n)$, $n = 1,2$; and the colour image is expressed as $f(x, y, z, n)$, $z = 1,2,3$, $n = 1,2$ (RGB channels).

In HDMR expansion for a multivariate function, the fixed term is defined as the sum of functions of one variable, functions of two variables, and others.

$$f(x_1, x_2, \dots, x_N) = f_0 + \sum_{i_1=1}^{N} f_{i_1}(x_{i_1}) + \sum_{i_1, i_2=1}^{N} f_{i_1 i_2}(x_{i_1}, x_{i_2}) + \dots + f_{12\dots N}(x_1, x_2, \dots, x_N) \tag{8.4}$$

In general, functions are represented up to univariate or bivariate functions, and the remainder is ignored as an approximation error. The function created for greyscale images can be fully represented when it is extended up to two-variable functions in the HDMR expansion [11]. The researchers presented a representation by making up to three variable functions for in-painting with colour images and aimed to estimate the lost pixels by applying them with Lagrange interpolation.

In another study, Karaca and Tunga [12] also studied the in-painting of a rectangular area using the same method. They tried 5×5, 10×10, and 20×20 pixel dimensions on different images for the in-painting area.

The algorithms used for in-painting in the literature have also been used for noise removal. Jassim [13] tried the Kriging algorithm for salt and pepper noise reduction. For noise of varying intensity, it first *detects* noise using an 8×8 pixel filter on the image and then applies Kriging interpolation to correct incorrect pixel values in this area.

8.3 MATERIALS AND METHODS

8.3.1 Materials

The internally stained images obtained from the methods used in the studies and their originals were presented with peak signal-to-noise ratio (PSNR) and SSIM criteria. In addition, although the masks used in the studies seem similar to each other, they have differences. This is a limitation to the exact comparison of algorithms. In order to make a full comparison of the algorithms proposed in this part of the study, we compared the results obtained by using the same masks and images and each of the methods.

256×256 greyscale and colour images were used to compare the algorithms. The images used were selected from the images commonly used in image processing research: Lena, Mandril, Peppers, Jetplane, and House. The masks applied to the images are created as follows:

- Mask 1: simple curve, drawn with a four-pixel-thick pencil
- Mask 2: A non-condensed font consisting of several lines between 12 and 19 fonts
- Mask 3: Intense font created with 12 font letters
- Mask 4: Intense scratches of oblique, horizontal, and vertical lines drawn with a four-pixel-thick pencil
- Mask 5: Frame created with a size of 40×40 pixels.

In addition, to test the noise reduction efficiency of interpolation methods, masks created with salt and pepper noise were applied. Their density levels are Noise 1: 10%, Noise 2: 30%, Noise 3: 50%, Noise 4: 70%, and Noise 5: 90%, respectively.For instance, the presentation of the greyscale Lena image with the masks to be used is demonstrated in Table 8.1.

In all the algorithms, corrected images were obtained by calculating only unknown pixel values, and the obtained images were compared with the original image using PSNR and SSIM criteria. In order to determine the unknown pixel values, the difference between the pixel values in the original image and the false image was taken, and the non-zero pixel values were estimated by the algorithms. In the comparison of the results with the actual image, the MSE, which is a part of the SSIM and PSNR criteria, is indicated.

8.3.2 Method

The evaluation presented in this study employs a number of various interpolation techniques: two-dimensional cubic interpolation, Kriging interpolation, interpolation with radial-based functions (RBF), and interpolation using HDMR. These techniques are detailed in the following sub-sections.

8.3.2.1 Two-Dimensional Cubic Interpolation

For this method, x, y: row, column coordinate values of known pixels, v: pixel values, x_q, y_q: coordinate values of the desired pixels, cubic: to be used as input parameters for the interpolation. After the information of the 2-D image is put into the form that the function can use, the function is executed, and the new pixel values are updated.

8.3.2.2 Kriging Interpolation

For this method, Kriging (x_i, y_i, z_i, x, y) and auxiliary functions presented by Schwanghart and Kuhn [14] were used (x_i, y_i: row, column coordinate values of

TABLE 8.1
Display of the Masks Used in the Study on a Sample Image

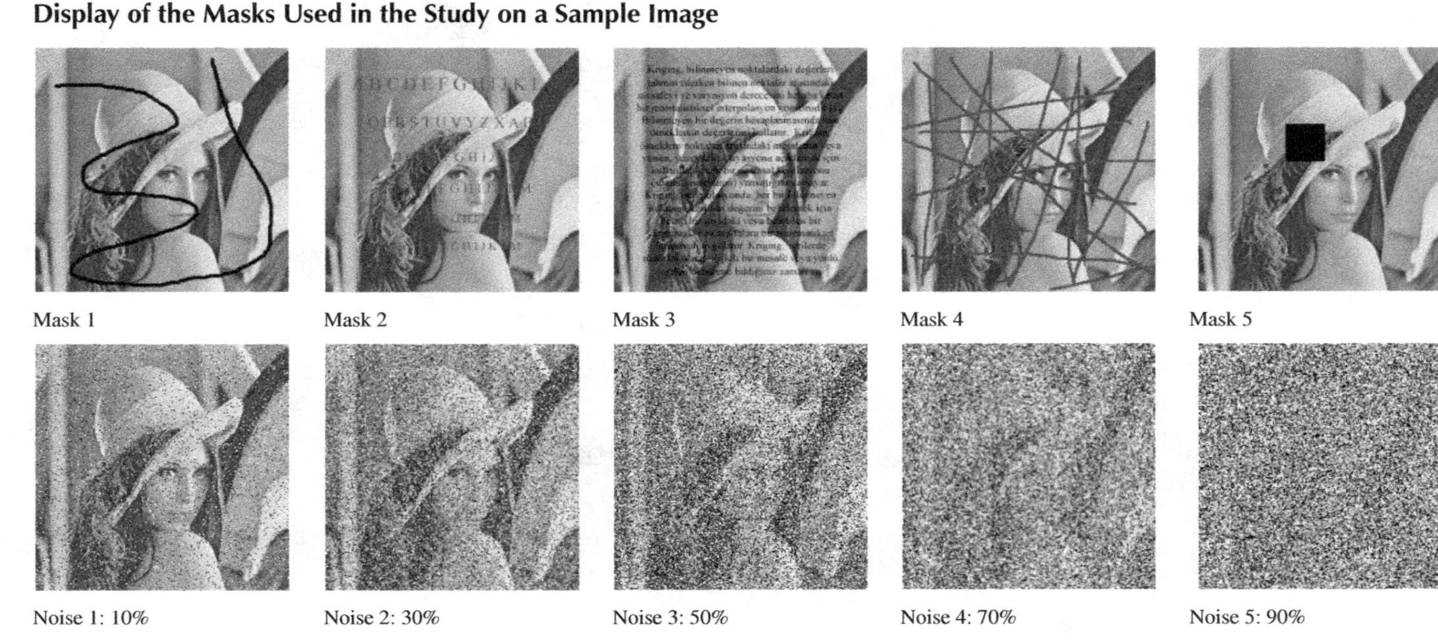

Mask 1	Mask 2	Mask 3	Mask 4	Mask 5
Noise 1: 10%	Noise 2: 30%	Noise 3: 50%	Noise 4: 70%	Noise 5: 90%

known pixels, z_i: pixel values, x, y: to be the value of the coordinate values of the desired pixels); 256×256 pixel images were used along with masks of size 16×16 pixels and noise densities defined in the neighbourhood of 8×8 pixels. The unknown values in each sub-image were calculated using the Kriging function.

In this way, it is ensured that the unknown points in the sub-images are related to all pixel values in the sub-images. The 8×8 pixel size has enough data for noise reduction. However, the Kriging algorithm did not work for Noise5 with a 90% density (see Table 8.1). Therefore, only 16×16 pixels are used for Noise5. On the other hand, although the calculation method by sub-images is suitable for scattered errors, it will not be suitable for an internal painting problem as in Mask5. Therefore, for the case in Mask5, the image in-painting is reduced to a 90×90 pixel image size that will take the centre of the lost frame area (40×40 pixels), and the function is run on this 90×90 pixel neighbourhood.

8.3.2.3 Interpolation with Radial-Based Functions

The approach of Foster[15]was adopted in order to compute the RBF function in a similar fashion to the Kriging interpolation where sub-images are used for the computation.

8.3.2.4 Interpolation Using High-Dimensional Model Representation

For this method, following the approach of Tunga and Koçanoğulları's [16] study, constant, one-variable, and two-variable functions represented greyscale images. The fixed, one-, two-, and three-variable functions representing the coloured images were found, and the missing areas on these functions were corrected by an interpolation approach. In this section, spline for interpolation of one-dimensional functions and cubic interpolation for interpolation of two-dimensional functions were implemented.

These methods can be applied directly to greyscale images, which can be expressed as two-variable functions. Interpolation of coloured images was obtained by applying these interpolations separately to the three layers of the images. At this point, there is no difference from the method applied to greyscale images. For the calculations, the same algorithms were applied to the RGB layers three times and then combined to produce corrected colour images.

8.4 RESULTS

In-painting results were obtained using the various interpolation methods discussed earlier. All the results obtained for a sample are presented in Figures 8.1 (in greyscale) and 8.2 (in colour).

The numerical comparison results obtained from all images are summarized in Table 8.2 for greyscale and in Table 8.3 for colour images. The results of 2-D cubic interpolation, Kriging interpolation, RBF interpolation, and YBMG interpolation are shown in four large blocks. Orange, blue, green intracellular staining was used to compare the outputs of the four methods. For example, in the PSNR values of four different methods used to remove Mask1 in the colour Lena image, 2-D cubic

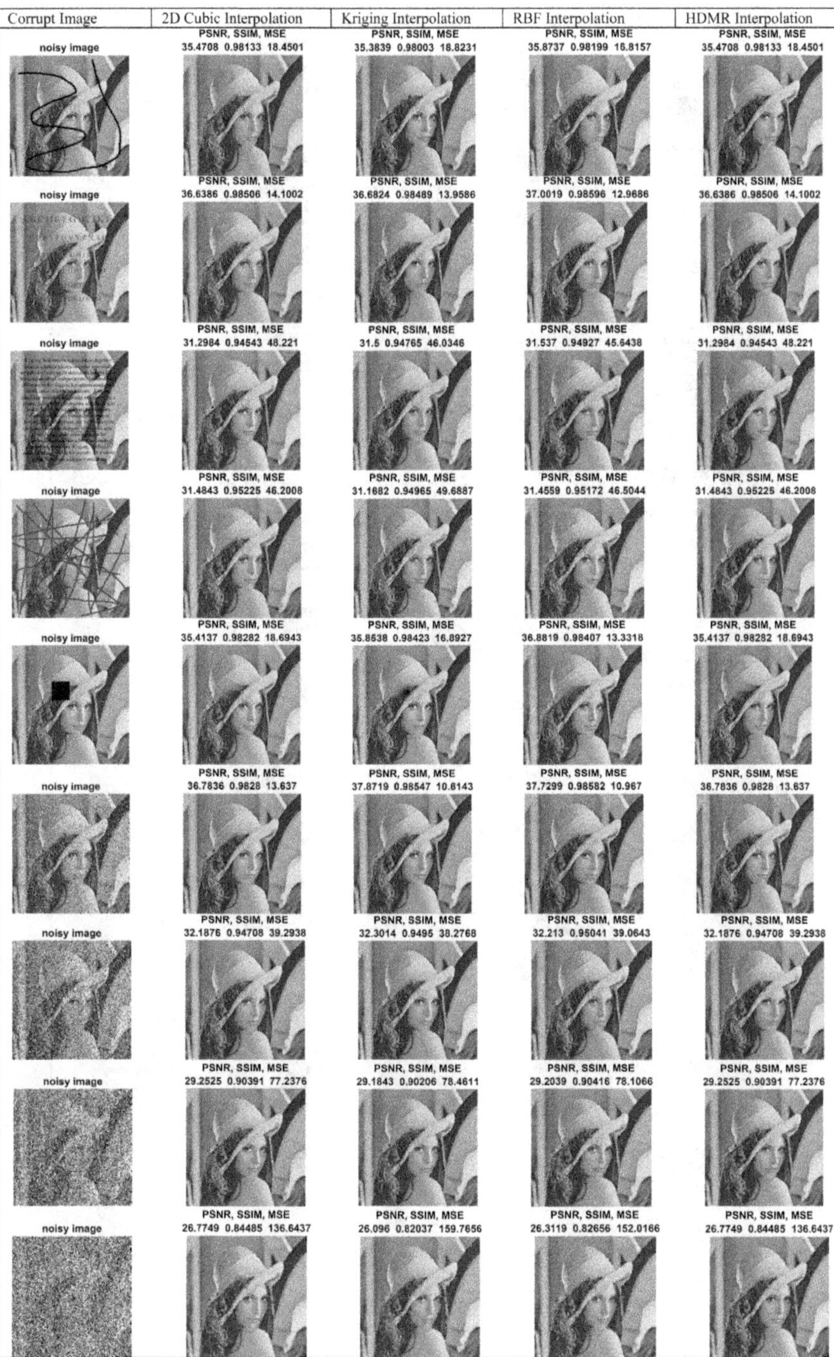

FIGURE 8.1 Greyscale sample image with in-painting.

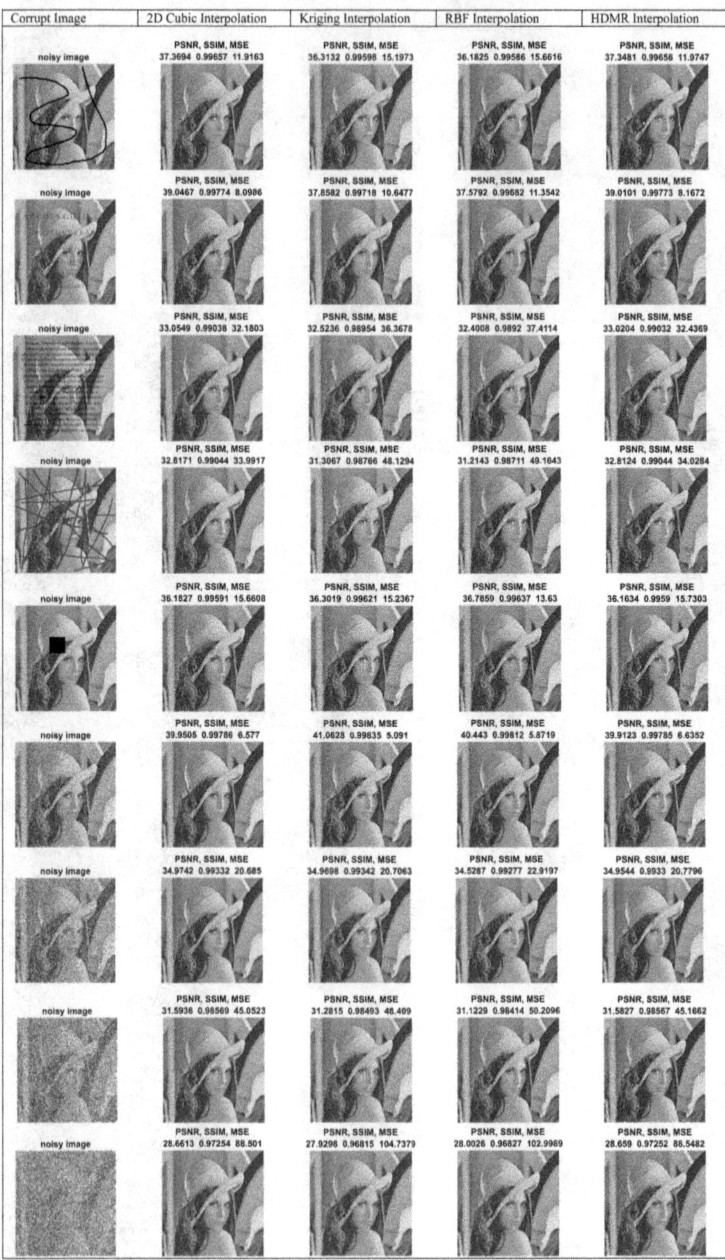

FIGURE 8.2 Colour sample image with in-painting.

TABLE 8.2
Comparison of All Images, Masks, and Methods: Greyscale

		2-D Cubic Interpolation			Kriging Interpolation			RBF Interpolation			HDMR Interpolation		
		PSNR	SSIM	MSE	PSNR	SSIM	MSE	PSNR	SSIM	MSE	PSNR	SSIM	MSE
Lena	Mask1	35.471	0.98133	18.45	35.384	0.98003	18.823	35.874	0.98199	16.816	35.471	0.98133	18.45
	Mask2	36.639	0.98506	14.1	36.682	0.98489	13.959	37.002	0.98596	12.969	36.639	0.98506	14.1
	Mask3	31.298	0.94543	48.221	31.5	0.94765	46.035	31.537	0.94927	45.644	31.298	0.94543	48.221
	Mask4	31.484	0.95225	46.201	31.168	0.94965	49.689	31.456	0.95172	46.504	31.484	0.95225	46.201
	Mask5	35.414	0.98282	18.694	35.854	0.98423	16.893	36.882	0.98407	13.332	35.414	0.98282	18.694
	Noise1	36.784	0.9828	13.637	37.872	0.98547	10.614	37.73	0.98582	10.967	36.784	0.9828	13.637
	Noise2	32.188	0.94708	39.294	32.301	0.9495	38.277	32.213	0.95041	39.064	32.188	0.94708	39.294
	Noise3	29.253	0.90391	77.238	29.184	0.90206	78.461	29.204	0.90416	78.107	29.253	0.90391	77.238
	Noise4	26.775	0.84485	136.64	26.096	0.82037	159.77	26.312	0.82656	152.02	26.775	0.84485	136.64
	Noise5	23.598	0.72678	283.94	23.019	0.69605	324.46	23.27	0.70052	306.23	23.598	0.72678	283.94
House	Mask1	40.928	0.98931	5.2512	40.359	0.9893	5.9872	39.515	0.9883	7.2715	40.928	0.98931	5.2512
	Mask2	43.898	0.99498	2.6504	43.479	0.99501	2.9186	42.548	0.99388	3.6162	43.898	0.99498	2.6504
	Mask3	37.635	0.97906	11.211	37.862	0.97944	10.64	36.753	0.97635	13.733	37.635	0.97906	11.211
	Mask4	35.842	0.97333	16.939	35.701	0.97317	17.497	34.682	0.96931	22.123	35.842	0.97333	16.939
	Mask5	38.951	0.98915	8.2782	40.742	0.99042	5.4816	43.857	0.99163	2.6756	38.951	0.98915	8.2782
	Noise1	47.819	0.99738	1.0744	47.855	0.99746	1.0657	45.775	0.99639	1.7201	47.819	0.99738	1.0744
	Noise2	37.905	0.9874	10.535	39.997	0.98683	6.5063	38.573	0.98345	9.0317	37.905	0.9874	10.535
	Noise3	36.898	0.97258	13.282	35.187	0.96611	19.697	34.445	0.96072	23.367	36.898	0.97258	13.282
	Noise4	32.53	0.934	36.314	30.709	0.91736	55.226	30.35	0.90886	59.994	32.53	0.934	36.314
	Noise5	25.212	0.82057	195.84	25.312	0.80812	191.38	25.289	0.80354	192.41	25.212	0.82057	195.84
Peppers	Mask1	39.189	0.99088	7.8374	39.631	0.99074	7.0791	39.416	0.99066	7.4377	39.189	0.99088	7.8374
	Mask2	42.063	0.99409	4.0435	42.656	0.99451	3.5274	41.299	0.99345	4.8212	42.063	0.99409	4.0435
	Mask3	35.572	0.97824	18.025	35.484	0.97721	18.396	35.288	0.97772	19.243	35.572	0.97824	18.025
	Mask4	32.549	0.96837	36.159	32.458	0.96667	36.926	31.916	0.96426	41.832	32.549	0.96837	36.159
	Mask5	32.701	0.98492	34.916	36.59	0.99072	14.257	36.018	0.99031	16.264	32.701	0.98492	34.916

(Continued)

TABLE 8.2 (CONTINUED)
Comparison of All Images, Masks, and Methods: Greyscale

		2-D Cubic Interpolation			Kriging Interpolation			RBF Interpolation			HDMR Interpolation		
		PSNR	SSIM	MSE	PSNR	SSIM	MSE	PSNR	SSIM	MSE	PSNR	SSIM	MSE
	Noise1	42.558	0.99442	3.6084	42.964	0.99472	3.286	41.813	0.99402	4.2836	42.558	0.99442	3.6084
	Noise2	36.44	0.9797	14.758	36.079	0.97823	16.037	35.309	0.97622	19.151	36.44	0.9797	14.758
	Noise3	32.927	0.96034	33.146	32.572	0.9524	35.961	32.196	0.95017	39.22	32.927	0.96034	33.146
	Noise4	28.73	0.91716	87.117	27.579	0.88713	113.55	27.886	0.88925	105.8	28.73	0.91716	87.117
	Noise5	24.524	0.80588	229.46	24.03	0.77226	257.07	24.092	0.77667	253.46	24.524	0.80588	229.46
Mandril	Mask1	33.607	0.9697	28.341	34.403	0.97272	23.591	34.308	0.97213	24.112	33.607	0.9697	28.341
	Mask2	35.45	0.98171	18.537	35.744	0.98254	17.325	35.505	0.98199	18.307	35.45	0.98171	18.537
	Mask3	29.277	0.92569	76.803	29.582	0.92778	71.59	29.385	0.92522	74.918	29.277	0.92569	76.803
	Mask4	29.17	0.92005	78.714	29.423	0.92242	74.256	29.178	0.91994	78.572	29.17	0.92005	78.714
	Mask5	37.325	0.98466	12.037	38.406	0.98865	9.3865	38.836	0.98899	8.5017	37.325	0.98466	12.037
	Noise1	35.096	0.98013	20.113	35.138	0.98042	19.921	34.633	0.97827	22.374	35.096	0.98013	20.113
	Noise2	29.509	0.92561	72.805	29.348	0.92245	75.555	29.118	0.91904	79.67	29.509	0.92561	72.805
	Noise3	26.259	0.84231	153.87	26.297	0.83673	152.53	26.035	0.83151	162.03	26.259	0.84231	153.87
	Noise4	23.592	0.71527	284.39	23.715	0.69501	276.4	23.381	0.69276	298.56	23.592	0.71527	284.39
	Noise5	21.045	0.48207	511.14	21.325	0.46271	479.25	21.189	0.473	494.53	21.045	0.48207	511.14
Jetplane	Mask1	35.763	0.98851	17.249	35.734	0.98738	17.363	35.419	0.98768	18.673	35.763	0.98851	17.249
	Mask2	36.561	0.99187	14.355	36.541	0.99196	14.422	35.943	0.9909	16.548	36.561	0.99187	14.355
	Mask3	31.514	0.96883	45.881	31.836	0.9692	42.605	31.48	0.96812	46.244	31.514	0.96883	45.881
	Mask4	30.23	0.96264	61.668	30.528	0.96125	57.586	30.098	0.96007	63.578	30.23	0.96264	61.668
	Mask5	34.77	0.99074	21.682	35.562	0.99056	18.067	35.175	0.99184	19.752	34.77	0.99074	21.682
	Noise1	39.482	0.99356	7.3265	39.906	0.99405	6.6453	38.988	0.99309	8.208	39.482	0.99356	7.3265
	Noise2	33.332	0.97603	30.188	33.393	0.97464	29.771	32.737	0.97209	34.625	33.332	0.97603	30.188
	Noise3	29.752	0.94846	68.848	29.742	0.94317	69.002	29.321	0.93934	76.032	29.752	0.94846	68.848
	Noise4	26.865	0.89929	133.82	25.929	0.87449	166.03	25.854	0.87406	168.91	26.865	0.89929	133.82
	Noise5	22.372	0.7664	376.59	22.087	0.73366	402.12	22.301	0.74699	382.77	22.372	0.7664	376.59

TABLE 8.3
Comparison of all Images, Masks, and Methods: Colour

		2D Cubic Interpolation			Kriging Interpolation			RBF Interpolation			HDMR Interpolation		
		PSNR	SSIM	MSE	PSNR	SSIM	MSE	PSNR	SSIM	MSE	PSNR	SSIM	MSE
Lena	Mask1	37.3690	0.9965	11.9160	36.3130	0.9959	15.1970	36.1820	0.9958	15.6620	37.3480	0.9965	11.9750
	Mask2	39.0470	0.9977	8.0986	37.8580	0.9971	10.6480	37.5790	0.9968	11.3540	39.0100	0.9977	8.1672
	Mask3	33.0550	0.9903	32.1800	32.5240	0.9895	36.3680	32.4010	0.9892	37.4110	33.0200	0.9903	32.4370
	Mask4	32.8170	0.9904	33.9920	31.3070	0.9876	48.1290	31.2140	0.9871	49.1640	32.8120	0.9904	34.0280
	Mask5	36.1827	0.9959	15.6608	36.3019	0.9962	15.2367	36.7859	0.9963	13.6300	36.1634	0.9959	15.7303
	Noise1	39.9510	0.9978	6.5770	41.0630	0.9983	5.0910	40.4430	0.9981	5.8719	39.9120	0.9978	6.6352
	Noise2	34.9740	0.9933	20.6850	34.9700	0.9934	20.7060	34.5290	0.9927	22.9200	34.9540	0.9933	20.7800
	Noise3	31.5940	0.9856	45.0520	31.28200	0.9849	48.4090	31.1230	0.9841	50.2100	31.5830	0.9856	45.1660
	Noise4	28.6610	0.9725	88.5010	27.9300	0.9681	104.7400	28.0030	0.9682	103.0000	28.6590	0.9725	88.5480
	Noise5	24.5101	0.9364	230.1814	24.1843	0.9302	248.1129	24.2096	0.9303	246.6749	24.4822	0.9359	231.6658
House	Mask1	40.4790	0.9959	5.8238	39.4160	0.9947	7.4390	39.0240	0.9945	8.1407	40.4790	0.9959	5.8238
	Mask2	41.5140	0.9972	4.5883	41.4500	0.9970	4.6565	41.0230	0.9968	5.1379	41.5140	0.9972	4.5885
	Mask3	35.6460	0.9892	17.7200	34.4480	0.9863	23.3500	34.0940	0.9858	25.3320	35.6460	0.9892	17.7200
	Mask4	33.9880	0.9850	25.9600	32.7490	0.9813	34.5320	32.3050	0.9801	38.2500	33.9880	0.9850	25.9590
	Mask5	40.0240	0.9977	6.4662	39.3290	0.9968	7.5886	41.6210	0.9986	4.4765	40.0240	0.9977	6.4662
	Noise1	42.5520	0.9973	3.6128	41.9370	0.9972	4.1625	41.3550	0.9968	4.7599	42.7410	0.9973	3.4590
	Noise2	35.5040	0.9889	18.3090	35.6730	0.9890	17.6100	35.3360	0.9880	19.0300	35.7270	0.9891	17.3920
	Noise3	32.6610	0.9779	35.2370	32.2310	0.9763	38.9060	32.0950	0.9752	40.1400	33.0250	0.9783	32.4040
	Noise4	29.3080	0.9565	76.2580	28.4140	0.9476	93.6950	28.5930	0.9482	89.9060	29.4980	0.9572	72.9990
	Noise5	24.8730	0.8977	211.7300	24.5450	0.8880	228.3600	24.5710	0.8889	226.9800	25.0030	0.8994	205.5100
Peppers	Mask1	36.7600	0.9964	13.7130	34.0110	0.9944	25.8200	35.1310	0.9955	19.9530	34.7740	0.9932	21.6610
	Mask2	39.0040	0.9973	8.1792	38.1130	0.9970	10.0420	38.1140	0.9972	10.0400	36.1120	0.9942	15.9170
	Mask3	32.7020	0.9898	34.9060	32.0700	0.9888	40.3730	32.0650	0.9892	40.4210	31.5940	0.9860	45.0530
	Mask4	30.5920	0.9852	56.7340	28.9100	0.9800	83.5830	29.1750	0.9812	78.6360	29.7810	0.9812	68.3920
	Mask5	32.510	0.9962	36.4840	35.3340	0.9976	19.0400	35.4160	0.9977	18.6830	31.7720	0.9934	43.2440

(Continued)

TABLE 8.3 (CONTINUED)
Comparison of all Images, Masks, and Methods: Colour

		2D Cubic Interpolation			Kriging Interpolation			RBF Interpolation			HDMR Interpolation		
		PSNR	SSIM	MSE	PSNR	SSIM	MSE	PSNR	SSIM	MSE	PSNR	SSIM	MSE
	Noise1	37.0490	0.9960	12.8290	37.3210	0.9963	12.0500	37.3460	0.9964	11.9800	35.0560	0.9929	20.3010
	Noise2	31.9010	0.9873	41.9780	31.9410	0.9876	41.5890	32.0020	0.9878	41.0100	30.7700	0.9830	54.4590
	Noise3	29.2230	0.9769	77.7570	28.9520	0.9758	82.7650	29.0920	0.9765	80.1390	28.4840	0.9716	92.1980
	Noise4	26.9580	0.9621	131.0100	26.0360	0.9550	161.9900	26.2500	0.9565	154.1800	26.0540	0.9523	161.3300
	Noise5	23.1160	0.9209	317.3300	22.8300	0.9129	338.9300	22.8610	0.9130	336.5100	22.2150	0.8988	390.4300
Mandril	Mask1	32.5990	0.9813	35.7410	32.1320	0.9806	39.8000	31.7410	0.9793	43.5520	32.5970	0.9813	35.7550
	Mask2	34.1960	0.9884	24.7460	34.2940	0.9887	24.1930	34.0950	0.9882	25.3270	34.1920	0.9884	24.7660
	Mask3	28.5570	0.9525	90.6530	28.4160	0.9498	93.6420	28.2680	0.9487	96.8830	28.5600	0.9525	90.5800
	Mask4	28.4080	0.9535	93.8250	28.0750	0.9526	101.3000	27.5810	0.9496	113.4800	28.4070	0.9535	93.8400
	Mask5	31.6180	0.9898	44.8000	30.8770	0.9922	53.1310	32.6760	0.9933	35.1160	31.6170	0.9898	44.8140
	Noise1	33.9730	0.9854	26.0490	34.0290	0.9852	25.7150	33.7550	0.9844	27.3870	33.9720	0.9854	26.0570
	Noise2	28.6800	0.9484	88.1250	28.7050	0.9475	87.6170	28.5010	0.9456	91.8380	28.6800	0.9484	88.1200
	Noise3	25.6260	0.8935	178.0400	25.6400	0.8903	177.4400	25.5230	0.8893	182.2900	25.6230	0.89354	178.1300
	Noise4	23.1580	0.8132	314.2800	23.0380	0.80268	323.0300	22.9140	0.8021	332.4000	23.1540	0.81314	314.5200
	Noise5	20.4720	0.6666	583.3300	20.6280	0.66629	562.7500	20.5690	0.6665	570.4300	20.4700	0.66662	583.5400
Jetplane	Mask1	37.1680	0.9923	12.4810	36.6330	0.9904	14.1190	35.7000	0.9897	17.5000	37.1600	0.99237	12.5040
	Mask2	38.4190	0.9942	9.3570	37.6990	0.99307	11.0440	37.4830	0.9928	11.6100	38.4120	0.99428	9.3736
	Mask3	32.2060	0.9772	39.1250	31.6920	0.97279	44.0430	31.4280	0.9729	46.8000	32.1880	0.97726	39.2910
	Mask4	30.5990	0.9713	56.6480	29.3490	0.96121	75.5430	29.1340	0.9606	79.3740	30.5970	0.97132	56.6740
	Mask5	35.4370	0.9909	18.5940	36.6620	0.99098	14.0250	36.6260	0.9922	14.1420	35.4320	0.99091	18.6170
	Noise1	39.8120	0.9950	6.7904	39.7660	0.99492	6.8626	38.9670	0.9943	8.2479	39.7950	0.99509	6.8170
	Noise2	33.7350	0.9817	27.5160	33.5000	0.97901	29.0470	33.0860	0.9781	31.9500	33.7190	0.98176	27.6190
	Noise3	30.4710	0.9604	58.3400	29.6470	0.94927	70.5260	29.5550	0.9501	72.0480	30.4650	0.96044	58.4200
	Noise4	27.0080	0.9208	129.5100	25.9770	0.8900	164.2200	25.8960	0.8922	167.2800	26.9900	0.9207	130.0300
	Noise5	22.3250	0.8036	380.7200	22.4040	0.76772	373.8400	22.4520	0.7792	369.7000	22.3390	0.80356	379.4600

interpolation has the highest value. This cell was stained with orange. The highest SSIM values of the four methods were painted with blue, while the lowest MSE values of the four methods were painted with green. By means of this staining, it is easier to observe which method gives better results in which situation.

In addition, the data of four different methods were compared with a one-way analysis of variance (ANOVA) test to determine whether there was a difference between the methods in terms of PSNR, MSE, and SSIM. There was no difference between the methods in terms of PSNR, MSE, and SSIM values in the evaluation ($p=0.997$, $p=0.998$, $p=0.986$ for grey images and $p=0.974$, $p=0.994$, $p=0.988$ for Grey images, respectively).

In addition, in order to compare the numerical results of the four methods, the difference between the maximum value and the smallest value of the PSNR and SSIM results obtained for each image and mask was observed as a percentage change. These ratios are presented in Figures 8.3 and 8.4.

PSNR and MSE values are directly related to each other. It is known that the MSE value is part of the PSNR criterion. The PSNR value is inversely proportional to the logarithm of the square root of MSE (Equation 8.5). Therefore, it is essential to have the lowest MSE when PSNR is highest.

$$\mathrm{PSNR} = 20\log_{10}(255/\sqrt{\mathrm{MSE}}) \tag{8.5}$$

In addition, the formula of MSE (Equation 8.6) is built on differences in density levels between the two images.

$$\mathrm{MSE} = \frac{1}{\mathrm{NM}}\sum_{x=1}^{N}\sum_{y=1}^{M}\left[f(x,y)-g(x,y)\right]^2 \tag{8.6}$$

In other words, the closer the pixel values in the same position in the two images, the closer the MSE value is to 0. We understand that the smaller the MSE value and the higher the PSNR, the better the results. On the other hand, the SSIM criterion was formulated to measure luminance, contrast, and structural correlation between the two images. The results obtained from functions comparing these three criteria separately are multiplied to find the SSIM value. Each function takes values in the range 0–1; as a result, SSIM has values from 0 to 1. The SSIM value approaches unity when the correlations are high, i.e. the images are similar. Although the PSNR and SSIM calculations do not seem to be alike, there are studies showing that these are analytically related [17]. Therefore, SSIM and PSNR values are expected to be higher in a well-obtained in-painting.

As a result of this evaluation, in Table 8.2 and Table 8.3, it will be natural for cells stained with orange, blue, and green to coexist. However, in some cases, it is observed that this association is not preserved, but this is due to very small differences that can be ignored. Therefore, the outputs we obtained do not contradict theoretical knowledge.

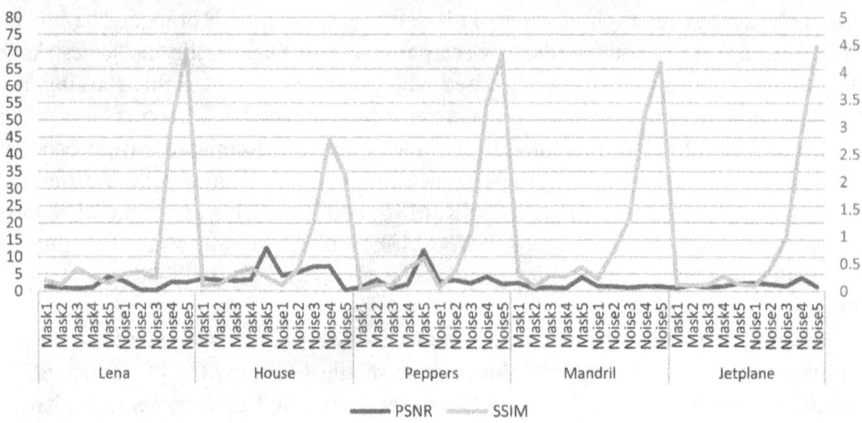

FIGURE 8.3 Percentage variation of the difference between the highest and lowest values of PSNR and SSIM scores obtained from greyscale images.

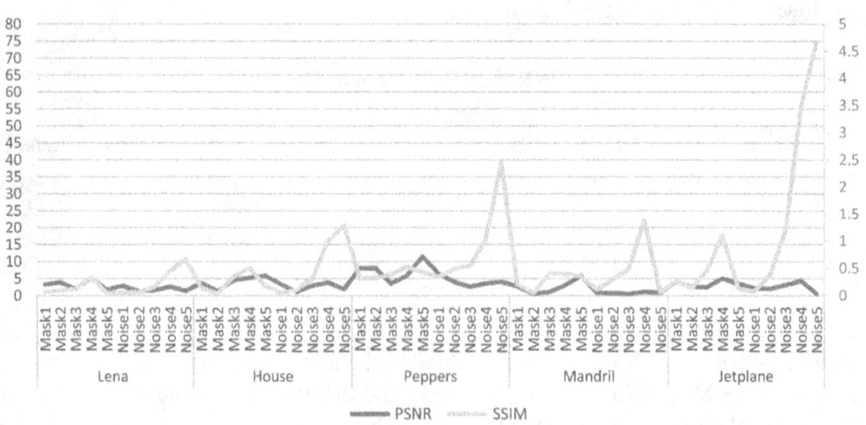

FIGURE 8.4 Percentage variation of the difference between the highest and lowest values of PSNR and SSIM scores obtained from colour images.

For each image and method in Figure 8.3 and Figure 8.4, we tried to observe the percentage change between the methods by the ratio of the highest and lowest obtained score differences to the lowest value. As can be observed here, the numerical comparison results for images obtained by in-painting are generally very close to each other. The difference in SSIM values does not exceed 5%. PSNR values are mostly within the 5% change band.

In addition, according to the statistics obtained from the one-way ANOVA test, it was confirmed that there was no difference between the methods in numerical evaluation scales. Therefore, it is observed that none of the methods has absolute superiority over the others in terms of in-painting results.

If the methods of application are taken into consideration, the in-painting problem in Mask5 will need to be evaluated separately. In other masks, the problem of internal painting is scattered throughout the image, where it focuses on a specific area. Therefore, in Kriging and RBF interpolation methods, we focused on the 90×90 pixel area in order to keep the lost area in the middle. It was considered that the selected area would provide sufficient data to fill the missing area.

In fact, in the first trials, the whole image was tested using the Kriging and RBF functions at once for both Mask5 and the four other masks. However, the functions needed during the operation of the array required mesh grid structures that need to be produced. When the results obtained for Mask5 are evaluated; it is observed that Kriging and RBF methods produce better outputs. On the other hand, this advantage will be determined in the evaluation made through observation. The results obtained in Mask5 can be compared visually with the study of Karaca and Tunga [11]. It is observed that the results obtained for staining of the quadratic region are much larger than the MSE values.

According to the results, the square area is still clearly visible and filled with diagonal lines. In our results, the square area (for Kriging and RBF interpolation) was less prominent, and no diagonal lines were observed in the faulty area. The difference between greyscale and colour images of the two studies may be inaccurate, but the interiors of in-painting characters can still be observed, although relatively, the results obtained in our study may be more visually appealing.

8.5 CONCLUSION

The literature lacks a detailed evaluation on the use of interpolation methods for the in-painting problem. Accordingly, this study aimed to evaluate state-of-the-art approaches, namely, 2-D, Kriging, RBF, and YBMG-based Lagrange interpolation methods, which have recently been studied in the literature.

A benchmark dataset was generated in order to perform a detailed evaluation of the in-painting algorithms. Several comprehensive experiments were conducted in order to make a fair assessment. According to the results obtained, it is observed that none of the methods has absolute superiority to the others in terms of in-painting results. However, Kriging and RBF interpolation produce better results for both numerical data and visual evaluation for image in-painting problems having large area losses. Studies on this subject have also yielded good results for large area in-painting, but Kriging and RBF interpolation outputs look better in terms of visual quality.

REFERENCES

1. M. Bartalmio, G. Sapiro, C. Ballester and C. Ballester, "Image Inpainting," in SIGGRAPH00: The 27th International Conference on Computer Graphics and Interactive Techniques Conference, Los Angeles, USA, 2000.
2. A. Bugeau, M. Bertalmío, V. Caselles and G. Sapiro, "A Comprehensive Framework for Image Inpainting," *IEEE Transactions on Image Processing*, vol. 19, no. 10, p. 2634–2645, 2010.

3. K. S. Amasidha, A. S. Awati, S. P. Deshpande and M. R. Patil, "Digital Image Inpainting: A Review," *International Research Journal of Engineering and Technology*, vol. 3, no. 1, p. 851–854, 2016.
4. L. Chang and Y. Chongxiu, "New Interpolation Algorithm for Image Inpainting," in International Conference on Physics Science and Technology (ICPST), Hong Kong, 2011.
5. F. Jassim, "Image Inpainting by Kriging Interpolation Technique," *World of Computer Science and Information Technology Journal*, vol. 3, no. 5, pp. 91–96, 2013.
6. M. S. Sapkal and P. Kadbe, "Kriging Interpolation Technique Basis Image Inpainting," *International Journal of Latest Trends in Engineering and Technology*, vol. 6, no. 3, p. 467–470, 2016.
7. M. S. Sapkal, P. K. Kadbe and B. H. Deokate, "Image Inpainting by Kriging Interpolation Technique For Mask Removal," in International Conference on Electrical, Electronics, and Optimization Techniques (ICEEOT), 2016.
8. A. Awati, S. P. Deshpande, P. Y. Belagal and M. R. Patil, "Digital Image Inpainting Using Modified Kriging Algorithm," in 2nd International Conference for Convergence in Technology (I2CT), Mumbai, India, 2017.
9. W. Wang and X. Qin, "An Image Inpainting Algorithm Based on CSRBF Interpolation," *Dianzi Yu Xinxi Xuebao/Journal of Electronics and Information Technology*, pp. 890–894, 2006.
10. E. Karaca and M. A. Tunga, "A Method for Inpainting Rectangular Missing Regions using High Dimensional Model Representation and Lagrange Interpolation," in 24th Signal Processing and Communication Application Conference (SIU), Zonguldak, TR, 2016.
11. E. M. Altın and B. Tunga, "High Dimensional Model Representation in Image Processing," in Proceedings of the 14th International Conference on Computational and Mathematical Methods in Science and Engineering, CMMSE, Cadiz, Spain, 2014.
12. E. Karaca and M. A. Tunga, "Interpolation-Based Image Inpainting in Colour Images Using High Dimensional Model Representation," in 24th European Signal Processing Conference (EUSIPCO), 2016.
13. F. A. Jassim, "Kriging Interpolation Filter to Reduce High Density Salt and Pepper Noise," *World of Computer Science and Information Technology Journal*, vol. 3, no. 1, pp. 8–14, 2013.
14. Y. Shen, W. Dai, and V. M. Richards, "A MATLAB toolbox for the efficient estimation of the psychometric function using the updated maximum-likelihood adaptive procedure," Behavior Research Methods. vol. 47, no. 1, pp. 13–26. 2015.
15. A. Horé and D. Ziou, "Image Quality Metrics: PSNR vs. SSIM," in 20th International Conference on Pattern Recognition, İstanbul, TR, 2010.
16. W. Schwanghart and N. J. Kuhn, "TopoToolbox: A Set of Matlab Functions for Topographic Analysis," *Environmental Modelling and Software*, vol. 25, no. 6, p. 770–781, 2010.
17. B. Tunga and A. Koçanaoğulları, "Digital Image Decomposition and Contrast Enhancement Using High-Dimensional Model Representation," *Signal, Image and Video Processing*, p. 299–306, 2018.

9 Real Time Density–Based Traffic Congestion Detection System Using Image Processing and Fuzzy Logic Controller

Anita Mohanty, Bhagyalaxmi Jena, and Subrat Kumar Mohanty

CONTENTS

9.1 INTRODUCTION

Traffic congestion in the present scenario is due to increased usage of vehicles in urban areas. Day by day, this problem is increasing because of a large number of vehicles with limited infrastructural development. The roads are the common path for traveling and daily activities. As traffic on these roads is common today, a lot of vehicle fuel and travel time is wasted, which induces impatience and irritation in drivers. Traffic jams are caused by countless reasons, like office hours, holidays, bad weather conditions, or unpredictable events such as road accidents or construction

DOI: 10.1201/9781003221333-9

works. Once we face a traffic jam, we may have to wait for some hours to come out of it. This plays a vital role in everybody's life as well as in society. A person in India spends 7% of their day commuting to work [1]. That's why an intelligent transportation system (ITS) [2,3] is an alternative. The most realistic solution to this is to use existing infrastructure intelligently and efficiently. In the real scenario, to regulate and control the traffic on roads, it is necessary to acquire and analyze the road traffic information. Several methods are employed to gather traffic facts, like magnetic inductive loop [4, 5], gain amplifier theory [6], congestion detection by wavelet analysis [7] and pattern recognition technique to detect congestion [8], microwave radar [9], etc.

In many developed countries, traffic is lane based. But Indian traffic is not lane based and is chaotic, so a vehicle sensing method is preferred to count the number of vehicles present. One of the methods to measure the traffic density on a road in the Indian scenario is using image processing techniques. This vision-based system has been used in traffic monitoring because of the advantages that make it unique with respect to other systems [10, 11]. Some unique characteristics are ease of installation, lower cost than other physical sensors, and large coverage area. This system can give a significant explanation of traffic status including speed and number of vehicles. These parameters can provide us with total traffic movement information, which meets the demand of ITS. Image processing techniques can be utilized to count the actual number of vehicles in all ambient conditions without any error. Once this vision sensor gives the number of vehicles, fuzzy control logic [12] is implemented to detect the level of congestion in that particular lane.

This chapter proposes an intelligent traffic system to extract the number of vehicles from video images captured by installed stationary roadside cameras. This number of vehicles is utilized in a fuzzy logic controller (FLC) with other parameters extracted from the fuzzy C-means (FCM) clustering technique implemented on parameters taken from the simulation of urban mobility (SUMO) simulator. The FCM clustering technique is used to minimize the traffic load by utilizing the vehicle parameters taken from the SUMO simulator and to extract the key features to be utilized by FLC to calculate the level of congestion. This chapter proposes a technique to quantify the level of congestion that incorporates congestion attributes such as vehicle count, speed, CO_2 emission, and fuel consumption.

The rest of this chapter is organized as follows. A literature review of the related work is presented in Section 9.2. The proposed model of a real time density–based traffic congestion detection system is discussed in Section 9.3. All the modules of this model are also discussed in detail here. The experimental analysis and the results of each and every module are presented in Section 9.4. Finally, Section 9.5 concludes the paper.

9.2 RELATED WORK

Recently, image processing techniques to control traffic signals have been examined by many researchers. Different image processing operations like segmentation, filtering, and object analysis are used on the input image to evaluate the traffic. This aforementioned technique is classified into two parts: classical machine vision methods and composite deep learning method. Classical methods utilize vehicle movement to detach it from a steady background image. These methods are classified into three types: (a) using background subtraction [1313], (b) using endless video frame difference [14], and (c) using optical flow [15]. Today, deep convolution networks are contributing a lot to detection of vehicles on the road. These methods have a distinct capability to study image features and can execute multiple associated works like classification and bounding box regression [16].

Atkociunas et al. [17] proposed an approach to monitor roads and traffic problems like tracking of vehicles, measurement of speed, detection of traffic jams, and recognition of number plates. Tracking based on contour eradication is used to track a vehicle by using the edge linking process. Finally, the geometric center is calculated and used for vehicle tracking. Then, the authors used a motion detection method to measure the speed of vehicles and neural network technology to detect number plates. Sharma et al. [18] used average filter in their proposed automated vehicle detection system to diminish the effect of noise and also examined differential morphological techniques for vehicle segmentation and shape detection. Thresholding amount is used to abolish undesirable objects besides the vehicles. Lastly, to derive the marked vehicle shape, index thresholding is used. Pejman [19] proposed to deliver an automated traffic control system for traffic on highways and streets based on contour extraction and detection of motion. The technique proposed is background subtraction. It clears away all background factors and detects the foreground. For image contrast adjustment, gamma correction is used. Secondary Hough Transform technique is used by other authors in their algorithms [20] for the discovery of road and environmental conditions as well as curved road positions. The transformation of the gray scale image into a black and white image is also discussed by the authors. Gupta et al. [21] elaborated a technique to calculate the traffic load by edge detection, histogram equalization, labeling, and removed the noise with the support of median filter. Median filter is used to get smooth image and sharp boundaries. Coifmana et al. [22] discussed a vehicle detection and tracking system which is planned to perform in different conditions (sunlight, sunset, and night-time situations). Instead of detecting the entire vehicle, some features of the vehicle are traced, which renders the whole system less susceptible to the dilemma of uncompleted occlusion. The authors also explained image segmentation techniques using active contour, which segments the two-dimensional gray scale image into foreground (object) and background regions. Changalasetty et al. [23] have automated the traffic monitoring system by performing the recognition and categorization of vehicles on the road. This system utilizes LABVIEW for processing of sample images of vehicles to get features like area, perimeter, width, and length. The neural network classification technique of data mining is used by the authors for the categorization of vehicles in terms of big or small. Gunasekaran and Padmavathi [24] proposed techniques to detect vehicles during the night time. To modify the contrast of images, White Top Hat Transform method is used. To detect vehicles for further analysis, methods such as multilevel thresholding, support vector machine classification, rule based component analysis, and symmetric based identification are used. Dharani and Anitha [25] proposed an image cropping method to estimate the number of vehicles on the road and also manage the traffic without considering the whole image. Prutha and Anuradha [26] authors proposed a real time vehicle identification method by considering background differencing, morphological operations, and edge detection. An effective and efficient approach based on the SIFT feature extraction algorithm, which can detect and track vehicles in the night time, is presented by Shao-Ping and Xiao-Ping [27].

After detecting vehicles on roads by using an image processing technique, it can be further processed to detect the intensity of traffic jams. For the detection of congestion, the number of messages collected from all the vehicles is higher. Because of limited availability of bandwidth, the key parameters from this large number of messages are extracted using FCM. The FCM technique is preferable here because of the dynamic nature of vehicle motion. These key parameters with the vehicle count extracted from the image processing technique are used in FLC to quantify the level of congestion.

The main contributions of the proposed work are as follows:

1. The proposed work uses an image processing technique to count the number of vehicles.
2. Real world vehicle data such as speed, CO_2 emission, and fuel consumption are extracted using the SUMO simulator.
3. It uses the FCM technique to extract the key features.
4. These key features are utilized with the number of vehicles extracted from the image processing technique to quantify the level of congestion.
5. This proposed work is tested to quantify the level of congestion in Bhubaneswar city, which is an administrative, educational, information technology, and tourism center in Odisha, India.

9.3 PROPOSED SYSTEM MODEL

The main objectives of this work are the detection and counting of vehicles, gathering of vehicle parameters, extraction of key parameters, and estimation of level of traffic congestion. The architecture of the proposed model is shown in Figure 9.1. This system has four major units, referred to in this work as the moving vehicle detection and counting system, SUMO simulator, FCM clustering unit, and FLC.

9.3.1 Moving Vehicle Detection and Counting System

On a lane, the traffic video is recorded through a color camera. This video is affected by noise due to illumination, shadow, camera imaging errors, and camera jitters. This single video is a series of frames of color images. To improve the images, the denoising of the original image and extraction of relevant information are required. This helps in improvement of operation accuracy and reduction of computation time of subsequent modules.

The block diagram of this moving vehicle detection and counting system is shown in Figure 9.2. This actually works on the reloaded video as an input. This system takes a frame of the image as an input, which is extracted from the video sequence.

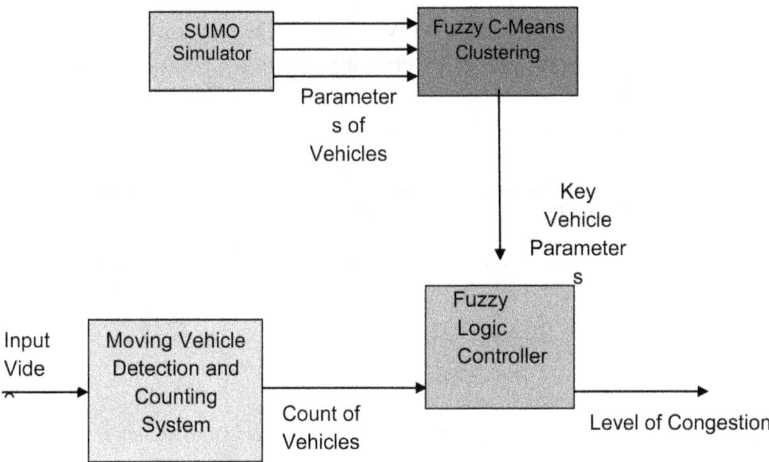

FIGURE 9.1 Block diagram of real time density–based traffic congestion detection system.

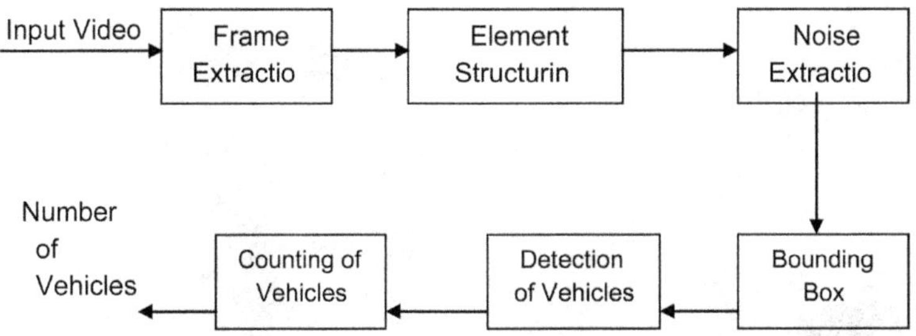

FIGURE 9.2 Moving vehicle detection and counting system.

The structuring element identifies the pixel in the frame of the image and defines the neighborhood used in the processing of each pixel. Actually, it is used to detect an object of the same shape and size as the structuring element in the frame. After elimination of noise, these object elements are put in a box, and the count of those boxes is the count of vehicles in a frame. Then, these frames are sequentially grouped to create the same video again, and the total count of vehicles in each frame is gathered to give an idea of the number of vehicles in that video.

9.3.2 Parameter Extraction Using SUMO Simulator

The SUMO simulator is taken to set up a traffic scenario from which the vehicle parameters are taken to identify the traffic congestion in a particular lane. The flow diagram for getting the actual data from the vehicles moving in the scenario is shown in Figure 9.3. The actual road map of Infocity Square, the most crowded area of Bhubaneswar, is collected from Open Street Map and is provided to SUMO for real time simulation.

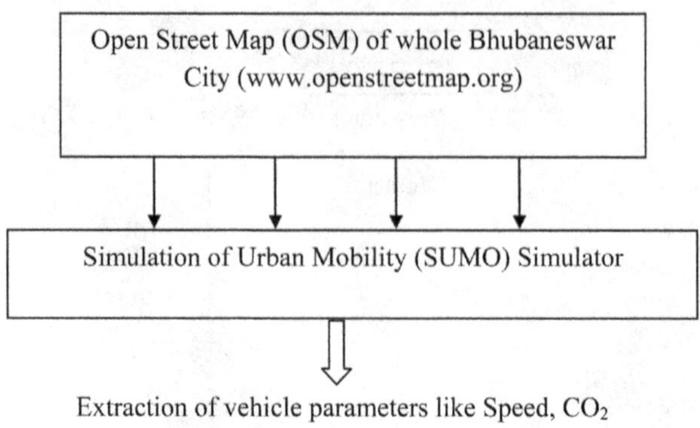

FIGURE 9.3 Flow diagram for getting real time parameters from vehicles.

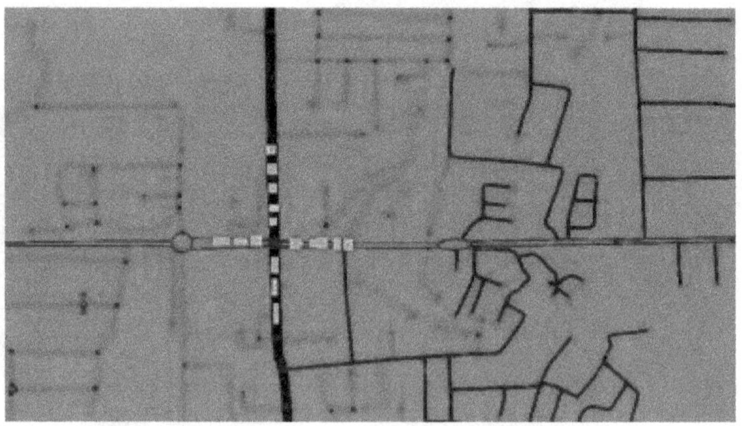

FIGURE 9.4 Congestion created at Infocity Square, Bhubaneswar.

At Infocity Square, traffic congestion is generated in SUMO as shown in Figure 9.4. Then, the rough outputs, like speed, CO_2 emission, and fuel consumption, etc., are collected from it to perform the simulation.

9.3.3 Key Features Extraction using Fuzzy C-Means Clustering

Actually, a number of various on-board sensors installed in vehicles produce abundant messages, which create a channel clash issue and disable the available bandwidth. The on-board units are set up in vehicles to gather output from sensors, like the speed of vehicles, CO_2 emission, and consumption of fuel, and transmit these messages to roadside infrastructure, where the key parameters are extracted using the clustering technique. In place of a number of sensors, the SUMO simulator is used here to generate vehicle parameters such as speed, CO_2 emission, and fuel consumption. Using this clustering technique, the data set including vehicle parameters is precisely grouped into a single cluster to generate the key parameters.

The unsupervised FCM clustering creates *k* number of clusters by considering the data points with a greater belongingness to that particular cluster [28,29]. The FCM algorithm is mentioned in Annexure 9.3.

The data set D_{SUMO}, which indicates the messages with features of speed (km/h), CO_2 (mg/s) emission, and fuel consumption (ml/s), is collected from the SUMO simulator and is listed in Table 1 in Annexure 9.1. For the extraction of key parameters, 10 vehicles are captured to generate the data set D_{SUMO}. The key parameters are mean values of the centers of the clusters.

Let $X_1, X_2, X_3, \ldots, X_n$ be the centers of n number of clusters, then the mean values of speed, CO_2 emission, and fuel consumption are estimated for a set of data by using Equations 9.1, 9.2, and 9.3.

$$S_m = \frac{\sum_{i=1}^{n} S_i}{n} \tag{9.1}$$

$$C_m = \frac{\sum_{i=1}^{n} C_i}{n} \tag{9.2}$$

$$F_m = \frac{\sum_{i=1}^{n} F_i}{n} \tag{9.3}$$

where S_i, C_i, and F_i are the speed, CO_2 emission, and fuel consumption of ith center X_i and S_m, C_m, and F_m are the mean values of speed, CO_2 emission, and fuel consumption of all centers.

9.3.4 Traffic Congestion Level Estimation Using Fuzzy Logic Controller

The level of traffic congestion is estimated using a Mamdani FLC [30,31] based on fuzzy logic decision support. The detection mechanism consists of four linguistic inputs, the vehicle count, speed, CO_2 emission, and fuel consumption, that mostly affect the congestion, and the linguistic output is the traffic congestion level. Each of these variables has different ranges to create different possible combinations of inputs to evaluate the congestion level. The proposed FLC inference engine as shown in Figure 9.5 is designed with 81 rules for congestion. It has three main blocks: fuzzifier, fuzzy rules, and defuzzifier. In this work, triangular membership functions are considered.

In fuzzifier, the linguistic variable speed could have a value that is one member of this set {low, medium, high}. As shown in Figure 9.6, the labels send a message by correlating with everyone a fuzzy subgroup of a few universes of discourse.

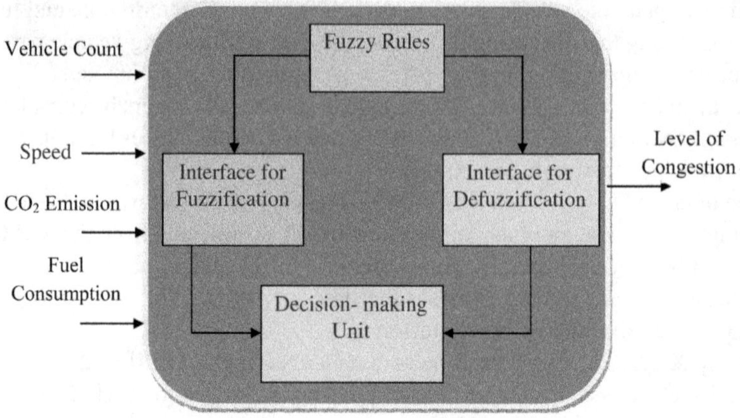

FIGURE 9.5 Architecture of fuzzy logic controller for estimating the level of congestion.

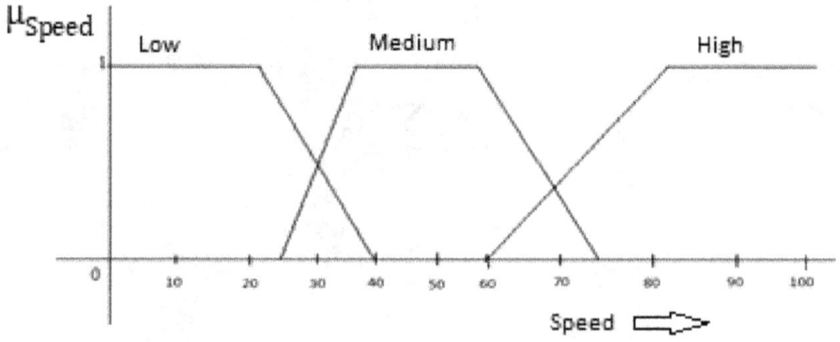

FIGURE 9.6 Linguistic variable vehicle speed.

The fuzzy rules of FLC to control are declared in the mode of a fuzzy if-then conditional statement as in Equation 9.4.

$$R_i : \text{If a is } M_i \text{ And}\ldots\text{and b is } N_i\text{, THEN c is } P_i \tag{9.4}$$

where a, b, and c are variables of fuzzy, and M_i, N_i, and P_i are subsets of fuzzy in the universal set X, Y, and Z. R_i, the fuzzy rule, can be viewed as a relation in fuzzy from the universal set (X and … and Y) to universal set Z. For p rules, the set of rules may be defined by a union of all those rules: $R = R_1 \cup R_2 \cup R_3 \ldots\ldots R_p$. The resultant set c can be created by the use of compositional rule of inference c = (a And …..And b)○R, where ○ stands for the compositional rule of inference.

In general, the most commonly used fuzzy inference techniques in fuzzy control are Mamdani's fuzzy implication method as in Equation 9.5.

$$\mu_C(c) = \left\{\min\left\{\mu_M(a), \ldots \mu_N(b), \mu_R(a, \ldots b, c)\right\}\right\} \tag{9.5}$$

If the variables of fuzzy a_1, b_2, and c_3 are fuzzy singletons, then the complete outputs of both the implication methods will be as given in Equation 9.6.

$$c_{0_i} = \min\left\{\mu_{M_i}\left(a_0\right),\ldots\mu_{N_i}\left(b_0\right)\right\}\cdot P_i \tag{9.6}$$

where a_0, b_0, and c_0 are fuzzy singletons, and P_i is the singleton value of c using the ith rule.

Actually, there is no such systematic process or method to select a defuzzification approach in FLC. The most commonly used is the MAXimum criterion method (MAX). This method generates the mark where the control action membership function arrives at a maximum value. The mean value of maxima is calculated using the mean-of-maximum (MOM) method. The Center-Of-Gravity (COG) method produces the center of gravity of the membership function area of the control action as shown in Figure 9.7.

The fuzzy rules for detection of congestion taking vehicle count, speed, CO_2 emission, and fuel consumption are listed in Table 2 in Annexure 9.2.

9.4 EXPERIMENTAL ANALYSIS AND RESULTS

To test the efficiency and accuracy of the moving vehicle detection and counting system, 10 videos are tested. The different steps to count the number of vehicles in a frame are shown in Figure 9.8.

The output of vehicle count from all these 10 videos is listed in Table 9.1.

The overall accuracy of the system is calculated to be 87.56%. After detection of the number of vehicles, the count is fed to the FLC. To test and design the FLC, the vehicle parameters are extracted from the SUMO simulator. The parameters extracted from SUMO simulator are listed in Table 1 in Annexure 9.1.

The information obtained from vehicles travelling in a crowded lane was used to extract parameters in our experiment. For video 1, a maximum of 10 pieces of

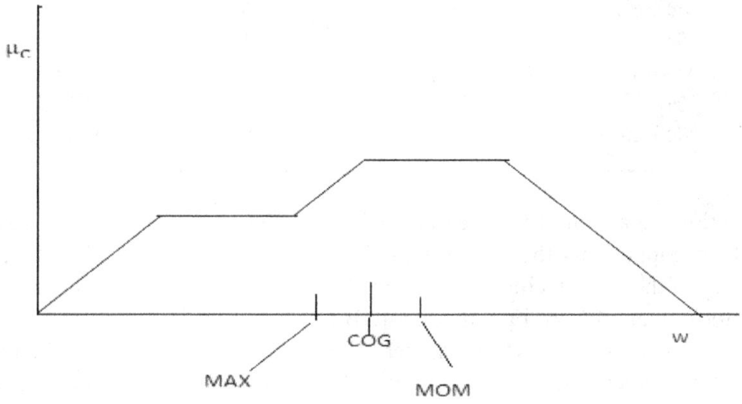

FIGURE 9.7 Strategy in defuzzification.

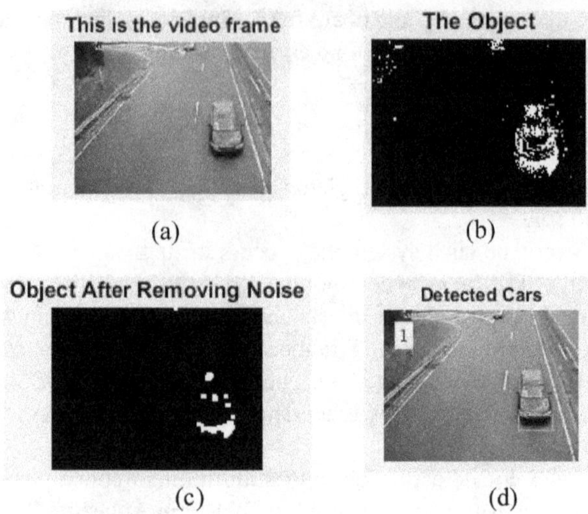

FIGURE 9.8 (a) Extracted video frame. (b) Detection of object from first frame. (c) Detection of object from first frame after removal of noise. (d) Count of vehicles from first frame.

TABLE 9.1
Vehicle Count

Name of the Video	Actual Number of Vehicles	Detected Number of Vehicles	Accuracy (%)
Video 1	10	9	90
Video 2	8	6	75
Video 3	12	10	83.33
Video 4	6	5	83.33
Video 5	16	14	87.50
Video 6	5	5	100
Video 7	1	1	100
Video 8	21	18	85.71
Video 9	18	16	88.89
Video 10	11	9	81.81

information are acquired with features of speed, CO_2 emission, and fuel consumption. That implies that this data set carries 10 data points. In MATLAB [32], the efficiency of the FCM technique is tested. All the calculations are achieved on Asus Vivo Book Intel 14 Core i3 10th Gen-4GB CPU @ 2.1 GHz with Turbo Boost up to 4.1 GHz with 4 GB RAM computer. The FCM technique clusters the different information gathered from the whole variety of vehicles in a lane. The FCM function works on the data set generated and completed by information coming from the vehicles, and finally, some clusters are generated.

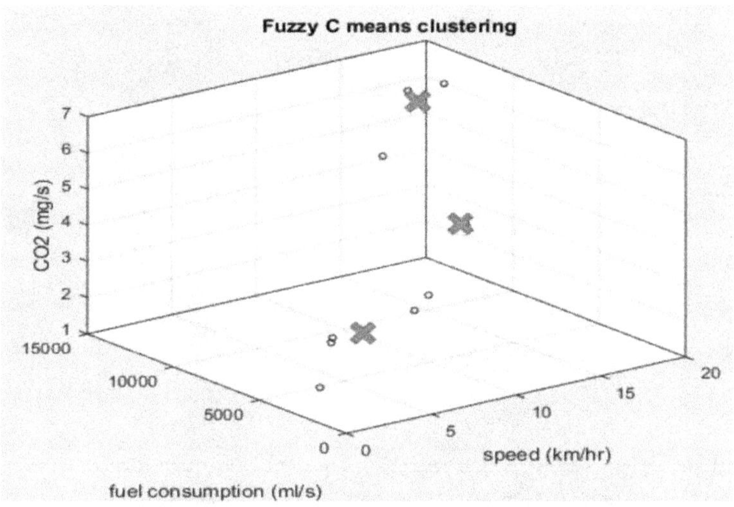

FIGURE 9.9 Formation of clusters taking the data messages using fuzzy C-means clustering.

TABLE 9.2
Centers and Mean of the Clusters

Clusters	Speed	CO_2 Emission	Fuel Consumption
Cluster 1	15.3	8,433.9	3.6
Cluster 2	18	13,489	6
Cluster 3	6.2	5,071.9	2.2
Mean value	13.167	8,998.267	3.933

A three-dimensional plot formed with three attributes, speed, CO_2 emission, and fuel consumption, for all the vehicles is shown in Figure 9.9. The X marks show the centers of the clusters. The three cluster centers and their mean value are shown in Table 9.2.

The mean value of the cluster centers is the key message from this group of data. After obtaining these key parameters, FLC is used in conjunction with the vehicle count. This Mamdani FLC works with four input variables: vehicle count, speed, CO_2 emission, and fuel consumption. This controller uses one output variable: traffic congestion level. The fuzzy rules for detection of traffic congestion level are listed in Table 2 in Annexure 9.2. Figure 9.10a to 10d depicts the membership functions of different inputs. Figure 9.11 depicts the membership functions of output variables.

Rule Viewer defines the whole fuzzy inference process at once. It reveals how the shape of certain membership functions impacts the overall result. As it displays every part of every rule, it performs well for small number of inputs and outputs. Figure 9.12 shows the response of Traffic Congestion Level FLC using Rule Viewer.

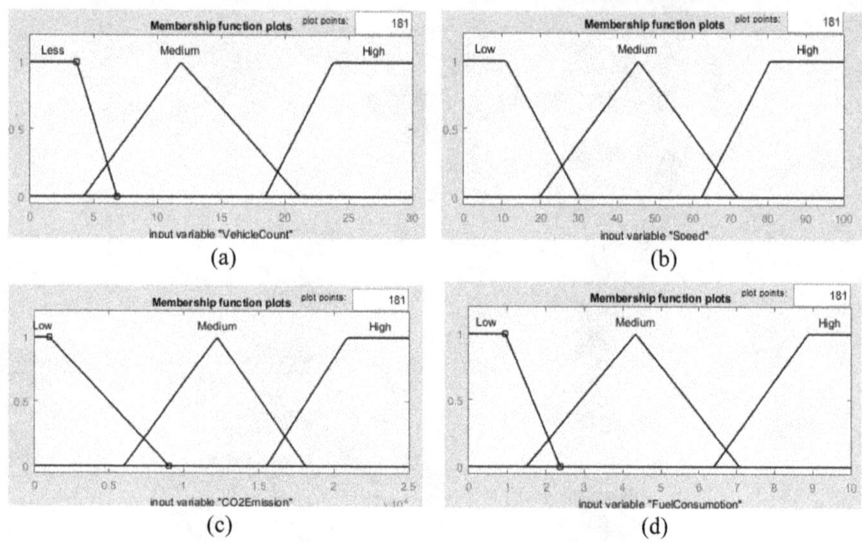

FIGURE 9.10 (a) Membership function (vehicle count). (b) Membership function (vehicle speed). (c) Membership function (CO_2 emission). (d) Membership function (fuel consumption).

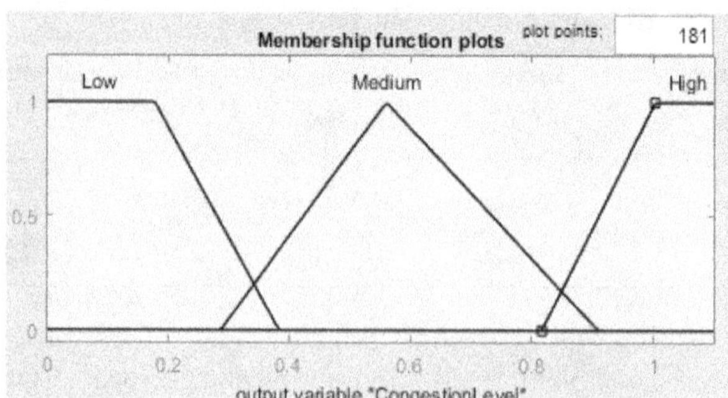

FIGURE 9.11 Membership function (congestion level).

For video 1, where the number of vehicles counted is 9, the congestion level is detected. The other vehicle outputs are the FCM outputs: speed is 13.167 km/h, CO_2 emission (mg/s) is 8998.267, and fuel consumption is 3.933 ml/s. With these parameters, the traffic congestion level is detected to be 53.4%.

9.5 CONCLUSION

A dynamic and efficient real time density–based traffic congestion detection system is presented, which specifies an excellent traffic detection technique without waste of time. It is also a great way of identifying the existence of vehicles in the lane, as it uses image data.

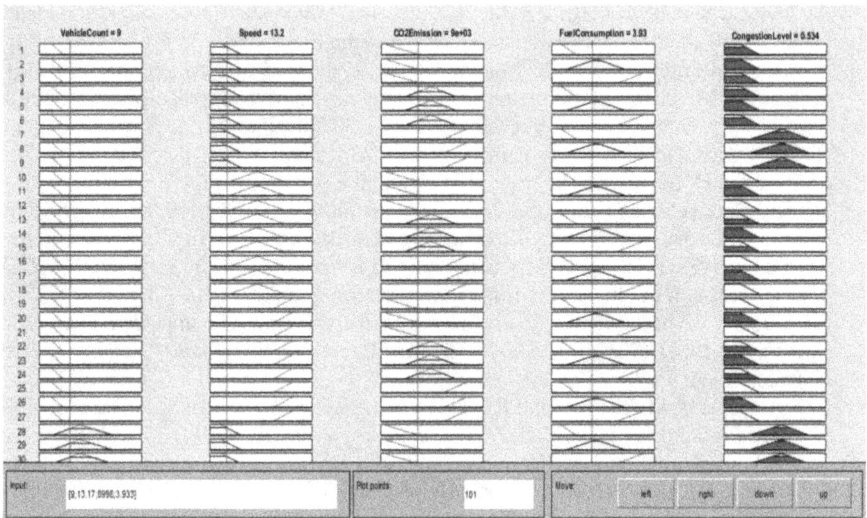

FIGURE 9.12 Rule viewer of congestion level.

Here, this image processing technique detects vehicles with an accuracy of 87.56% and also overcomes the drawbacks of every conventional approach of traffic detection. It excludes the requirement for additional hardware and sensors. With the image processing technique, it also uses FCM and FLC. This model is able to detect the congestion level for a particular video with some particular parameters at an accuracy of 53.4%. To further enhance the learning and for other prospective associated research, to depict the parameter more precisely, one can conclusively describe the membership functions range values. Also, a more strong and decisive vehicle detection algorithm can be adopted to increase the accuracy of the parameter values that will be supplied to the fuzzy logic system.

REFERENCES

1. Indians spend 7% of their day getting to their office, *The Economic Times* (indiatimes .com).
2. R. Bauza, J. Gozalvez, J. Sanchez-Soriano, "Road traffic congestion detection through cooperative vehicle-to-vehicle communications," in Proceedings of the IEEE Conference on Local Computer Networks, LCN'10, Denver, CO, pp. 606–612, 2010.
3. F. Terroso-Saenz, M. Valdes-Vela, C. Sotomayor-Martinez, R. Toledo-Moreo, A. Gomez-Skarmeta, "A cooperative approach to traffic congestion detection with complex event processing and VANET," *IEEE Transactions on Intelligent Transportation Systems*, vol. 13, no. 2, pp. 914–929, 2012.
4. M. Wright, R. Horowitz, "Fusing loop and GPS probe measurements to estimate freeway density," *IEEE Transactions on Intelligent Transportation Systems*, vol. 17, no.12, pp. 3577–3590, 2016.
5. D. Guibert, S.-S. Leng, et al., "Robust blind deconvolution process for vehicle reidentification by an inductive loop detector," *IEEE Sensors Journal*, vol. 14, no.12, pp. 4315–4322, 2014.

6. G. Jiang, S. Niu, A. Chang, Z. Meng, C. Zhang, "Automatic traffic congestion identification method of expressway based on gain amplifier theory," in Proceedings of the Advanced Computer Control, Vol. 2, ICACC'10, Shenyang, China, pp. 648–651, 2010.
7. W. Zhu, M. Barth, "Vehicle trajectory-based road type and congestion recognition using wavelet analysis", in Proceedings of the IEEE Intelligent Transportation Systems Conference, ITSC'06, Paris, France, pp. 879–884, 2006.
8. S. Vaqar, O. Basir, "Traffic pattern detection in a partially deployed vehicular ad hoc network of vehicles", *IEEE Wireless Communications*, vol. 16, no. 6, pp. 40–46, 2009.
9. K. Mandal, A. Sen, et al., "Road traffic congestion monitoring and measurement using active RFID and GSM technology," in Proceedings of the International IEEE Conference on Intelligent Transportation Systems, pp. 1375–1379, 2011.
10. D. Beymer, P. McLauchlan, B. Coifman, J. Malik, "A real-time computer vision system for measuring traffic parameters," IEEE Conference on Computer Vision and Pattern Recognition, pp. 495–501, 1997.
11. R. Cucchiara, M. Piccardi and P. Mello, "Image analysis and rule-based reasoning for a traffic monitoring system," *IEEE Transactions on Intelligent Transportation Systems*, vol. 1, no. 2, pp. 119–130, 2000.
12. Z. Yang, H. Liu and C. Du, "Study of fuzzy control for intersections with traffic intensity being priority", *Computer Engineering and Applications*, vol. 45, no. 36, 2009.
13. M. Radhakrishnan, "Video object extraction by using background subtraction techniques for sports applications," *Digital Image Processing*, vol. 5, no. 9, pp. 91–97, 2013.
14. L. I. Qiu-Lin, H. E. Jia-Feng, "Vehicles detection based on three-frame-difference method and cross-entropy threshold method," *Computer Engineering*, vol. 37, no. 4, pp. 172–174, 2011.
15. Y. Liu, L. Yao, Q. Shi, J. Ding, "Optical flow based urban road vehicle tracking," In 2013 9th International Conference on Computational Intelligence and Security, 2014. IEEE. https://doi.org/10.1109/cis.2013.89
16. Z. Q. Zhao, P. Zheng, S. T. Xu, X. Wu, "Object detection with deep learning," A review. arXiv e-prints, arXiv:1807.05511, 2018.
17. E. Atkociunas, R. Blake, A. Juozapavicius, M. Kazimianec, "Image processing in road traffic analysis," *Nonlinear Analysis: Modeling and Control*, vol. 10, no. 4, pp. 315–332, 2005.
18. B. Sharma, V. K. Katiyar, A. K. Gupta, and A. Singh, "The automated vehicle detection of highway traffic images by differential morphological profile," *Journal of Transportation Technologies*, vol.4, pp. 150–156, 2014.
19. P. Niksaz, "Automatic traffic estimation using image Processing," *International Journal of Signal Processing, Image Processing and Pattern Recognition*, vol. 5, no. 4, December, 2012.
20. J. G. Haran, J. Dillenburg, P. Nelson, "Real-time image processing algorithms for the detection of road and environmental conditions," in 9th International Conference on Applications of Advanced Technology in Transportation (AATT), 2006.
21. P. Gupta, G. N. Purohit, A. Gupta, "Traffic load computation using corner detection technique in Matlab Simulink model," *International Journal of Computer Applications*, vol. 72, no. 6, June 2013.
22. B. Coifmana, D. Beymerb, P. McLauchlanb, J. Malik, "A real-time computer vision system for vehicle tracking and traffic surveillance," *Transportation Research Part C*, vol.6, pp. 271–288, 1998.
23. S. B. Changalasetty, A. S. Badawy, W. Ghribi, H. I. Ashwi, "Identification and classification of moving vehicles on road," *Computer Engineering and Intelligent Systems*, vol.4, no. 8, 2013.

24. K. Gunasekaran, S. Padmavathi, "Night time detection for real time traffic monitoring systems: A review," *Int. J. Computer Technology & Applications*, vol. 5, no. 2, pp. 451–456, 2014.
25. S. J. Dharani, V. Anitha, "Traffic density count by optical flow algorithm using image processing," *Automative Parts system and Application*, vol. 3, no. 2, April 2014.
26. Y. M. Prutha, S. G. Anuradha, "Morphological image processing approach of vehicle detection for real-time traffic analysis", *International Journal of Engineering Research and Technology*, vol. 3, no. 5, May 2014.
27. Z. Shao-Ping, F. Xiao-Ping, "Nighttime motion vehicle detection based on MIL boost," *Sensors and Transducers*, vol. 171, no. 5, pp. 93–98, May 2014.
28. Y. Yong, Z. Chongxun, L. Pan, "A novel fuzzy C-means clustering algorithm for image thresholding," *Measurement Science Review*, vol. 4, no. 1, 2004.
29. A. Mohanty, S. Mahapatra, U. Bhanja, "Traffic congestion detection in a city using clustering techniques in VANETs," *Indonesian Journal of Electrical Engineering and Computer Science*, vol. 5, no. 3, pp. 401–408, March 2017.
30. R. Bauza, J. Gozalvez, "Traffic congestion detection in large-scale scenarios using vehicle-to-vehicle communications," *Journal of Network and Computer Applications*, vol. 36, pp. 1295–1307, 2013.
31. A. Mohanty, U. Bhanja, S. Mahapatra, "Intelligent traffic quantification system," Proceedings of the IOP Conference Series: *Material Science and Engineering*, vol. 225, 2017.
32. Math works http://www.mathworks.com.

ANNEXURE 9.1

TABLE 1
D_{SUMO}: Data Set

Attributes → ↓ Identification of Vehicle	Speed (km/h)	CO_2 Emission (mg/s)	Fuel Consumption (ml/s)
1	4.25	5,006.74	2.15
2	14.10	11,690.13	5.03
3	6.18	5,236.77	2.25
4	15.54	8,364.15	3.6
5	1.89	3,340.78	1.44
6	17.94	14,013.07	6.02
7	9.62	5,504.7	2.37
8	4.45	5,175.93	2.22
9	20.00	13,938.78	5.99
10	10.97	5,993.99	2.58

ANNEXURE 9.2

TABLE 2
Some of the Rules for Detection of Congestion

	Linguistic Inputs				Linguistic Output
Rule No.	Vehicle Count	Speed	Fuel Consumption	CO_2 Emission	Level of Congestion
1	Le	Lo	Lo	Lo	Lo
2	Le	Lo	Lo	Me	Lo
3	Le	Lo	Lo	Hi	Me
4	Le	Lo	Me	Lo	Me
5	Le	Lo	Me	Me	Me
6	Le	Lo	Me	Hi	Se
7	Le	Lo	Hi	L	Me
8	Le	Lo	Hi	Me	Hi
9	Le	Lo	Hi	Hi	Hi
10	Le	Me	Lo	Lo	Lo
11	Le	Me	Lo	Me	Lo
12	Me	Me	Lo	Hi	Me
13	Me	Me	Me	Lo	Me
14	Me	Me	Me	Me	Me
15	Me	Me	Me	Hi	Me
16	Me	Me	Hi	Lo	Me
17	Me	Me	Hi	Me	Hi
18	Me	Me	Hi	Hi	Lo
19	Me	Hi	Lo	Lo	Lo
20	Me	Hi	Lo	Me	Me
21	Me	Hi	Lo	Hi	Lo
22	Hi	Hi	Me	Lo	Lo
23	Hi	Hi	Me	Me	Lo
24	Hi	Hi	Me	Hi	Me
25	Hi	Hi	Hi	Lo	Me
26	Hi	Hi	Hi	Me	Me
27	Hi	Hi	Hi	Hi	Lo
28	Hi	Hi	Lo	Lo	Lo
29	Me	Lo	Me	Lo	Lo
30	Me	Lo	Lo	Lo	Lo

Here Lo is Low, Le is Less, Hi is High, Me is medium, and Se is Severe.

ANNEXURE 9.3

ALGORITHM: FUZZY C-MEANS CLUSTERING [28,29]

Input:

- k: the number of clusters
- X: the data set containing objects
- f: the parameter in objective function
- ϵ: a threshold for the convergence criteria

Output:

- A set of clusters

Method:

1. Initialize V= $\{v_1, v_2 \ldots, v_k\}$
2. $V^{previous} \leftarrow V$
3. Compute the membership function $\mu_{C_i}(x)$ using Step (2).
4. Update the prototype v_i in V using Step (3) till $\sum_{i=1}^{C} v_i^{previous} - v_i \leq \varepsilon$; otherwise, return to Step (2).

10 Fundamentals of Face Recognition with IoT

Payal Parekh and Hina J. Chokshi

CONTENTS

10.1 INTRODUCTION

In the preceding few years, face recognition has been a working research field. At first, a branch of AI (artificial intelligence) was integrated with robotics to enable robots to have visual perception. But today, it is part of a broader and larger discipline of computer vision and pattern recognition. Furthermore, many engineers and scientists all around the globe have been hard at work developing increasingly powerful and precise approaches and algorithms for these kinds of processes and their applications in everyday life. In order to protect any personal data, various forms of security measures have become necessary. Assigning a password is the most popular way to protect personal information. Several systems have started to utilise various biometric standards to identify tasks as security algorithms and information technology have advanced [1].

The physiological or behavioural characteristics of people can be easily identified by these biometric factors. Many other advantages are also realised, such as the fact that a person's presence in front of the sensor is sufficient to identify them, eliminating the need to remember complicated passwords and sensitive pass codes.

DOI: 10.1201/9781003221333-10

Many recognition systems based on biometric characteristics, including fingerprints [2], iris, face, and voice [3] have been employed in the last few decades in this area.

Many simple systems are gaining popularity and appeal as a result of their ability to identify persons based on their biological traits. As a result, given its potential in a wide range of applications and industries, this has become one of the most extensively utilised biometric authentication methods in recent years (surveillance, home security, border control, and so on) [4].

Different robust and discriminating face detection and identification strategies, such as local, subspace, and hybrid approaches, are proposed [5].

Due to various difficulties, such as facial expressions, lighting circumstances, and head orientation, there are still issues that need to be properly addressed. Reliable face recognition and identification systems have been developed to deal with these issues. These systems are extremely complicated, necessitating greater processing time and system memory consumption. Because of its vast potential applications and theoretical importance, face recognition is a prominent research issue in the disciplines of computer vision and image processing. Because it integrates two modern technologies, namely, face recognition and the Internet of Things (IoT), it may provide a high level of security. These two are quickly evolving industrial technologies, which scientists are still studying, and have had a tremendous impact on the development of security systems. As a result, data security is doubled. The usage of IoT and face recognition also allows remote control and monitoring. This feature makes hacking nearly impossible.

Face biometrics are being applied in many areas in manufacturing, industries, construction, disrupting design, and healthcare.

- **Payment**: Postmodern purchasing includes online shopping and the use of cards, to name a few examples. Facial recognition (FR) is already in use in businesses, but the next step is to make it possible to pay for it online, e.g. applications like Alibaba and Alipay affiliate payment software [6].
- **Access and security**: A face biometric can be integrated into the physical device and object used for payment confirmation. Apple, Samsung, and Xiaomi Crop phones all have FaceTech, which is used for access through the owner's facial features instead of using a password or pattern. Advanced face security can be especially useful for companies or organizations that manage sensitive information and have tight control over who enters the facility [6].
- **Criminal identification**: The FBI now maintains a record that has facts on half of the population of the United States. This is an easy method to keep track of crime statistics around the country. Cameras equipped with AI to identify criminals in the UK have also been tested [6].
- Advertising: Marketers and advertising companies are now able to collect personal information from their target markets because the data is readily available. OptimEyes screens can be installed in public places. If your consumer is a man between the ages of 15 and 25, for example, the current FIFA games may be advertised on the screen [6].

- **Healthcare**: Instead of recognising an individual through Face Tech, medical professionals can look at patient characteristics to identify illnesses. In the medical profession, face biometrics is used to safeguard patient data by utilising a unique patient picture instead of a password and login to protect patient data [6].
- **Smart homes**: Home locking devices used currently follow the traditional way of locking the doors using locks. Face recognition systems can be integrated with IoT for face recognition. A Raspberry Pi is installed to perform the desired functionalities, along with enabling the owner to add family members or known faces to the database [7].
- **Surveillance systems**: Surveillance face detection and recognition systems are capable of recognising faces captured in video frames captured by Internet protocol (IP) cameras. These systems are extremely expensive. To keep systems cost-effective, IP cameras can be prepared utilising low-price cloud computing services and IoT devices [8].

This chapter first introduces the applications and basics of face recognition. Every step is a second theme that provides a process of the overall flow of face recognition described in the method used at each stage. This is the third theme, which describes the IoT system architecture using facial applications. Fourth, we present a comparison table related to facial applications used in IoT. Fifth, the challenges and limitations of IoT systems comprise the final conclusion of the chapter.

10.2 PROCESS OF FACE RECOGNITION

It is essential to give a brief explanation of the challenges that need to be addressed and overcome in order to accurately complete the face identification task before detailing the approaches employed. A facial recognition system has the following features: it can process video and photos in real time, it is robust under varied lighting circumstances, it is independent of humans (regardless of hair, ethnicity, or gender), and it can handle faces from different angles [9]. To obtain data, different types of sensors are used: for example, electroencephalography (EEG), and wearable inertial sensors. Using these sensors provides extra information that enables the FR systems to recognise both the static human two-dimensional (2-D) face and video sequences. Also, there are three different types of sensors that help in improving the accuracy and reliability of an FR system by dealing with the challenges of head pose, illumination variation, and facial expression in video and 2-D image processing. The first, most common type is the non-visual sensors, namely, depth, EEG, and audio sensors, that provide extra data along with the visual dimension in order to improve the image recognition reliability. The second type of sensor is the detailed face sensors that help to detect small changes in face components. Examples include eye-trackers, which may help to differentiate the face image from the background noise. The last type of sensor is target-focused sensors, which facilitate the FR system to differentiate the least required visual contents, which helps in dealing with illumination variation (Figure 10.1).

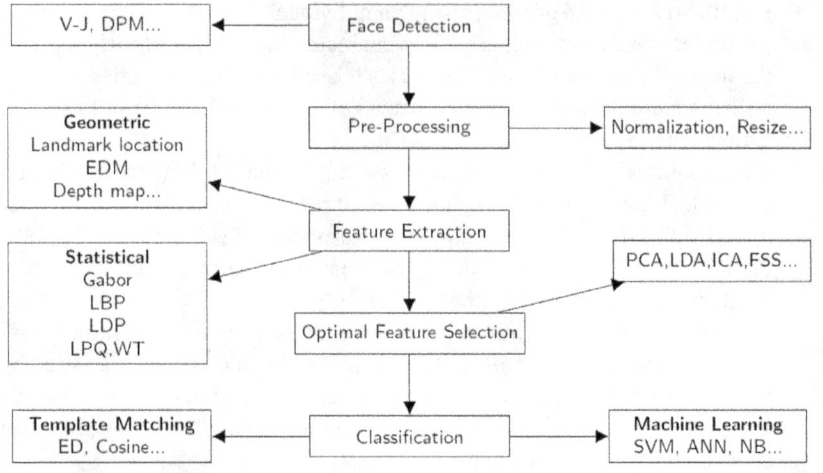

FIGURE 10.1 Processing flow of face recognition. (Payal, P. and Goyani, M.M., *The Imaging Science Journal*, 68, 114–127, 2020.)

10.2.1 Fundamentals of Face Recognition Steps

10.2.1.1 Face Detection

In the above-mentioned face recognition system, the initial stage is to localise a person's face in a specific image, where the photos carry out a person's face or merely a fatch face part. Many differences in facial expression and lighting can make accurate face identification difficult. A pre-processing phase is conducted in the design of the FR system to make the process more resilient.

Face detection progress is largely due to the cascade detector described by Viola and Jones (2001) [11] and Jones (2004) [12]. Fanaee et al. (2019) [13] have developed a face detection system that is both fast and reliable. The AdaBoost-based method is used for nonlinear classification, where the cascade detector leads to real-time detection. To achieve higher detection accuracies, various improvements have been proposed whereby training and testing time is reduced. Elastic bunch-graph templates, soft cascade method, modified cascade detector, and FloatBoost are examples of discrete AdaBoost versions.

10.2.1.2 Pre-Processing Image

The purpose of the pre-processing algorithm is to remove the impacts of lighting, position fluctuations, and occlusion factors that can degrade an FR system's performance. A possible concept for the pre-processing stage is to convert the photos from RGB to greyscale. Then, it ensures that all the faces are pointing in the same direction and that the features are taken from the same semantic regions. To achieve good results, contrast and illumination adjustments will be made. In order to get an appropriate contrast, histogram equalisation can be used to the input face image. As stated, to improve the colour contrast and brightness value of an image, a new approach for appropriate choice of discrete wavelet transform using contrast limited adaptive histogram equalisation (E-CLAHE) parameters was developed by Thamizharasi Ayyavoo et al.(2018) [14]. As the brightness and contrast were significantly boosted, the recognition accuracy of discrete wavelet transform E-CLAHE in databases was achieved extremely well. For pre-processing, Xinfang Cui et al. (2017) [15] utilised a Gaussian Kernel, whereas Zhihua Xie et al. (2017) [16] used a Gaussian filter and a normalised 2-D face image, and Farnaz et al. (2018) [13] used a highpass filter.

10.2.1.3 Feature Extraction

In an FR system, the most important function is through face feature face will be recognised or pre-processing images. This stage depicts a face by detailing the distinguishing characteristics of the face picture, such as the lips, eyes, and nose as well as their geometric distribution [17]. Structure, shape, and size are features of each face that identify the face. A large number of methods are employed for identifying the face using the size and distance parameters by extracting features like eyes, mouth shape, and nose [10] Global feature extraction techniques are used in global techniques for selecting a complete set of features, as stated in the paper by Xie et al. (2017) [16] and by Yang et al. (2018) [18]. Local methods are used for unusable information and another advantage of local approaches is that they are unaffected by facial geometry, ageing, changing position, or face rotation, as described in the articles by Weng et al., Sariyanidi et al., and Jabid et al. [19–21]. For balancing the tradeoffs, both these techniques can also be combined together as in Fanaee et al., Ayyavoo et al., Goyani and Patel, Koteswara Rao et al., and Gajendrasinh et al. [13, 14, 22–24]. Local Binary Pattern Histograms are used for person face identification using real-time components [25].

10.2.1.4 Optimal Feature Selection and Reduction

For combining different features, a huge dimension would be required. More computing power is needed for processing the big vectors, which in turn leads to more power consumption and slowing down the process. Due to this drawback, this method is considered inappropriate for various FR applications. The feature reduction technique can be applied in two different ways: feature selection and dimensionality reduction. Dimensionality reduction techniques include principal component analysis (PCA), linear discriminant analysis (Fisherfaces), dynamic link architecture (DLA), modular eigenfaces, elastic bunch-graph matching (EBGM), independent component analysis (ICA), and many more.

10.2.1.5 Classification

The next phase of classification is face identification and categorisation into classes making use of the optimal feature set. The two popular classification and categorisation techniques are template matching and matching learning.

Another name for the template-based classifier is the distance measure, which consists of city-block distance method, chi-square statistic, quadratic form distance, Hausdorff distance, Canberra distance, and squared Euclidean distance classification

[26]. The different methods available for machine learning are artificial neural network (ANN), support vector machine (SVM), K-nearest neighbour (KNN), and Naive Bayes algorithm (NBetc).

10.3 SYSTEM ARCHITECTURE OF IOT AND FACE APPLICATION

The face recognition system imports the input into photos and stream videos. We need to get the subject (face image) to check the proof output. Some methodologies characterize the face recognition framework in five steps.

Figure 10.2 depicts the system-level architecture of the device that is used for the implementation of the proposed project. The IoT device consists of components such as the operating system (OS) Kernel, container, bin/lib/modules, and the application. An OS like Raspberry Pi divides the kernel and the user namespace. Furthermore, the IoT device is installed on the OS with the supported binaries to run the device engine to host containers. To provide isolation from other dependencies, the AI model FR application on the IoT platform is bundled inside the container. The application finally authenticates the person through face recognition confidence level. The authenticity of the person is based on the confidentiality level, i.e. the higher the confidence level, the higher the authenticity.

As shown in Figure 10.1, an application-level architecture is represented, consisting of five steps for an FR system. The five steps of FR are face pre-processing, face detection, feature extraction, optimal feature selection, and classification. To make the FR system more secure, it makes use of the system-level architecture and the application-level architecture to deliver the two latest technologies.

Bhatia (2018) made use of the Raspberry Pi 3 microprocessor, external web camera, speaker, and stepper motor hardware for implementing the system [25]. Raspberry Pi was used as an IoT device for storing the facial information, as mentioned by Pawar et al. [27]. A combination of edge-based AI, cloud computing and rk3288 was proposed by Zeng et al. [28]. Priyan Malarvizhi Kumar and her group have proposed a new image recognition and navigation system, which provides accurate and rapid messages in the form of a voice for easy navigation by visually challenged people [29].

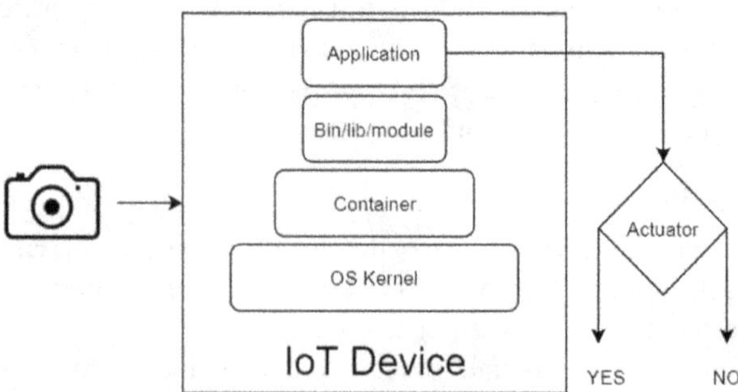

FIGURE 10.2 IoT system-level architecture.

Aydin used computer vision in the IoT. This allows the use of the Pi camera and sensor. Real-time home monitoring is required for security. It also provides a security system to implement real-time face detection [30].

It becomes difficult to apply the FR techniques on an embedded platform for Visual IoT due to shortage of computational resources, which are based on huge deep neural networks, as mentioned by Seon ho Oh. The Compact Deep neural networks-based FR system turned out to be a solution for this issue. Making use of the deep neural network, a new methodology was proposed. This method a minimal model complexity, allowing it to be efficiently implemented in an embedded context, and it was regarded as one of the most powerful ways of dealing with posture and illumination changes [31]. Making use of the cloudlet, face detection and extraction techniques are implemented, so that the need for large data transfers to remote processing centres can be minimised. Also, the paper proposes a face matching technique, which was implemented on the face characteristics vector within a private cloud environment. The cloudlets are positioned next to CCTV cameras [32]. A scheme for detecting and locating missing persons was proposed by the researcher Amin in 2016, stating the effectiveness of the procedure [32].

The Face in the Video Identification Framework (FivR) was proposed by Xuan Qi (2018). This takes the real-time key frame from IoT edge devices and performs FR on the cloud backend using the retrieved key frames. This reduces the volume of data and, as a result, lowers the cloud backend processing load [33].

As stated by Othman, for detecting movement in a specific area, passive infrared (PIR) sensors are used (2018), and the photos are captured by Raspberry Pi, detected, and recognised. Later, using the Telegram application, the photos and notifications are sent to a smartphone-based IoT. Overall, the resulting system is real time, fast, and low cost [34]. Kone Srikrishnaswetha in 2019 introduced an electronic voting machine having powerful secured features like unique aadhar card number and face recognition through biometric IoT [35–38].

10.4 TABLE OF COMPARISON

Table 10.1 provides a summary of the studied methods.

10.5 CHALLENGES AND LIMITATIONS

To date, many different devices with IoT embedded are used for face detection. These devices use specific algorithms, and these algorithms also have many variations that can be applied for feature extraction to get more specific and accurate results. These variant algorithms can be used in various phases of an FR system to get the desired outcomes. This also results in reducing the volume of data. The major challenge faced here is the low computation power for processing the image data. Sensors and networks are an indispensable part of the IoT. But, not all machines are equipped with advanced sensors and network functions to effectively communicate and share data. In addition, sensors with different power consumption capabilities and safety standards built into traditional machines may not provide the same

TABLE 10.1
Comparison of State-of-the-Art Face Recognition with IoT

Paper	Pre-Process	FD	FE	S/D	FS	Classifier	Database	Validation	Device	Result Accuracy
[25]	Resize	–	LBP-histogram	S	N	–	Own data	–	Raspberry Pi 3 microprocessor	90%
[27]	Pi camera module	Voila-Jones	Pi camera	S/D	N	-	LFW	–	Raspberry Pi	Face identify
[28]	Background removal	LBP eigenvalue	Edge AI	D	N	–	Own data	–	rk3288, cloud	Reduce the delay%
[29]	–	Open CV	Open CV	D	N	Neural network	Own Data	Euclidean distance	ardiuono, ultrasoni	95%
[30]	Greyscale	Haar cascade	–	S/D	N	AdaBoost training, cascade classifier	Own data	Euclidean distance	Raspberry Pi 3, smartphone, SD card memory, Raspberry Pi camera	Detection rate 100%
[31]	Greyscale	HOG-based detector	–	S/D	N	Deep neural network	LFW	–	Visual IoT	93%
[32]	Surveillance camera	Cloudlet	Cloudlet	S/D	N	Private cloud	Real data	–	Private cloud, cloudlet	Find missing person %
[33]	Face Quality Assessment	Key-frame extraction engine (KFE)		D	N	CNN	Corridor, hallway	–	Cloud, Open CV	95.00%, 96.60%
[34]	Greyscale, Normalisation	Haar-like feature AdaBoost	LBPH, PCA, LDA	D	N	KNN	Real-time image/ own data	8*8	PIR sensor	90.90%, 93.08%
[35]		Haar-like feature AdaBoost	–	D	N	–	Real-time image/ own data	–	Electronic voting machine, Raspberry	Secured authentication

Abbreviations: CNN: Convolutional Neural Network, D: Dynamic face (3-D), FD: face detection, FE: feature extraction, FS: feature selection, S: static face (2-D), HOG: Histogram of Oriented Gradients, LBP: Local Binary Patterns, LDA: Linear Discriminant Analysis, LFW: Labeled Faces in the Wild.

results. For IoT applications, lack of proper integration can lead to abnormal functions and efficiencies.

10.6 CONCLUSIONS

This chapter proposes a complete FR process flow. The study concludes that along with implementing the existing feature extraction algorithms, we can also implement variations of those algorithms in feature extraction, yielding a more concrete and accurate result. Also, by applying machine learning algorithms like KNN and CNN, we can get the desired results. Combining this methodology with IoT will enable more secured and concrete data extraction. We expect this chapter to encourage researchers in future to place more emphasis on the use of local approaches for FR techniques combined with the IoT applications.

REFERENCES

1. Yousri Ouerhani, Maher Jridi, and Ayman Alfalou. "Fast face recognition approach using a graphical processing unit 'GPU'". In: IEEE International Conference on Imaging Systems and Techniques. IEEE, 2010, pp. 80–84.
2. Wencheng Yang et al. "A fingerprint and finger-vein based cancelable multi-biometric system". *Pattern Recognition* 78 (2018), pp. 242–251.
3. Nitesh Purushottam Patel and Ashwini Kale. "Optimize approach to voice recognition using iot". In: International Conference On Advances in Communication and Computing Technology (ICACCT). IEEE, 2018, pp. 251–256.
4. Chunhui Zhao, Xiaocui Li, and Yan Cang. "Bisecting k-means clustering based face recognition using block-based bag of words model". *Optik-International Journal for Light and Electron Optics* 126.19 (2015), pp. 1761–1766.
5. Yassin Kortli et al. "A comparative study of CFs, LBP, HOG, SIFT, SURF, and BRIEF techniques for face recognition". In: *Pattern Recognition and Tracking XXIX* Vol. 10649. International Society for Optics and Photonics, 2018, p. 106490M.
6. URL: https://disruptionhub.com/5-applications-facial-recognition-technology/.
7. Pallavi Gupta and Manisha Rajoriya. "Face recognition based home security system using IoT". *Journal of Critical Reviews* 7.10 (2020), pp. 1001–1006.
8. Asif Ahmed et al. "Real-time face recognition based on IoT: A comparative study between IoT platforms and cloud infrastructures". *Journal of High Speed Networks* 26.2 (2020), pp. 155–168.
9. H. D. Supreetha Gowda, G. Hemantha Kumar, and Mohammad Imran. "Multimodal Biometric Recognition System Based on Nonparametric Classifiers". In: *Data Analytics and Learning*. Springer, 2019, pp. 269–278.
10. Parekh Payal and Mahesh M. Goyani. "A comprehensive study on face recognition: methods and challenges". *The Imaging Science Journal* 68.2 (2020), pp. 114–127.
11. Paul Viola and Michael Jones. "Rapid object detection using a boosted cascade of simple features". IEEE Computer Society Conference *on* Computer Vision *and* Pattern Recognition 1 (2001), pp. I–I. DOI: 10.1109/CVPR.2001.990517.
12. Paul Viola and Michael J. Jones. "Robust real-time face detection". *International Journal of Computer Vision* 57.2 (2004), pp. 137–154. DOI: https://doi.org/10.1023/B:VISI.0000013087.49260.fb.
13. Farnaz Fanaee, Mehran Yazdi, and Mohammad Faghihi. "Face image super-resolution via sparse representation and wavelet transform". *Signal, Image and Video Processing* 13.1 (2019), pp. 79–86. DOI: https://doi.org/10.1007/s11760-018- 1330-9.

14. Thamizharasi Ayyavoo and Jayasudha John Suseela. "Illumination pre-processing method for face recognition using 2D DWT and CLAHE". *IET Biometrics* 7.4 (2018), pp. 380–390. DOI: 10.1049/iet-bmt.2016.0092.
15. Xinfang Cui, Peng Zhou, and Wankou Yang. "Local dominant orientation feature histograms for face recognition". *Applied Informatics* 4. SpringerOpen (2017), p. 14. DOI: 10.1186/s40535-017-0043-4.
16. Zhihua Xie, Peng Jiang, and Shuai Zhang. "Fusion of LBP and HOG using multiple kernel learning for infrared face recognition". In: IEEE/ACIS 16th International Conference on Computer and Information Science. IEEE, 2017, pp. 81–84. DOI: 10.1109/ICIS.2017.7959973.
17. Fethi Smach et al. "An FPGA-based accelerator for Fourier descriptors computing for color object recognition using SVM". *Journal of Real-Time Image Processing* 2.4 (2007), pp. 249–258.
18. Fuwei Yang et al. "Discriminative multidimensional scaling for low-resolution face recognition". *IEEE Signal Processing Letters* 25.3 (2018), pp. 388–392. DOI: 10.1109/LSP.2017.2746658.
19. Renliang Weng, Jiwen Lu, and Yap-Peng Tan. "Robust point set matching for partial face recognition". In: *IEEE Transactions on Image Processing* 25.3 (2016), pp. 1163–1176. DOI: 10.1109/TIP.2016.2515987.
20. Evangelos Sariyanidi, Hatice Gunes, and Andrea Cavallaro. "Automatic analysis of facial affect: A survey of registration, representation, and recognition". In: *IEEE transactions on pattern analysis and machine intelligence* 37.6 (2015), pp. 1113–1133. DOI: 10.1109/TPAMI.2014.2366127.
21. Taskeed Jabid, Md Hasanul Kabir, and Oksam Chae. "Local directional pattern (LDP)–A robust image descriptor for object recognition". In: Seventh IEEE International Conference on Advanced Video and Signal Based Surveillance. IEEE, 2010, pp. 482–487. DOI: 10.1109/AVSS.2010.17.
22. Mahesh M. Goyani and Narendra M. Patel. "Robust facial expression recognition using local haar mean binary pattern". *Journal of Information Science and Engineering* 16 (1 2018), pp. 54–67.
23. L. Koteswara Rao, P. Rohini, and L. Pratap Reddy. "Local color oppugnant quantized extrema patterns for image retrieval". *Multidimensional Systems and Signal Processing* (2018), pp. 1–23. DOI: 10.1186/s40064-016-2664-9.
24. Rathod Dharmendrasinh Gajendrasinh and Mohammed Husain Bohara. "Sentiment analysis for Feature extraction using dependency tree and named entities". International Conference on Innovations in information Embedded and Communication Systems (ICIIECS). Vol. 1, 2017, pp. 587–592.
25. Prayag Bhatia et al. "IOT based facial recognition system for home security using LBPH algorithm". In: 3rd International Conference on Inventive Computation Technologies (ICICT). IEEE, 2018, pp. 191–193.
26. Vipin Tyagi. *Content-Based Image Retrieval: Ideas, Influences, and Current Trends.* Springer, 2018.
27. Suraj Pawar et al. "Smart home security using IoT and face recognition". In: Fourth International Conference on Computing Communication Control and Automation (ICCUBEA). IEEE, 2018, pp. 1–6.
28. Junjie Zeng, Cheng Li, and Liang-Jie Zhang. "A face recognition system based on cloud computing and AI edge for IOT". In: International Conference on Edge Computing. Springer, 2018, pp. 91–98.
29. Priyan Malarvizhi Kumar et al. "Intelligent face recognition and navigation system using neural learning for smart security in Internet of Things". In: *Cluster Computing* 22.4 (2019), pp. 7733–7744.

30. Ilhan Aydin and Nashwan Adnan Othman. “A new IoT combined face detection of people by using computer vision for security application”. In: International Artificial Intelligence and Data Processing Symposium (IDAP). IEEE, 2017, pp. 1–6.
31. Seon Ho Oh, Geon-Woo Kim, and Kyung-Soo Lim. “Compact deep learned feature-based face recognition for visual internet of things”. *The Journal of Supercomputing* 74.12 (2018), pp. 6729–6741.
32. Anang Hudaya Muhamad Amin, Nazrul Muhaimin Ahmad, and Afiq Muzakkir Mat Ali. “Decentralized face recognition scheme for distributed video surveillance in IoT-cloud infrastructure”. In: *IEEE Region 10 Symposium (TENSYMP)*. IEEE, 2016, pp. 119–124.
33. Xuan Qi, Chen Liu, and Stephanie Schuckers. “IoT edge device based key frame extraction for face in video recognition”. In: 18th IEEE/ACM International Symposium on Cluster, Cloud and Grid Computing (CCGRID). IEEE, 2018, pp. 641–644.
34. Nashwan Adnan Othman and Ilhan Aydin. “A face recognition method in the Internet of Things for security applications in smart homes and cities”. In: 6th International Istanbul Smart Grids and Cities Congress and Fair (ICSG). IEEE, 2018, pp. 20–24.
35. Kone Srikrishnaswetha, Sandeep Kumar, and Md Rashid Mahmood. “A study on smart electronics voting machine using face recognition and aadhar verification with IoT”. In: *Innovations in Electronics and Communication Engineering*. Springer, 2019, pp. 87–95.
36. URL: https://www.amicusint.org/articles/2015/10/13/face-recognition.
37. URL: https://deepai.org/machine-learning-glossary-and-terms/feature-extraction.
38. URL: https://www.facefirst.com/blog/brief-history-of-face-recognition-software/.

11 IoT for Health Monitoring

Anshu Saxena, Asmita A. Moghe, and Ratish Agarwal

CONTENTS

11.1 INTRODUCTION

Internet of Things (IoT) devices may use a single sensor or multiple sensors. They may be wearable or standalone devices. These devices can be worn by patients and doctors,who can use themto observe their patients remotely. This will also help withpoor health infrastructure in remote village areas, lack of medical professionals in remote areas, physical connectivity, etc.IoT-based healthcare systems can overcome some of these drawbacks to a great extent. In this system, we have incorporated an alert system to alert the doctor when a health parameter crosses a threshold. The

DOI: 10.1201/9781003221333-11

patient's health will be monitored every few seconds, and new data will be updated every few seconds. Therefore, in any emergency condition, patients can be observed. The system will be a remote monitoring system that will be portable with a power supply.

11.2 LITERATURE REVIEW

Gupta et al. [1] have developed a healthcare monitoring system with a pulse sensor and a temperature sensor. In this system, the authors have used the Global System for Mobile Communications (GSM) system with Short Message Service (SMS) to alert the doctor and caretaker. It usesATmega 328 and ArduinoUno, and a Li battery is used to power the system. The pulse rate is observed to see if whether is normal or higher or lower than the threshold values. If it goes above the threshold, it could be an indication of major or minor heart attack, depending on pulse rate. In the case of extreme abnormal values of these parameters, an SMS will be sent to concerned doctors.SMS can also be sent to close members of the family or any caregiver. In addition to this, updated vital parameters will be constantly updated to the server so that any person who has authority can log in and check the details when desired. Furthermore, the use of GSM also helps to send emergency signals such as an "SOS" message to loved ones/caregiver/doctor. Analysis was carried out on several patients in the age group of 21 to 50 based on some evaluation parameters (before going to sleep, after waking up, after exercise, when at rest, or tired, or at work, etc.)such as temperature and beats per minute(bpm) using the free heart rate application. Observations show that heart rate increases after exercise but stays normal when resting or tired or before sleeping. However, there is a small increase in heart rate above the normal level after getting up.

Jaiswal et al. [2] developed an .0automated system using Raspberry Pi by collecting patients' statistics from medical devices fitted with sensors and further processing this data by uploading it to the medical centre's cloud. They also used docker containers for storage distribution, etc. Data through sensor nodes is collected, encoded, and transferred over a wireless channel to the server through the necessary software. Data (patient's temperature, blood pressure, etc.) is input into Raspberry Pi before being pushed to the server. The docker container and the local database in the server are used for further processing of collected data and providing it to the doctor, nursing staff, and hospital to monitor and diagnose health problems. Proper authentication and data encryption are employed among the docker container users and the rest of the components. As well as data storage, the system also provides services like gathering information from doctors, users, hospitals, etc. Instead of using virtual machines, in this system, the docker container and server are loaded in Raspberry Pi. The container receives data from the sensors. Containers in the network receive data and also send it to the local docker server for processing. The local server helps in emergency situations by sending an alert message to the doctor/caregiver. At the same time, it also communicates with the remote server by forwarding to it the sensed data for processing and distribution at the global level to health experts for advice and diagnosis.

An Android app [3] again measures the temperature and pulse rate of a patient, which are basic indications of change in the health of the patient. These sensors send input to Raspberry Pi, and the data collected from this is sent to the cloud. For measuring temperature, ds18b20 is used. Raspberry Pi continuously monitors temperature and pulse rate. For measuring heartbeat, sensor 11574 is used. These measured values are sent to the cloud (ThingSpeak) for continuous monitoring. This data can also be viewed by the doctor or caregiver, and if there is any variation, they can take appropriate measures. This data is available on the Android app, and the doctor or caregiver can take further necessary actions based on these observations. In the case of any emergency, they can call for help in the app calls to an emergency number. There is a facility to send text messages like "patient is not well, need help" to the doctor.

Another health monitoring system, developed by Sasidharan et al. [4], uses three non-contact sensors, which consecutively measure heart rate, respiration rate, SpO_2, and temperature. The body sensor network uses signal processing and data transmission modules through which the collected data can be transmitted wirelessly via the internet. Sensor input is sent toArduino Uno. SpO_2indicated the oxygen level in the blood. It is defined as the ratio of oxygenated haemoglobin to deoxygenated haemoglobin in RBC (red blood cells). With change in intensity of red light, the SpO_2 values change. Its normal range is from 95% to 100% for a healthy individual. Temperature is measured using an LM series temperature sensor. Respiration is also used for diagnosing other respiratory disorders.

A data mining smartphone application is proposed in Saranya et al. [5]. Here, sensors are again used for a health disease prediction and diagnosis system. This system measures various parameters of the patient, like pressure, heartbeat, and temperature, using wireless communication. In this project, again, Arduinois connected to various sensors. Three sensors (1, 2, and 3) are worn by the subject, which measure the heart rate, body temperature, and pressure level of the human body, respectively. The corresponding measurements will be shown on liquid crystal displays (LCD). These parameters are used to predict the disease of the patient intelligently. If all the three parameters that are sensed are between the threshold values, then the patient is considered healthy. Patient data is stored in the cloud, so healthcare professionals can monitor their patients from a remote location at any time. Deviations in data values are collected from wearable sensor nodes that process time alert messages, which are transmitted to concerned smartphones. Apart from monitoring, disease prediction and diagnosis can also be carried out with this system by the data mining smartphone application.

Another Raspberry Pi-based health monitoring system employing wireless communications and having low power consumption with maximum functionality with multiple sensors and freedom of movement to users is described by Sambhram [6]. Data is collected from ECG (electrocardiogram), BP (blood pressure), heart rate, and temperature sensors using a RaspberryPi module, and then analysed, and the fog computing technique is used to decide whether the values read from the patient are normal or abnormal. These are sent to cloud storage and can be retrieved on a smartphone by connecting to the internet. In the case of abnormal values being read, there is the facility for an SMS notification, message, etc. to be sent to the user's smartphone so that he/she can take care of the patient in critical emergency situations.

An automatic system to monitor a patient's body temperature, heart rate, body movements, and blood pressure is developed in Banka et al. [7], which can also predict whether the patient is suffering from any chronic disorder or disease using the various health parameters and many other symptoms. In level1, using a vibration sensor, a BP sensor, and pulse and temperature sensors, data is collected and stored in the cloud. It is further filtered and classified to get useful information. Further analysis or prediction is done using data mining techniques. Raspberry Pi reads the information of different vitals of patients and updates it to MySQL DB. This is used to display information via email, message, etc. to concerned doctors, caregivers, etc. in the case of deviation of vitals from threshold values. As well as collected data via sensors, patients were also questioned regarding the symptoms, etc., and comparing this data with existing knowledge enabled the prediction of disease/disorder, thus creating an efficient expert system.

Again inRaseduzzamanRuman et al. [8], the same three sensors to measure body temperature, pulse rate, and heartbeat are used. The heartbeat sensor (ECG) sends signals to the microcontroller Arduino Mega, which is connected by a Wi-Fi module. Arduino sends this data to ThingSpeak online software. A heart rate of 121 bpm was recorded from the patient, who was monitored with an IBI of 1826 ms. Data is taken with the patient in both relaxed and excited states. The bpm was recorded as 80–90 in the relaxed position and 120 under stress or anxiety. There is an increase in the temperature of the patient when stressed, and the temperature of the surroundings also plays a role in the increase in temperature. It was observed that when the patient comes to rest, the body temperature drops to its optimum value.

With the objective of effective communication between patient and doctor, another healthcare monitoring system is suggested inJagannadhaSwamy and Murthy [9]. To provide a better healthcare facility in rural areas, with basic health check-ups, a more technology-oriented system is developed. Such an intelligent system will have the facility for patient accounts that will maintain patient data collected using some hardware. This data can be read by patients also through an app on their mobiles using Bluetooth module HC-05, and the same data will also be uploaded to the fusion table's patient database at the same time. The doctor's app will also show individual patient accounts having unique ID. This unique patient ID is stored in the doctor's fusion table. The doctor app will use respective patient information from the doctor's fusion table, and by linking this information to the corresponding patient's fusion table, it will display the various measured parameters, like heart rate, temperature, and oxygen levels in the patient's blood, on the doctor app. Communication with hardware is established via HC-05 by selecting Bluetooth from a list of paired devices by pressing the connect button of Arduino. The text of this button is changed according to the connection status of the device. Also, the user can select the desired parameter to measure. This activates the hardware, and corresponding parameter values are displayed. As well as this, sensor data is gathered by the application at periodic intervals with the help of a clock timer. Data updating and uploading to fusion table are done constantly. The initial screen will call the fusion table with doctor details when the doctor logs in. The fusion table will carry out the process of linking patient ID and other details and update patient data. Such an automated system has considerable advantages.

Another health monitoring system built using ArduinoESP 32, heart rate, and temperature sensor, using Android Studio, Firebase, and ThingSpeak [10], uses an Android application with a measuring unit which can be carried. The Arduino ESP32 microcontroller board is connected to DS18B20 and pulse sensors. Values of temperature and heart rateare input through sensors, displayed on LCD, and also sent to the ThingSpeakplatform in real time via Wi-Fi. The platform maintains historical data. In the case of deviation in readings from desired values, the system sends a notification via the Line application in real time. Details of registered members, like their name, respective heart rate, temperature, measuring devices, etc., are maintained using an Android application in the Firebase Real-time Database. User authentication is done with the Firebase Authentication application. The caregiver is the admin user, and members are the other category of users. Heart rate and temperature information can be updated/monitored by admin users using Web View. The web page displays embedded ThingSpeak graphics for heart rate and temperature. Information on vitals is also visible to member users.

Tripathi and Shakeel [11] describe the use of the IoThNet topology, which is a network of various sensors, computing, and storage capability taken together in a heterogeneous system. It integrates ECG, BP, SpO_2, temperature measurements with mobiles, servers, laptops, optical instruments, etc. Apple Watch can measure calorie count, distance, step count, BP, etc. using a large number of sensors based on infrared and visible light. A near field communication (NFC) antenna establishes communication between two wearable devices and a mobile phone. Data is stored on the IoT server. The app is used to navigate health data on a smartphone.

An IoT-based healthcare system has also been applied to determine cases of heatstroke, hypothermia, bradycardia, and tachycardia in Kamal and Ghosal [12]. The system uses Raspberry Pi along with DS18B20, ADC1015, ADXL345, and a heartbeat sensor. Automatic patient monitoring, real-time data transmission and storage of medical data, etc. are its key features. The key component here is a decision-making model to bring the person's health status into the normal range in the case of an emergency condition. A stepper motor model is used for this. Communication of Raspberry Pi with sensors is through general-purpose input/output (GPIO). The processing model processes collected samples. The data are stored in a buffer for probabilistic analysis and live graphs. The probabilistic model's output is given to the decision-making model to make a decision about hardware in an emergency situation.

Use of ATmega 328 can also be seen in Antonio [13], Braam et al., and Maria et al. [14, 15] to measure the temperature of patients, alerting the caregiver/doctor along with pulse sensor data and pushing data to the cloud. Similar work is also achieved using esp 8266 inIrmansyah et al. [16]. A more specific health monitoring system with IoT is seen with BAN (body area network) developed in Vijay Kumar et al. [17] to collect information like heat, circulatory strain, and the pulse rate of patient and to observe these parameters using a specialized application.

Rajanna et al. [18] used (TI) MSP430F5529, which is a microcontroller. Algorithms for detection of peak ECG signals are implemented using microcontroller unit (MCU) to measure the heart rate, bpm, and IBI metrics. The json payload obtained from PubNub channels was published to freeboard.io subscription. The freeboard.io dashboard was used for visualisation. It displays variation in IBI/Heart rate variability (HRV). The peak threshold determined is simple and accurate. Performance with moving average algorithm is better than with IBI. A similar heart monitoring system and database management system is proposed in Ren and Lyu [19].

11.3 PROPOSED METHODOLOGY

We have developed a system in which patients' vitals will be detected and this data is sent to the cloud; therefore, the data will be accessible not only to the patients but also to the doctor. This will create - two-way communication between doctor and patients. In this system, we have incorporated an alert system to alert the doctor when a health parameter crosses a threshold. The patient's health will be monitored every few seconds, and new data will be updated every few seconds. Therefore, patients can be observed in any emergency condition. The system will be a remote monitoring system, which will be portable with a power supply. The system has a heartbeat detection mechanism, temperature detection, and a panic alert mechanism. A caregiver, health expert, or nursing staff can utilize this system to observe remotely crucial health parameters of the patient or person in need. The main idea is to make healthcare facilities more accessible, cost effective, and convenient for people who live in villages and remote locations so that they don't have to come to the city for every problem they have and can be treated and cured in their own place. This will help in making healthcare facilities available fast and will save a lot of time for patients as well as the doctor. By doing so, it will save many lives, and in turn, it can also save paying huge sums for transportation and medical assistants. It can make healthcare more accessible to poor people. This will benefit many critical patients who need constant observation even after being discharged from hospital after surgeries and treatment. They can also use this system, and doctors can monitor them remotely.

11.4 HARDWARE AND SOFTWARE SPECIFICATION

11.4.1 Arduino Uno

Arduino Uno [20] is a microcontroller board created on the ATmega328P (datasheet). Figure 11.1 shows the input/output pins, facility for USB connection, power supply reset, etc. The board has all the required support for the microcontroller.

11.4.2 Temperature Sensor

The LM35 series [21] are precision integrated-circuit temperature devices whose output voltage is linearly proportional to the temperature in degrees centigrade. This

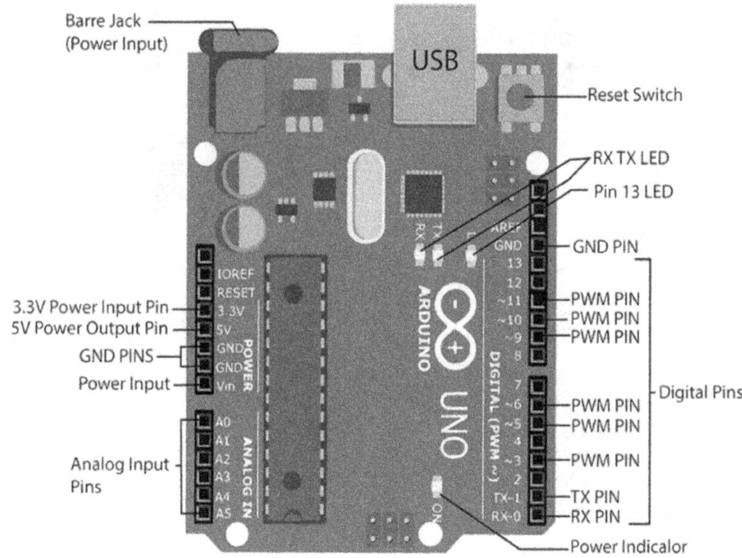

FIGURE 11.1 Arduino Uno.

sensor is comparable to a linear temperature sensor that gives readings in kelvin. Here, there is no need to subtract a certain constant voltage from the output to get temperature in degrees Celsius. Additional calibration is not required. Accuracies of ±¼ and ±¾ °C cover the full range from −55 to 150 °C (Figure 11.2).

11.4.3 LCD

LCDs are used to display various parameters [22]. The LCD shown in Figure 11.3 has 16 pins with2 rows accommodating 16 characters each. These are often used in 4-bit mode or 8-bit mode. It has eight data lines and three control lines, which will be used for control purposes.

11.4.4 ESP8266

The ESP8266shown in Figure 11.4 is a low-cost Wi-Fi microchip, with a full Transmission Control Protocol/Internet Protocol (TCP/IP) stack and microcontroller capability [23]. It allows microcontrollers to connect to a Wi-Fi network and make simple TCP/IP connections. The ESP8285 has1 MiB of built-in flash, which enables building single-chip devices capable of connecting to Wi-Fi.

11.4.5 Power Supply

The adapter used here provides 9 volt 1 ampere direct current (dc) output [24]. In this work, we have used the power supply to give 12 V power to the hardware to

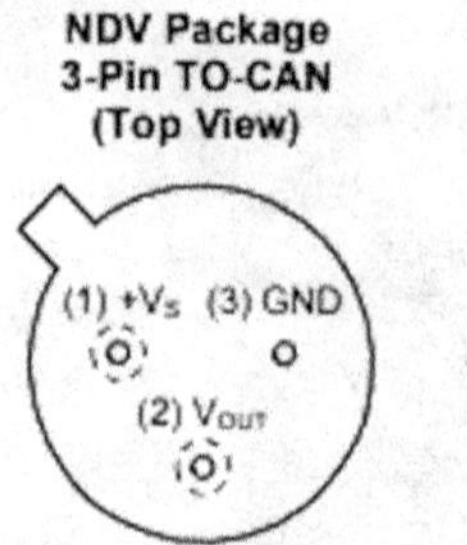

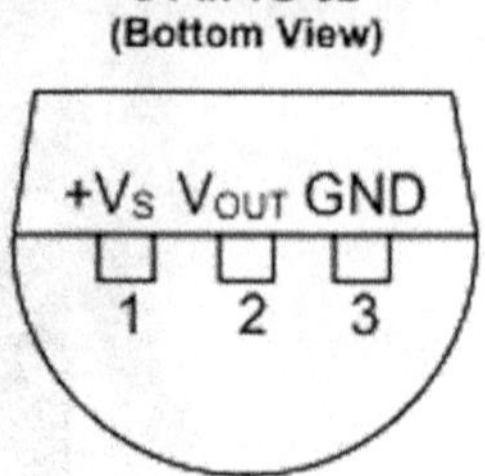

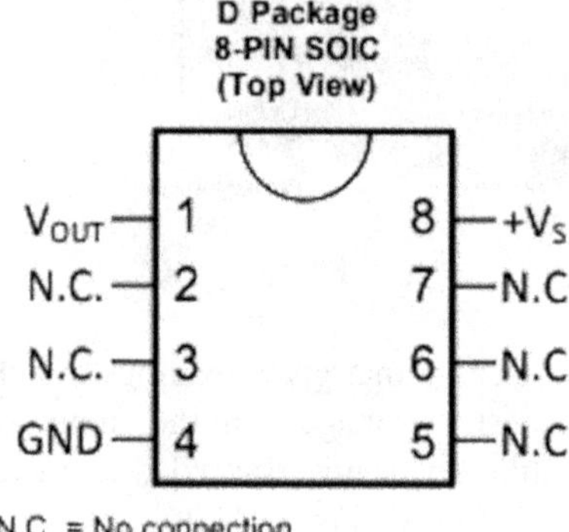

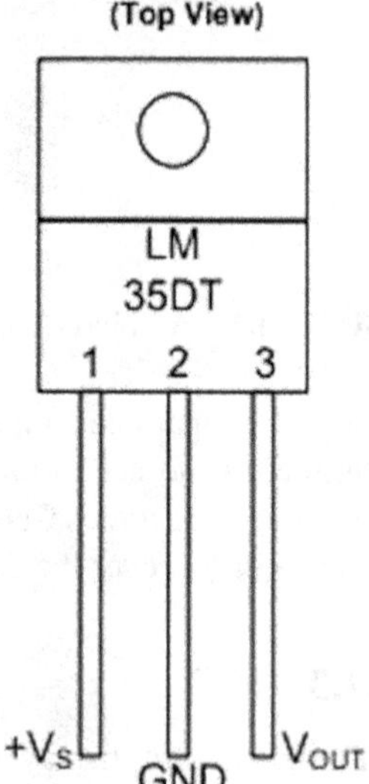

FIGURE 11.2 LM 35 temperature sensor.

function properly and provide the necessary power to all the components of the system (Figure 11.5).

Features:

- Incredibly low fault rates
- High reliability
- Regulated stable voltage
- Stable output
- Efficient and consumes less energy.

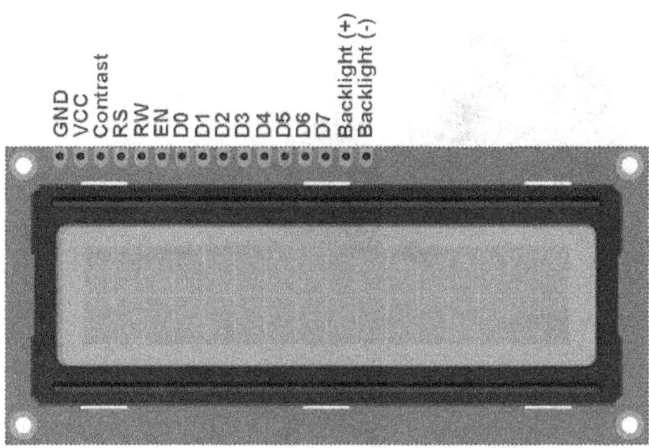

FIGURE 11.3 LCD with its pin out.

FIGURE 11.4 ESP8266.

11.4.6 Pulse Sensors

These are plug and play devices used for sensing heart rate with Arduino [25]. The sensor is connected to Arduino or plugged to the finger or ear lobe using jumper wires. The front side, having the guts logo, touches the skin. A light-emitting diode (LED)shines through the small hole. The square is an ambient light sensor, which

FIGURE 11.5 12 V adapter.

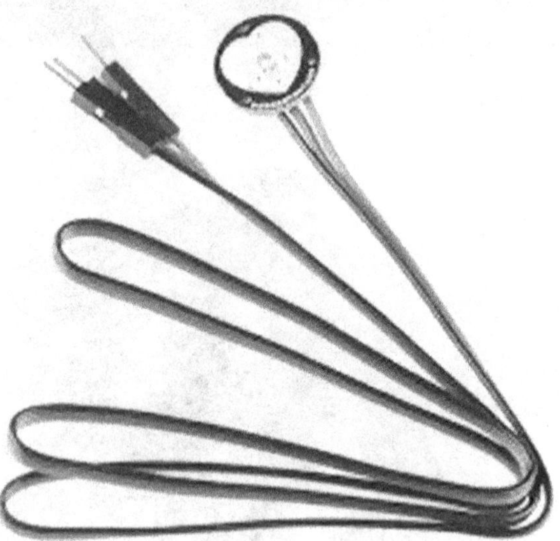

FIGURE 11.6 Pulse sensor.

regulates the brightness of the screen under various light conditions. When light emitted from the LED falls on the fingertip or earlobe or any other capillary tissue, it bounces back, and this is read by the sensor. It uses a 24-inch flat ribbon cable.

The black wire resembles ground, the signal is given to the next wire, and +3 to +5V supply is connected to the last wire (Figure 11.6).

11.5 SOFTWARE SPECIFICATION

11.5.1 Arduino IDE

The Arduino Integrated Development Environment (IDE) [26] is written using functions from c and c++. It is a cross-platform application. Using third party cores and development board made by other vendors, it may also be used to upload programs to Arduino-compatible boards.

11.5.2 ThingSpeak (API)

Data collected from various sensors can be stored and retrieved using HTTP AND MQTT protocol via the local area network (LAN)by use of the ThingSpeakapplication programming interface (API). Various applications related to logging of sensors, tracking location, etc. can be created.ThingSpeakis used to analyse and visualize uploaded data using MATLAB.MQTT is the most widely used machine to machine (M2M)/IoT protocol [27] and has numerous applications [28], including fetching real-time data [29].

11.6 RESULTS AND DISCUSSION

In this project, we have used the ThingSpeak cloud platform, in which patients' vitals are calculated and represented in graphical form. The patient's vitals are calculated in a few seconds, and the data is updated to the cloud, so that the doctor can login to the cloud at any given time and view all the details of the patient's vitals. This is done in three phases, involving the collection and measurement of the patient's vitals (temperature, heart rate, and pressing of the panic button). In this work, measurements are carried out on a normal person for experimental purposes.

11.6.1 Phases 1 and 2: Patient'sVitals Are CollectedandPushed to the Cloud, Where They Are Graphically Analysed

To measure a person's body temperature and heart rate, an LM35 sensor and a pulse sensor are connected to the analogue pins of Arduino Uno. Temperature and heartbeat information in degrees Celsius and bpm, respectively, are displayed on the LCD screen connected to Arduino Uno, as shown in Figure 11.7. In order to upload data

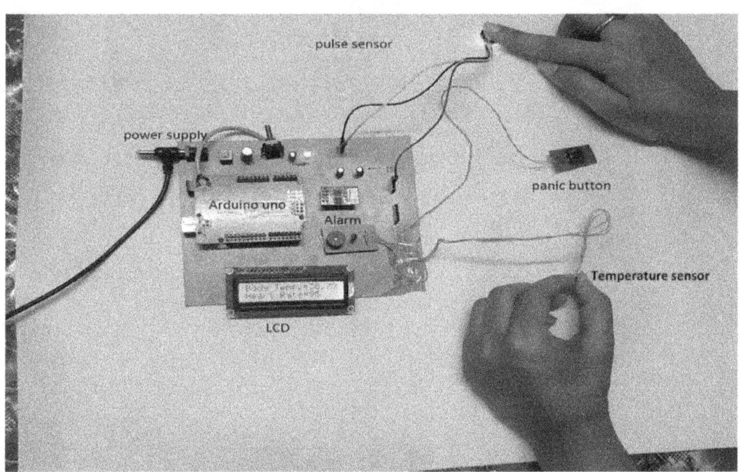

FIGURE 11.7 Temperature and heart rate displayed on LCD.

to ThingSpeak, Wi-Fi connectivity is achieved by connecting esp8266 to the output of the Arduino Uno board by using the appropriate authentication key and API. Temperature and heartbeat data is continuously uploaded to ThingSpeak, which updates the patient's body temperature every few seconds. A graphical representation of body temperature on ThingSpeaktaken on 1 March and 2 March at specific times for short durations is shown in Figure 11.8. In order to increase temperature, a candle was brought close to the sensor, and the temperature was seen to increase to 40 degrees Celsius momentarily, as shown in the figure.

Measured temperature data can be downloaded into a csv file as shown in Figure 11.9.

Heartbeat measured for the same person is presented graphically as observed on ThingSpeak in Figure 11.10. These values have been depicted for 1 and 2 March, respectively. ThingSpeakprovides a facility for taking 16 feeds as input from 16 sensors. Here, we have three feeds (represented as feeds) of temperature, heart rate, and

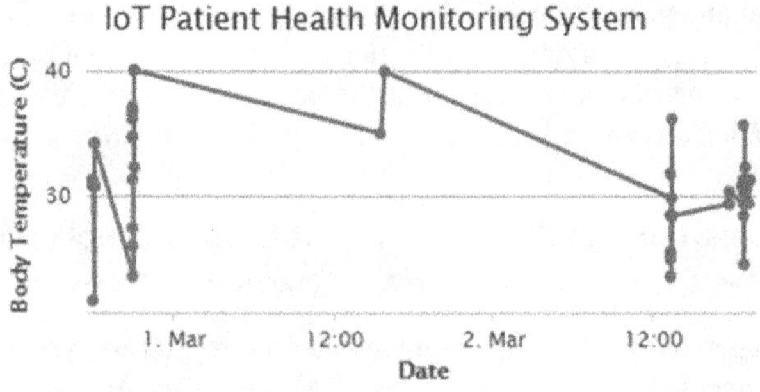

FIGURE 11.8 Measured temperature in degrees Celsius.

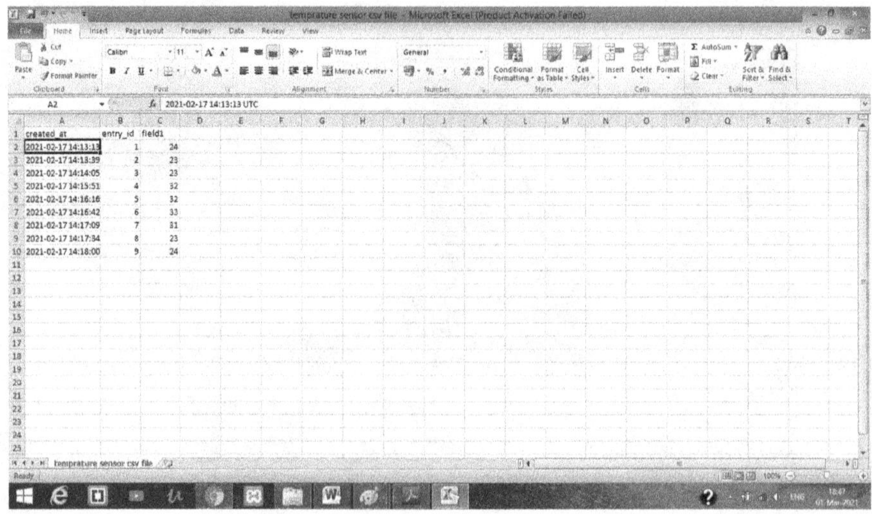

created_at	entry_id	field1
2021-02-17 14:13:13	1	24
2021-02-17 14:13:39	2	23
2021-02-17 14:14:05	3	23
2021-02-17 14:15:51	4	32
2021-02-17 14:16:16	5	32
2021-02-17 14:16:42	6	33
2021-02-17 14:17:09	7	31
2021-02-17 14:17:34	8	23
2021-02-17 14:18:00	9	24

FIGURE 11.9 csv file for measured temperature.

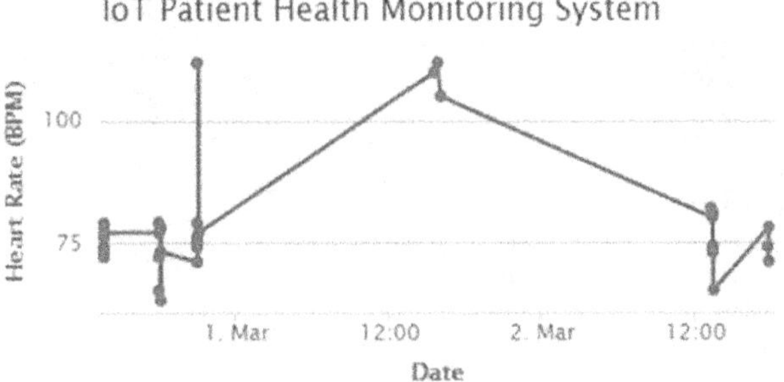

FIGURE 11.10 Measured heart rate in bpm.

panic button. This data is recorded in fields 1,2, and 3, respectively, as observed from the csv file in Figure 11.11. This represents the normal situation when the button is not pressed, and 0 represents the situation when the panic button is pressed.

If at any point in time the patient feels uncomfortable and needs emergency observation and critical care, then for such a situation, we need to include an alert mechanism, which will alert nearby caregivers or nursing staff. For this purpose, we have included a panic button, which will be available in this healthcare monitoring system. This button, when pressed by the patient for about 2 seconds, will generate an alarm that will alert the nursing staff immediately.

In the graph in Figure 11.12, we can see that 0 represents the normal situation when the button is not pressed, and 1 represents the state when the panic button is pressed.

11.6.2 MATLAB Analysis of 3-Day Body Temperature of Patients

In this section, we have also included 3 days' temperature comparison and MATLAB analysis of the patient's body temperature and found the minimum temperature of the patient during the last 3 days. The graph in Figure 11.13 is the temperature analysis of day 1, day 2, and day 3.

11.6.3 ThingSpeak Dashboard with All the Vital Parameters and Their Graphical Representation

This is the interface of this healthcare monitoring system that will be presented to the doctor, caregiver, and nursing staff once they are logged into the system using login credentials so that only authorized persons can view the patient's history, which will preserve the privacy of patients (Figure 11.14).

11.6.4 Phase 3: IFTTIntegration of Data from ThingSpeak to Generate Triggers at Particular Threshold Value

In the event of a panic situation being generated due to extreme variation in body temperature/heartbeat/pressing of panic button, the IFTT (if this then that) service

created_at	entry_id	field1	field2	field3	latitude	longitude	elevation	status
2021-02-28 08:09:57 UTC	1	31.28	73					
2021-02-28 08:10:15 UTC	2	23.46	74					
2021-02-28 08:10:33 UTC	3	23.46	78					
2021-02-28 08:11:07 UTC	4	34.7	76					
2021-02-28 08:11:25 UTC	5	28.35	79					
2021-02-28 08:12:06 UTC	6	33.24	74					
2021-02-28 08:12:23 UTC	7	29.33	72					
2021-02-28 08:12:41 UTC	8	31.77	77					
2021-02-28 12:25:54 UTC	9	25.42	77	1				
2021-02-28 12:26:12 UTC	10	28.35	79	1				
2021-02-28 12:26:34 UTC	11			0				
2021-02-28 12:27:43 UTC	12	24.93	79	1				
2021-02-28 12:28:05 UTC	13			0				
2021-02-28 12:29:01 UTC	14	31.28	72	1				
2021-02-28 12:29:19 UTC	15	30.79	65	0				
2021-02-28 12:33:01 UTC	16	21.51	63	1				
2021-02-28 12:33:18 UTC	17			0				
2021-02-28 12:34:02 UTC	18	30.79	78	1				
2021-02-28 12:34:20 UTC	19	30.79	73	1				
2021-02-28 12:34:38 UTC	20	34.21	73	1				
2021-02-28 15:27:44 UTC	21	23.46	71	1				
2021-02-28 15:28:02 UTC	22	25.9	79	1				
2021-02-28 15:28:19 UTC	23	27.37	75	1				
2021-02-28 15:28:38 UTC	24			0				
2021-02-28 15:29:23 UTC	25	36.17	71	1				
2021-02-28 15:29:40 UTC	26	37.15	74	1				
2021-02-28 15:29:58 UTC	27	36.66	75	1				
2021-02-28 15:31:19 UTC	28	31.28	75	1				
2021-02-28 15:31:37 UTC	29	34.7	75	1				
2021-02-28 15:31:55 UTC	30	34.7	76	1				
2021-02-28 15:32:12 UTC	31	32.26	112	1				
2021-02-28 15:32:30 UTC	32	40.08	76	1				
2021-02-28 15:33:03 UTC	33	40.08	77	1				
2021-03-01 10:01:07 UTC	34	35	110	1				
2021-03-01 10:17:17 UTC	35	40	112	1				
2021-03-01 10:31:49 UTC	36		105					
2021-03-02 07:50:16 UTC	37	29.81	80	1				
2021-03-02 07:50:34 UTC	38	23.46	82	1				
2021-03-02 07:51:21 UTC	39	24.93	81	1				
2021-03-02 07:51:38 UTC	40			0				
2021-03-02 07:52:46 UTC	41	28.35	74	1				
2021-03-02 07:53:04 UTC	42	24.93	73	1				
2021-03-02 07:53:22 UTC	43	31.77	73	1				
2021-03-02 07:53:39 UTC	44	25.42	74	1				
2021-03-02 07:54:46 UTC	45	36.17	74	1				
2021-03-02 07:55:04 UTC	46	28.35	65	1				
2021-03-02 12:13:06 UTC	47	29.33	78	1				
2021-03-02 12:13:23 UTC	48	29.33	74	1				
2021-03-02 12:13:41 UTC	49	30.3	71	1				

FIGURE 11.11 csv file for measured vitals.

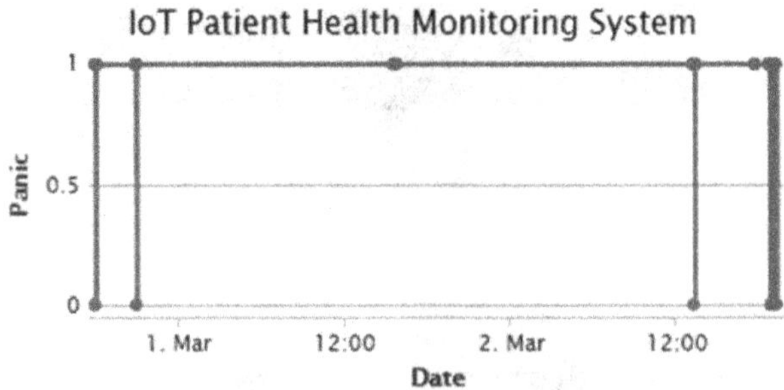

FIGURE 11.12 Graph of patient's panic.

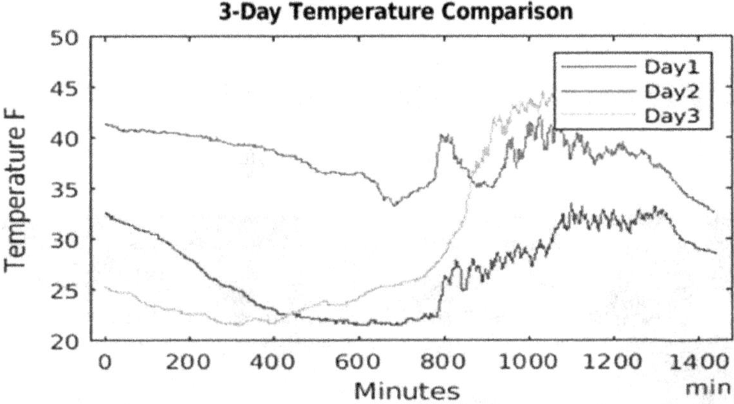

FIGURE 11.13 Graphical representation of MATLAB analysis of patient's temperature over 3 days.

is used to alert the doctor and/or caregiver. To implement this, a definite threshold is preset (110 bpm and 37 degrees Celsius), so that whenever the threshold value is exceeded, the IFTT service will send an email notification to the doctor's registered email address, and the doctor will be able to see the notification immediately. He/she can then check that patient, who needs emergency help. Figure 11.15 is a snapshot of the email notification generated on a distant mobile (caregiver/doctor) on 1 March 2021 at 15.31 hours when the heart rate of the patient exceeded 100 bpm.

Figure 11.16 is a snapshot of the email notification generated on a distant mobile (caregiver/doctor) on 28 February 2021 at 15.31 hours when the temperature of the patient exceeded 40 degrees Celsius.

Similarly, when the patient presses the panic button in the case of panic/emergency, then too, an email alert is generated. Figure 11.17 is a snapshot of the email notification generated on a distant mobile (caregiver/doctor) on 2 March 2021 at 9.18 hours in a panic/emergency situation [30].

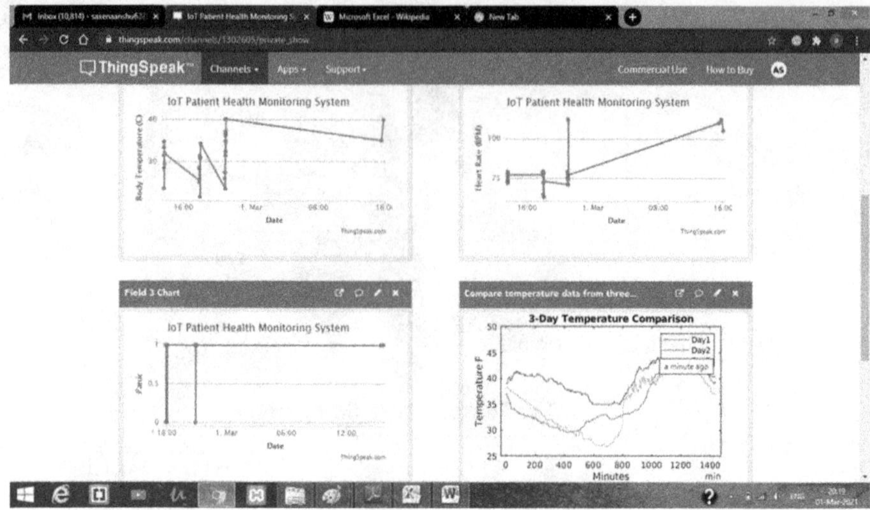

FIGURE 11.14 Dashboard with patient's vitals and its analysis.

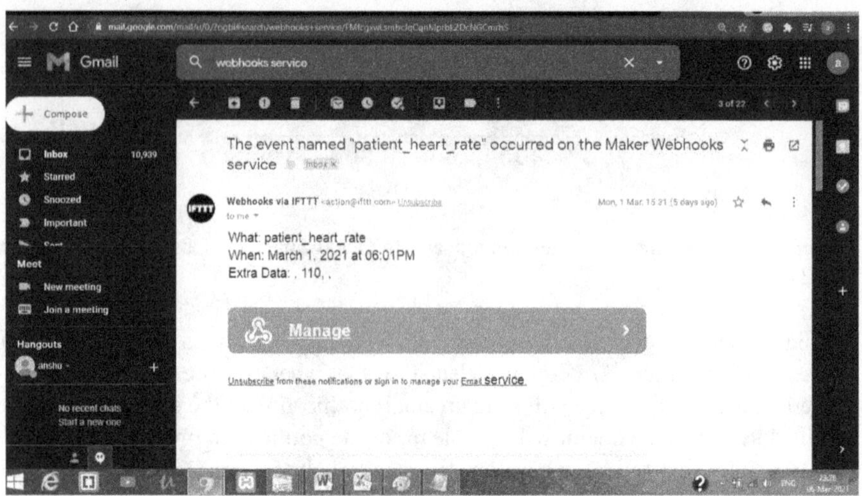

FIGURE 11.15 Email notification of patient in case of abnormal heart rate.

11.7 CONCLUSION AND FUTURE WORK

The remote patient monitoring system was developed with a vision to make healthcare more affordable and accessible to ordinary people. The patient's vitals, such as temperature and heart rate, are collected. The readings are collected in the cloud (after being taken via different sensors), from where they can be easily viewed by the doctor or caregiver remotely from any part of the world. This data can be useful in analysing patient health and can be used in medical investigations. The data can also

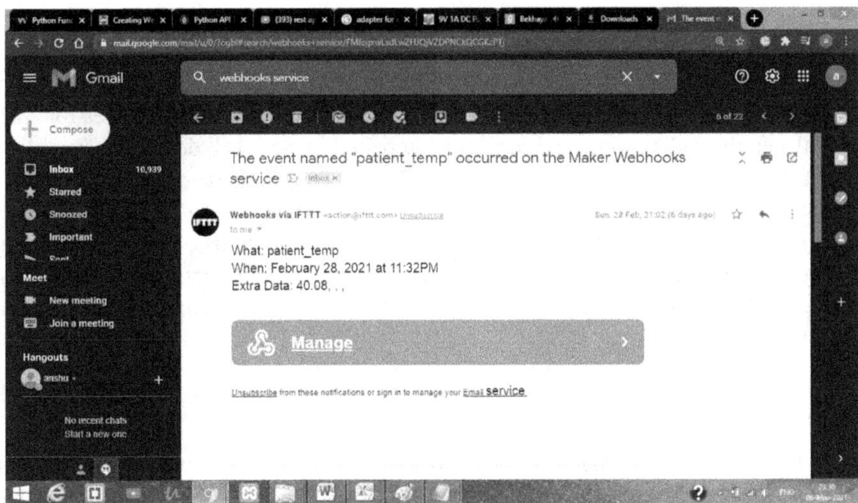

FIGURE 11.16 Email alert of patient's/person's temperature (abnormal).

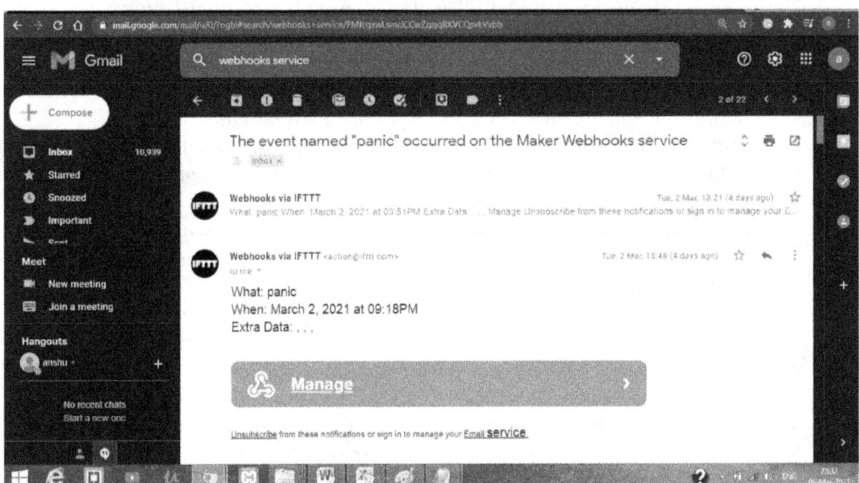

FIGURE 11.17 Email alert of patient for panic alert.

be used in research studies on medical issues that are affecting the chronically ill or elderly. The desired objective was attained. Heartbeat, temperature sensing, remote monitoring of patients' vitals, etc. gave the desired results. In emergency or critical situations, like deviation of temperature or heart rate from desired values or the patient being unwell and calling for help through the press of a button, alert signals to doctor/caregivers can help save lives by taking necessary actions. Also, routine monitoring of patient details is possible.

Microcontrollers today have high speeds, have become miniaturized, and are power efficient. They can be easily incorporated into healthcare devices. So, healthcare professionals are adopting the use of embedded systems for healthcare monitoring. In most developing countries, people are now using their smartphones for such useful health applications, and are also interested in new IoT devices that will be useful in daily life. This indicates that people will adopt the use of IoTin the future. A remote healthcare system promises to be useful to people and give them a better quality of life. In future, better sensors may address more vitals of a person, making the system more efficient and more accurate and useful in predicting some diseases. Also, an application can be designed so that it would be more convenient to use the system more efficiently.

REFERENCES

1. N. Gupta, H. Saeed, S. Jha, M. Chahande, S. Pandey, "Study and Implementation of IoT based Smart Healthcare System", in International Conference on Trends in Electronics and Informatics ICEI, 2017.
2. K. Jaiswal, S. Sobhanayak, B. K. Mohanta, D. Jena, "IoT-Cloud based framework for patient's data collection in smart healthcare system using Raspberry-pi", in International Conference on Electrical and Computing Technologies and Applications (ICECTA), 2017.
3. N. Reddy, A. M. Marks, S. R. S. Prabaharan, S. Muthulakshmi, *IoT Augmented Health Monitoring System*, IEEE, 2017.
4. P. Sasidharan, T. Rajalakshmi, U. Snekhalatha, "Wearable Cardiorespiratory Monitoring Device for Heart Attack Prediction", in International Conference on Communication and Signal Processing, April 4–6, 2019, India.
5. E. Saranya, T. Maheswaran, "IOT Based Disease Prediction and Diagnosis System for Healthcare", *International Journal of Engineering Development and Research*, Volume 7, Issue 2 (2019). ISSN: 2321-9939.
6. M. B. G. Sambhram, "Early Prediction System of Cardiovascular Disease through IoT", *International Journal of Combined Research & Development*, Volume 5, Issue 12 (December 2016). eISSN:2321-225X;pISSN:2321-2241
7. S. Banka, I. Madan, S. S. Saranya, "Smart Healthcare Monitoring using IoT," *International Journal of Applied Engineering Research*, Volume 13, Issue 15 (2018) pp. 11984–11989. ISSN 0973-4562. © Research India Publications. http://www.ripublication.com
8. M. R. Ruman, A. Barua, W. Rahman, "IoT Based Emergency Health Monitoring System", in International Conference on Industry 4.0 Technology (I4Tech), Feb. 13–15, 2020, Vishwakarma Institute of Technology, Pune, India.
9. T. J. Swamy, T. N. Murthy, "eSmart: An IoT based Intelligent Health Monitoring and Management System for Mankind", in International Conference on Computer Communication and Informatics (ICCCI –2019), Jan. 23 – 25, 2019, Coimbatore, India.
10. S. Nookhao, V. Thananant, T. Khunkhao, "Development of IoT Heartbeat and Body Temperature Monitoring System for Community Health Volunteer", in Joint International Conference on Digital Arts, Media and Technology with ECTI Northern Section Conference on Electrical, Electronics, Computer and Telecommunications Engineering (ECTI DAMT & NCON), 2020.
11. V. Tripathi, F. Shakeel, "Monitoring Health Care System using Internet of Things: An Immaculate Pairing", in International Conference on Next Generation Computing and Information Systems (ICNGCIS), 2017.

12. N. Kamal, P. Ghosal, "Three Tier Architecture for IoT Driven Health Monitoring System Using Raspberry Pi", in IEEE International Symposium on Smart Electronic Systems (iSES) (Formerly iNiS), 2018.
13. P. O. Antonio, "Heat Stroke Detection System Based in IoT".
14. K. Braam, T.-C. Huangy, C.-H. Cheny, E. Montgomery, S. Vo, R. Beausoleily, "Wristband Vital: A Wearable Multi-Sensor Microsystemfor Real-time Assistance via Low-Power Bluetooth Link".
15. A. R. Maria, P. Sever, S. George, "MIoT Applications for Wearable Technologies Used for Health Monitoring", in International Conference on Electronics, Computers and Artificial Intelligence ECAI, 2018.
16. I. Muhammad, E. Madona, N. Anggara, P. Roni, "Low Cost Heart Rate Portable Device for Risk Patients with loT and Warning System", in International Conference on Applied Information Technology and Innovation (ICAITI), 2018.
17. G. V. Kumar, A. Bharadwaja, N. N. Sai, "Temperature and Heart Beat Monitoring System Using IOT", in International Conference on Trends in Electronics and Informatics ICEI, 2017.
18. R. R. Rajanna, S. Natarajan, P. R. Vittal, "An IoT Wi-Fi Connected Sensor For Real Time Heart Rate Variability Monitoring", In IEEE 3rd International Conference on Circuits, Control, Communication and Computing, 2018.
19. N. L. Yanpin Ren, "A Pulse Measurement and Data Management System Based on Arduino Platform and Android Device", in Proceedings of 2016 IEEE 13th International Conference on Networking, Sensing, and Control Mexico City, Mexico, April 28–30, 2016.
20. https://store.arduino.cc/usa/arduino-uno-rev3
21. https://www.ti.com/lit/ds/symlink/lm35.pdf
22. https://www.electronicwings.com/arduino/lcd-16x2-interfacing-with-arduino-uno
23. https://en.wikipedia.org/wiki/ESP8266
24. https://www.electronicscomp.com/9v-1amp-dc-adaptor
25. https://pulsesensor.com/pages/pulsesensor-playground-toolbox
26. https://en.wikipedia.org/wiki/Arduino_IDE
27. https://en.wikipedia.org/wiki/ThingSpeak
28. B. Mishra et al., "The Use of MQTT in M2M and IoT Systems: A Survey", *IEEE Access*, volume 8, pp. 201071–201086.
29. D. Z. Maria Salagean, "IoT Applications based on MQTT Protocol", in International Symposium on Electronics and Telecommunications (ISETC), 5–6 Nov 2020.
30. R. A. Atmoko et al., "IoT real time data acquisition using MQTT protocol", *Journal of Physics: Conference Series*, May 2017.

12 Human Behavior Detection using Image Processing and IoT

Kamini Solanki, Abhishek Mehta, Dharmendrasinh Rathod, and Priya Swaminarayan

CONTENTS

DOI: 10.1201/9781003221333-12

12.1 INTRODUCTION

One of the most popular and challenging applications of computer vision is the analysis of visual scenarios involving humans. Face recognition, gesture recognition, and whole-body tracking are some of the tasks that fall within this domain. The motivation originates from a desire to better human–computer interaction, which is one of artificial intelligence's general goals. Human detection in still photos is a very recent field. Because of the requirement to segment frequently changing landscapes in real situations, this subject is both rich and demanding. Technological advancements in real-time image capture, transport, and analysis have given the movement a boost. A smart surveillance system's basic capability today is to recognize whether humans are present in the acquired frames/images. This study is about spotting humans in photographs who are in a pretty upright stance. Due to their varied appearance and vast range of stances, detecting humans in photographs is a difficult undertaking. As a result, a powerful feature set is required that allows the human shape to be readily distinguished even amid busy backgrounds and dim lighting. In the following are some examples of applications that demonstrate the necessity for a strong human detector based on Canny Operator. In this chapter, a technique was employed to detect various human beings. To detect humans, the system is initially trained with a variety of databases. This chapter employs a shape-based detection technique.

12.1.1 What Is Computer Vision?

In such systems, as well as human detection, computer vision techniques are applied. Many applications, such as human–robot interaction (HRI), video surveillance, human motion tracking, gesture recognition, and human behavior analysis, require human detection and identification. We are particularly interested in the topic of HRI, which has a wide range of applications. Because intelligent robots should live with humans in a human-friendly environment, they must be aware of and detect humans in their vicinity. Because of its low cost and ease of use, a single static camera is frequently utilized for person detection. However, because the robot (camera) and the human are moving relative to one other, and the lighting conditions and backgrounds are changing, a single camera is not practicable for human detection by a mobile robot. As a result, depth cues from a stereo camera can be used to successfully recognize and identify humans in mobile robot applications. In this research, stereo-based vision is used to recognize and identify humans.

12.1.2 Background of the Research

Human detection in real-world scenarios is a difficult task. Intelligent vehicles are cars, lorries, buses, and other vehicles with sensors and control systems that aid in driving. Several techniques have been developed in recent years, with outstanding results reported on a range of databases.

12.1.3 Objective of the Project

An intelligent machine must be able to recognize different types of humans. After training with multiple databases, it will be able to detect many types of objects. The major goal of the project is to recognize specific objects in still photographs.

The following are the major components:

1. Conversion from gray scale to color
2. Recognition of shapes
3. Comparisons of images
4. Detection of humans

12.1.4 Scope of the Project

This research describes a unique method for detecting and tracking moving vehicles in real time. The project's scope is flexible. Human detection in photographs has gained popularity in the field of computer vision, with applications in a variety of fields.

12.1.5 Overview of Proposed System

This chapter provides a boundary-based human detection system based on computer vision. The project begins with capturing real-time photos with a camera. The system finds the item after using a segmentation method to recognize moving objects and removing noise. It can detect humans by comparing the border with predetermined templates. This system's methodology has advantages over existing human detection systems in terms of speed, simplicity, learning, capability, and robustness to small image changes. The technology is capable of detecting not only humans but also any other thing we desire. This concept may be used to train with a variety of moving objects.

12.1.6 Project Organization

There are five sections in this chapter. The second section discusses earlier relevant studies as well as their successes and limitations. The third section delves into the specifics of the detection methods we employed to address the problem of human detection, including the experimental setup, feature vector generating process, datasets used, evaluation methodology, and feature descriptor configuration settings. The procedures taken to select the optimal feature descriptor for compressed images are detailed in the fourth section. Finally, the last section discusses the findings and finishes the study by discussing the chapter's main contributions, shortcomings, and future research.

12.2 LITERATURE REVIEW

Human detection is closely related to general object recognition techniques. It involves two steps. The first is feature extraction and training a classifier, as shown in Figure 12.1.

Figure 12.1 describes the techniques that are most relevant for object detection or classification while also being resistant to changes in illumination, viewpoint, and object contour alterations. Points [1] and [2], blobs (Laplacian of Gaussian [3] or Difference of Gaussian [4]), intensities [5], gradients [6] and [7], color, texture, or combinations of some or all of these [8] can be used to create such features. The resulting descriptors must adequately define the image for the detection and classification operation. The numerous approaches to descriptor selection will be divided into two basic categories. Local descriptors of relevant local picture regions are used to create sparse representations. Key point detectors, image fragments, and parts detectors can all be used to pick regions. Dense representations, on the other hand, are based on picture intensities, gradients, or higher-order differential operators. Picture characteristics are frequently extracted densely (typically pixel-by-pixel) throughout an entire image or detection window and then compiled into a high-dimensional descriptor vector that can be utilized for discriminative image classification or labeling the window as object or non-object.

12.2.1 Local Shape-Based Human Detection

Simulate human body configurations using local shape context to represent body part templates. They use normalized cuts segmentation and form, shading, and focus signals to retrieve body components in later work. construct a global human

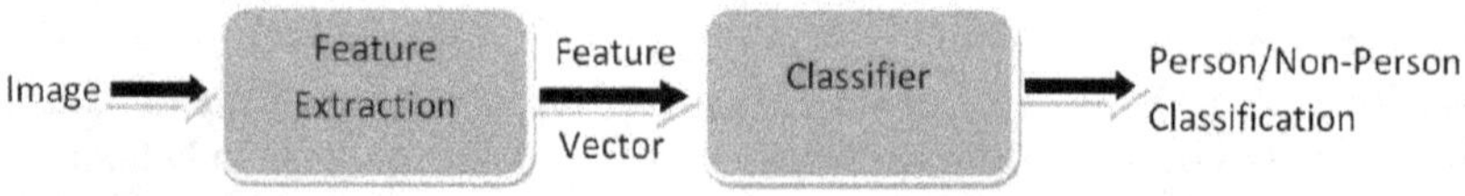

FIGURE 12.1 Components of human detection system.

model using Haar wavelet coefficients. Different shape-based person detection algorithms were investigated by Edgar Seeman, Bastian Leibe, and others [Leibe, 2003]. Shape-based detection techniques are divided into two categories: global and local approaches.

12.2.2 Global Approach

The global chamfer matching technique [Gavrila, 2000] is a global approach, while the Implicit Shape Model (ISM) is a local approach [Harris, 1998]. The global chamfer matching technique is used to match object shape silhouettes to image structure. A silhouette is shifted over the picture for this purpose, and a distance Dchamfer (T,L) between a silhouette T and the edge image is determined at each image location L.

12.2.3 Local Approach: Implicit Shape Model

The local strategy is broken down into three sections. Model training, hypothesis creation, segmentation, and verification are just a few examples. Local features extracted from training photos are used to train the ISM [Leibe, 2000]. Then, based on the spatial distribution of their occurrence on the item, their spatial occurrence is modelled. For each training image, an Interest Point Detector is used. A hypothesis is developed after training with the photos, and then, segmentation and verification are performed to detect the human. Local features are calculated using the Interest Point Detector. Interest Point Detector approaches include the Harris detector (Harris, 1988), the Difference of Gaussian detector (Lowe, 2004), and others, such as the Hessian–Laplace [Mikolajczyk, 2004] and Harris–Laplace detectors [Mikolajczyk, 2001].

12.2.4 Dense Descriptors of Image Regions

The Eigen faces approach of Sirovitch and Kirby [9] is one of the most well-known works that uses simple picture intensities. Image gradient descriptors have been employed in Dalal and Triggs [7], where histograms of gradients have been used. The Census algorithm [10] converts the intensity space to an order space, from which a bit pattern is generated by comparing the orders of a particular pixel with those of its neighbors. Mittal and Visvanathan [11] have often used an enhanced version of this approach to address the issue of counting salt and pepper noise in a pixel. Singh et al. [12] presented a way for enhancing noise immunity by making the penalty for an order flip proportional to the intensity difference between the two flipped pixels. Finally, Gupta and Mittal [13] describe a statistical method for adjusting the match measure to the underlying error process. All these methods assume that the pixel locations do not change between the two patches, making them unsuitable for a feature matching situation where the pixel locations may shift. Ojala et al. [14] create feature descriptors with Gaussian noise invariant point pairs. If the order of a point pair between the two patches changes, a penalty is applied, and the penalties for different pairings are added to determine the difference between the two features. In texture description, the Local Binary Patterns (LBP) descriptor [15], which is a version of the Census technique [10], has shown potential [16]. In Gupta et al. [17], a center-symmetric LBP (CS-LBP) that only compares center-symmetric pairings of pixels was investigated for feature description because the LBP operator produces a relatively high-dimensional histogram and is thus difficult to use in the context of an area descriptor. More recently, Papageorgiou and Poggio [18] proposed center-symmetric Local Ternary Patterns (CS-LTP) to make CS-LBP noise-resistant, as well as a global order-based descriptor (Histogram of Relative Intensities) that better handles saturation and illumination fluctuations.

12.2.5 Work in Human Detection

A pedestrian detector based on a polynomial support-vector machine (SVM) with rectified Haar wavelets as input descriptors is described in Mohan et al. [19], with a parts (subwindow)-based variation in Gavrila and Philomin [20]. However, we discovered that linear SVMs (weighted sums of corrected wavelet outputs) produce similar results and are far faster to compute. The findings for pedestrian, face, and automobile are shown in Mohan et al. [19]. Gavrila et al. [21] adopt a more direct technique, extracting edge pictures and utilizing chamfer distance to match them to a set of learned exemplars. This method has been implemented in a real-time pedestrian detection system [22]. Jones et al. [23] used AdaBoost [24] to train a series of progressively more complicated region rejection rules based on Haar-like wavelets and space-time differences to create an efficient moving person detector. In a dynamic programming framework similar to that of Krystian Mikolajczyk and Zisserman [25] and Wang et al. [26], Cordelia et al. [6] created an articulated body detector by combining SVM-based limb classifiers over first- and second-order Gaussian filters.

Lanitis et al. [27] build a parts-based technique with detectors for faces, heads, and front and side profiles of upper and lower body parts using a combination of orientation position histograms and binary-threshold gradient magnitudes. Panchal [28] combines Dalal and Triggs [7] and Ojala et al. [15] to create a more efficient description while also attempting to avoid occlusion.

12.2.6 Different Types of Edge Detector

The method of detecting and pinpointing sharp discontinuities in an image is known as edge detection. Discontinuities are sharp shifts in pixel intensity that define the edges of objects in a scene. Edge detection using traditional approaches entails convolving the image with an operator (a two-dimensional [2-D] filter) that is designed to be sensitive to big gradients in the image while returning zero values in uniform regions. There is a plethora of edge detection operators available, each of which is tuned to identify specific types of edges. Edge orientation, noise environment, and edge structure are all factors to consider when choosing an edge detection operator. The operator's shape determines a certain direction in which it is most sensitive to edges. The search for horizontal, vertical, or diagonal edges can be optimized using operators. In noisy photos, edge recognition is difficult because both the noise and the edges include high-frequency material. Efforts to decrease noise result in distorted and fuzzy edges. On noisy images, operators are often wider in scope, allowing them to average enough data to disregard localized noisy pixels. As a result, the detected edges are less accurately localized. Not every edge has a sharp change in intensity. Objects with boundaries defined by a gradual shift in intensity might be caused by refraction or inadequate focus [1].

In certain instances (see Figure 12.2), the operator must be selected to be responsive to such a progressive shift. As a result, there are issues such as erroneous edge detection, missing true edges, edge localization, excessive computing time, and noise-related issues, among others. As a result, the goal is to compare and contrast various edge detection techniques as well as examine their performance under various scenarios. Edge detection can be done using a variety of methods. The majority of diverse procedures, on the other hand, can be divided into two categories:

- Gradient-based edge detection:

The gradient approach finds edges by looking for maximum and minimum values in the image's first derivative.

- Laplacian-based edge detection:

To locate edges, the Laplacian approach looks for zero crossings in the image's second derivative. Calculating the image's derivative can reveal the location of an edge, which has the one-dimensional shape of a ramp.

The gradient of this signal (which is merely the first derivative with respect to t in one dimension) is as follows:

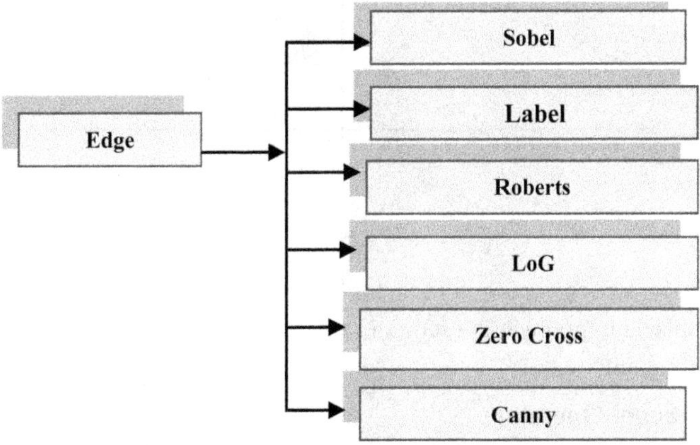

FIGURE 12.2 Different edge detectors.

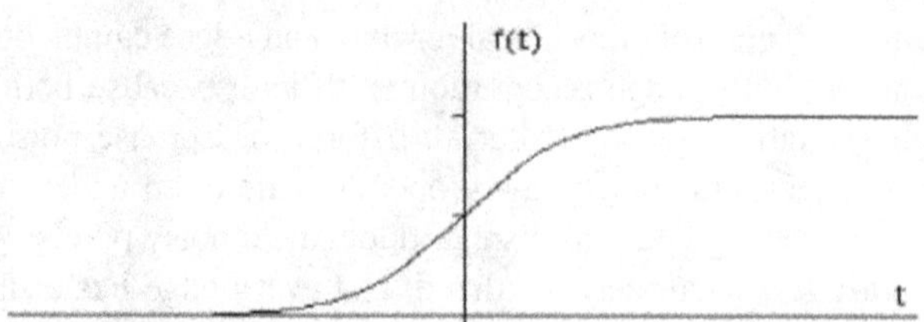

If we take the gradient of this signal (which, in one dimension, is just the first derivative with respect to t), we get the following:

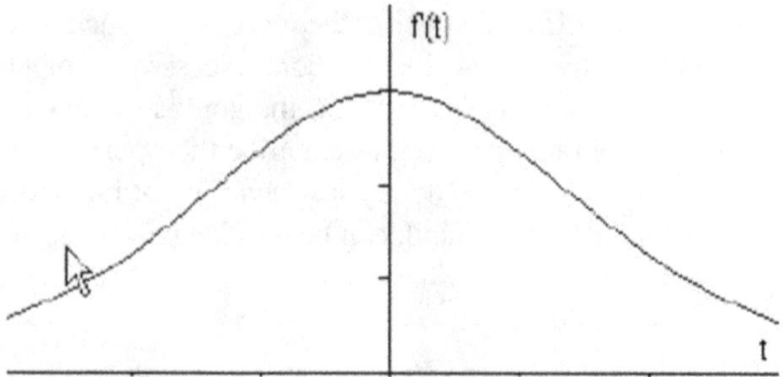

The derivative clearly reveals a maximum in the original signal at the middle of the edge. The Sobel method is part of the "gradient filter" family of edge detection filters, which includes this way of locating an edge. If the gradient value exceeds a certain threshold, a pixel position is deemed an edge location. As previously stated, edges will have higher pixel intensity values than those in their immediate vicinity. After you've set a threshold, you can compare the gradient value with the threshold value and identify an edge any time the threshold is surpassed.

Furthermore, the second derivative is zero when the first derivative is at its greatest. As a result, locating the zeros in the second derivative is another option for finding the position of an edge.

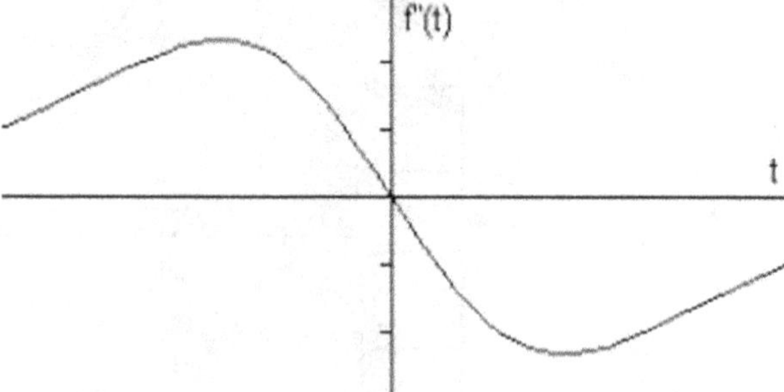

The Laplacian approach is used, and the second derivative of the signal is presented in the following.

12.2.6.1 Sobel Operator

As seen in Figure 12.3a, the operator is made up of a pair of 33 convolution kernels. One kernel is just 90 degrees rotated from the other.

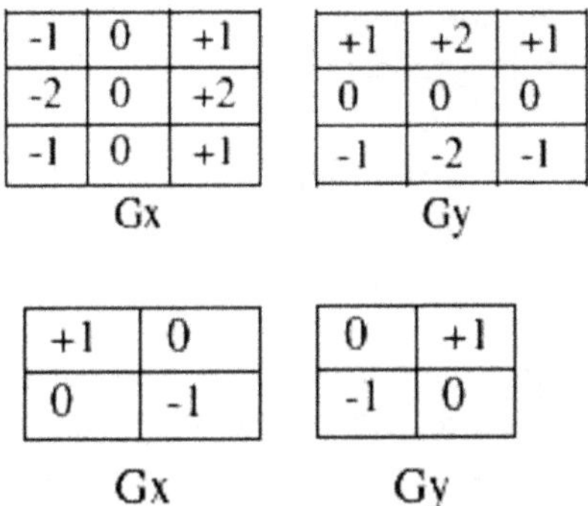

Gx

-1	0	+1
-2	0	+2
-1	0	+1

Gy

+1	+2	+1
0	0	0
-1	-2	-1

Gx

+1	0
0	-1

Gy

0	+1
-1	0

FIGURE 12.3 (a) Masks used by Sobel operator; (b) masks used for Robert operator.

These kernels, one for each of the two perpendicular orientations shown in Figure 12.3, are designed to respond maximally to edges traveling vertically and horizontally relative to the pixel grid. The kernels can be applied to the input image individually to provide separate gradient component measurements in each orientation (call these Gx and Gy). The absolute magnitude of the gradient at each site and its orientation can then be determined by combining these results. The magnitude of the gradient is calculated as follows.

Typically, an approximate magnitude is calculated using the following formula:

$$|G| = \sqrt{Gx^2 + Gy^2}$$

$$|G| = |Gx| + |Gy|$$

This takes a lot less time to compute. The spatial gradient is caused by the angle of orientation of the edge (relative to the pixel grid):

$$\theta = \arctan(Gy/Gx)$$

12.2.6.2 Roberts Cross Operator

The Roberts cross operator measures a 2-D spatial gradient on an image in a straightforward and quick way. The estimated absolute magnitude of the spatial gradient of the input image at that place is represented by pixel values at each position in the output. Figure 12.3b shows the operator, which is made up of a pair of 22 convolution kernels. One kernel is simply 90 degrees rotated from the other [4]. The Sobel operator is pretty similar to this.

These kernels, one for each of the two perpendicular orientations, are designed to respond maximally to edges running at 45° to the pixel grid. The kernels can be used individually to the input image, producing individual gradient component measurements in each orientation (call these Gx and Gy). The absolute magnitude of the gradient at each site and its orientation can then be determined by combining these results. The magnitude of the gradient is calculated as follows:

$$|G| = \sqrt{Gx^2 + Gy^2}$$

$$|G| = |Gx| + |Gy|$$

In most of the cases, an approximate magnitude is calculated, which is much faster. The angle of orientation of the edge that causes the spatial gradient (in relation to the pixel grid orientation) is

$$\theta = \arctan(Gy/Gx) - 3\pi/4$$

12.2.6.3 Prewitt's Operator

The Prewitt operator is used to detect vertical and horizontal edges in images and is comparable to the Sobel operator shown in Figure 12.4.

12.2.6.4 Laplacian of Gaussian

The Laplacian is a 2-D isotropic measure of an image's second spatial derivative. The Laplacian of an image is widely used for edge identification since it highlights regions of fast intensity change. To reduce susceptibility to noise, the Laplacian is frequently applied to an image that has first been smoothed with something resembling a Gaussian smoothing filter. Normally, the operator accepts a single gray level image as input and outputs another gray level image.

The Laplacian L(x,y) of a picture with I(x,y) pixel intensity values is calculated as follows:

$$L(x,y) = \frac{\partial^2 I}{\partial x^2} + \frac{\partial^2 I}{\partial y^2}$$

We must discover a discrete convolution kernel that can approximate the second derivatives in the definition of the Laplacian because the input image is represented as a series of discrete pixels. Figure 12.4 depicts three typically used tiny kernels.

Figure 12.5 describes that kernels are extremely sensitive to noise because they approximate a second derivative measurement on the image. To combat this, the image is frequently Gaussian smoothed before the Laplacian filter is applied. Prior

-1	0	+1
-1	0	+1
-1	0	+1

Gx

+1	+1	+1
0	0	0
-1	-1	-1

Gy

FIGURE 12.4 Masks for the Prewitt gradient edge detector.

to the differentiation stage, this pre-processing step eliminates the high-frequency noise components.

In fact, because convolution is an associative operation, we can first convolve the Gaussian smoothing filter with the Laplacian filter and then convolve this hybrid filter with the picture to get the desired result. There are two benefits to doing things this way. Because the Gaussian and Laplacian kernels are typically much smaller than the image, this approach takes far fewer arithmetic operations.

The LoG (Laplacian of Gaussian) kernel[6] can be pre-calculated in advance, requiring only one convolution on the picture at run-time.

The 2-D LoG function centered on zero and with Gaussian standard deviation has the form shown in Figure 12.6.

The LoG kernel becomes identical to the simple Laplacian kernels depicted in Figure 12.4 as the Gaussian becomes increasingly narrow. This is because on a discrete grid, smoothing with a very narrow Gaussian (0.5 pixels) has no effect. As a result, the simple Laplacian can be regarded as a limiting case of the LoG for narrow Gaussians on a discrete grid (Figure 12.7).

1	1	1
1	-8	1
1	1	1

-1	2	-1
2	-4	2
-1	2	-1

FIGURE 12.5 Three commonly used discrete approximations to the Laplacian filter.

$$LoG(x,y) = -1/\pi\sigma^4 \left[1 - \left(\frac{x^2+y^2}{2\sigma^2} \right) \right] e^{\frac{x^2+y^2}{2\sigma^2}}$$

and is shown

0	1	0
1	-4	1
0	1	0

in Figure 5.

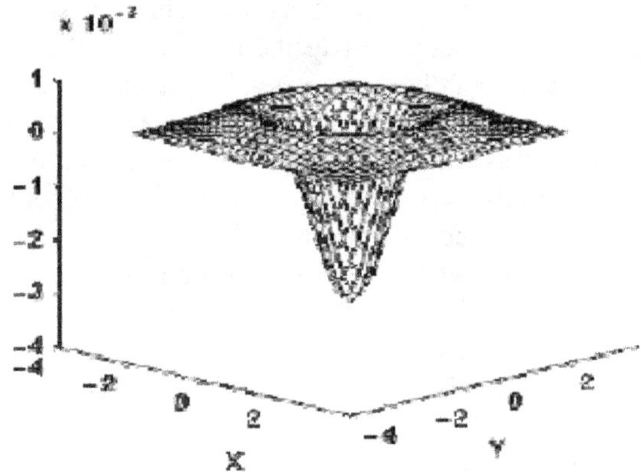

FIGURE 12.6 The 2-D Laplacian of Gaussian (LoG) function. The *x* and *y* axes are marked in standard deviations.

0	1	1	2	2	2	1	1	0
1	2	4	5	5	5	4	2	1
1	4	5	3	0	3	5	4	1
2	5	3	-12	-24	-12	3	5	2
2	5	0	-24	-40	-24	0	5	2
2	5	3	-12	-24	-12	3	5	2
1	4	5	3	0	3	5	4	1
1	2	4	5	5	5	4	2	1
0	1	1	2	2	2	1	1	0

FIGURE 12.7 Discrete approximation to LoG function with Gaussian S = 1.4.

12.2.7 Canny Edge Detection Algorithm

Many people consider the Canny edge detection algorithm to be the best. Canny's goal was to improve on the various edge detectors that were already on the market when he began his work. His concepts and methods can be found in his paper, "A Computational Approach to Edge Detection", which he published after completing his goal. In his research, he used a set of criteria to improve current edge detection systems. The first and most evident benefit is a low rate of error. It's critical that edges in images aren't overlooked and that there are no responses to non-edges. The edge points must be well localized as the second condition. In other words, the gap between the detector's edge pixels and the actual edge must be kept to a minimum. The presence of only one response to a single edge is a third criterion. Because the first two were insufficient to totally exclude the possibility of multiple replies to an edge, this was included. The Canny edge detector smooths the image first to remove noise based on these criteria. The picture gradient is then used to highlight places with high spatial derivatives. The program then moves over these zones, suppressing any pixels that are not at maximum brightness (non-maximum suppression). Hysteresis now reduces the gradient array much further. To follow along the remaining pixels that haven't been suppressed, hysteresis is used. Hysteresis employs two thresholds, with the first being set to zero if the magnitude falls below it (made a non-edge). It is designated as an edge if the magnitude exceeds the high threshold. If the magnitude falls between the two thresholds, it is set to zero unless a path exists between this pixel and a pixel with a gradient greater than T2.

Step 1: A set of steps must be followed in order to implement the Canny edge detector algorithm. Before attempting to locate and detect any edges, the first step is to filter out any noise in the original image. The Gaussian filter is used primarily in the Canny algorithm because it can be computed using a simple mask. Standard convolution methods can be used to accomplish Gaussian smoothing after a sufficient mask has been calculated. A convolution mask is usually a fraction of the size of the actual image. As a result, the mask is slid across the image, one square of pixels at a time, modifying it. The detector's sensitivity to noise decreases as the breadth of the Gaussian mask increases. As the Gaussian width is raised, the localization inaccuracy in the identified edges likewise increases slightly.

Step 2: After smoothing the image and removing the noise, the gradient of the image is used to determine the edge strength. On a picture, the Sobel operator provides a 2-D spatial gradient measurement. The magnitude of the approximate absolute gradient (edge strength) at each position can then be determined. The Sobel operator employs a pair of 3 × 3 convolution masks, one of which estimates the gradient in the x-direction (columns) and the other the gradient in the y-direction (rows).

-1	0	+1
-2	0	+2
-1	0	+1

Gx

+1	+2	+1
0	0	0
-1	-2	-1

Gy

The magnitude, or edge strength, of the gradient is then approximated using the formula:

$$|G| = |Gx| + |Gy|$$

Step 3: The direction of the edge is computed using the gradient in the x and y directions. However, an error will be generated when sumX is equal to zero. So, in the code, there has to be a restriction set whenever this takes place. Whenever the gradient in the x-direction is equal to zero, the edge direction has to be equal to 90 degrees or 0 degrees, depending on what the value of the gradient in the y-direction is equal to. If Gy has a value of zero, the edge direction will equal 0 degrees. Otherwise, the edge direction will equal 90 degrees. The formula for finding the edge direction is

$$\text{Theta} = \text{inv}\tan(Gy/Gx)$$

Step 4: Once the edge direction is known, the next step is to relate the edge direction to a direction that can be traced in an image. So if the pixels of a 5 × 5 image are aligned as follows:

X X X X X

X X X X X

X X a X X

X X X X X

X X X X X

Looking at pixel "a", it can be observed that there are only four viable orientations for characterizing the surrounding pixels: 0 degrees (horizontal), 45 degrees (positive diagonal), 90degrees (vertical), or 135 degrees (horizontal) (along the negative diagonal). So, depending on which of these four directions the edge orientation is closest to, it must now be resolved into one of these four directions (e.g. if the orientation angle is found to be 3 degrees, make it 0 degrees). Consider dividing a semicircle into five distinct areas.

Any edge direction that falls within the yellow range (0 to 22.5 and 157.5 to 180 degrees) is therefore set to 0 degrees. Any edge direction that falls between 22.5 and 67.5 degrees in the green range is set to 45 degrees. Any edge direction that falls between 67.5 and 112.5 degrees in the blue range is set to 90 degrees. Finally, every edge direction that falls between 112.5 and 157.5 degrees is set to 135 degrees.

Step 5: Now that the edge directions have been determined, non-maximum suppression must be applied. Non-maximum suppression is used to trace along the edge in the edge direction and suppress any pixel value that is not regarded as being an edge (makes it equal to 0). In the output image, this will result in a thin line.

Step 6: varying operator output above and below the threshold. When a single threshold, T1, is applied to an image, and an edge has an average strength equal to T1, noise will cause the edge to dip below the threshold in some cases. It will also extend above the threshold, giving the appearance of a dashed line. Hysteresis uses two thresholds, a high and a low, to avoid this. Any pixel in the image with a value larger than T1 is considered an edge pixel and is instantly designated as such. Then, any pixels with a value larger than T2 that are related to this edge pixel are likewise selected as edge pixels. If you want to follow an edge, you'll need a T2 gradient to begin, but you won't stop until you reach a T1 gradient.

12.2.8 Detection and Tracking Using Combination of Thermal and Visible Imaging

The previous method was useful indoors, but in the outdoor environment, it requires a high-resolution camera. In this case, a fusion of infrared camera and visible imaging is used. This significantly reduces the processing time and power required for detecting a human. The main processes of the technique are segmentation and classification. These are used in many detection techniques.

12.2.8.1 Segmentation

The initial step in any real-time person detection technique based on photos is to identify static objects in the background and ROI (region of interest). Segmentation is the term for this procedure. To create a background model, most methods leverage intensity, texture, and contrast attributes of the image over time. To account for slow changes in light, the background model is updated by averaging frames over time. The current image is pixel-by-pixel removed from the background model. The suggested system uses temperature information to segment the object. By observing heated objects, humans are segmented. Because not all hot objects are persons, this stage was also used to weed out small groups of hot objects that were unlikely to be individuals.

12.2.8.2 Classification

After segmentation, there is a group of heated objects, which must be classified as human or not. Straightforward shape analysis can be utilized. To determine the shape of the item, they employed vertical histogram projection. They discovered that individuals have a shape that is akin to a normal Gaussian curve, which distinguishes them from other hot items such as cars, buses, and other vehicles. They categorize persons based on their shapes.

12.2.8.3 Summary

Various methods of human detection are briefly explored in this section. The major goal of these studies is to use computers to recognize human photos that are more motionless. Because the approach has some limitations, researchers are working to enhance the techniques. Our proposed human detection methodology is presented in the following section.

12.3 PROPOSED HUMAN DETECTION METHODOLOGY

12.3.1 Introduction

We covered several human detection techniques, their successes, and their limitations in the preceding section. The proposed human detection system will be discussed in this section. The discussion is divided into three sections: image acquisition, edge detection, and human detection via shape analysis. The suggested system architecture is described in Section 12.3.2. The human detection technique is described in depth in Section 12.3.3.

12.3.2 Proposed System Architecture

Despite the fact that motion is a key indication for understanding activities, we can more or less interpret human actions in such images. This is especially evident in news or sports photos, as people are photographed in stylized positions that reflect an action. The photo was taken with a digital camera. The boundary of the segmented images is detected, and the detected boundary is compared with predetermined templates for matching (Figure 12.8).

12.3.3 Details of Human Detection

12.3.3.1 Human Detection

Human detection is strongly linked to object recognition systems in general. It consists of two stages: training or learning phases and classifier phases for human detection. The flow chart of the proposed human detection method is shown in Figure 12.9. The next subsections will go over each aspect of the project. The acquisition of photos is the subject of Section 12.3.3.1. The gray scale conversion from RGB images is the subject of Section 12.3.3.2. The shape/boundary detection approach is discussed in

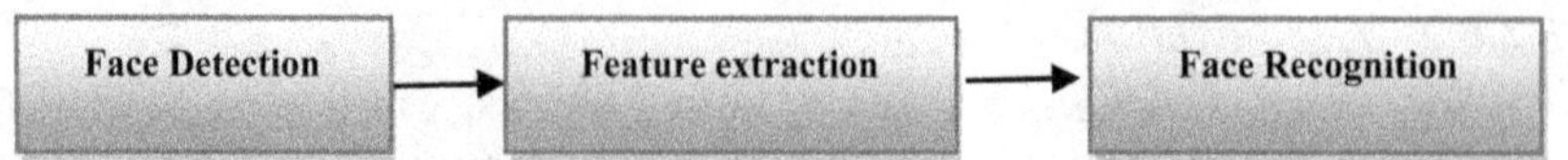

FIGURE 12.8 Framework for face recognition.

Section 12.3.3.3. Normalization techniques are discussed in Section 12.3.3.4, while pattern matching approaches are discussed in Section 12.3.3.5.

12.3.3.2 Image Acquisition

We can get images in various ways, such as digital camera and scanner, and store images on a computer hard disk.

12.3.3.3 Gray Scale Conversion

There are several methods for image conversion. In this chapter, we use the gray scale conversion for further processing to detected human areas by the traditional approach.

$$\text{New}_i = (R_i + G_i + B_i)/3) \tag{3.1}$$

First, input the original image; then, remove the background from the image and convert it into gray scale (Figures 12.4).

12.3.3.4 Edge Detection

Many people consider the Canny edge detection algorithm to be the best. Canny's goal was to improve on the various edge detectors that were already on the market when he began his work. His concepts and methods can be found in his paper, "A Computational Approach to Edge Detection", which he published after completing his goal. In his research, he used a set of criteria to improve current edge detection systems. The first and most evident benefit is a low rate of mistakes. It's critical that edges in images aren't overlooked and that there are no responses to non-edges. The

FIGURE 12.9 Original image.

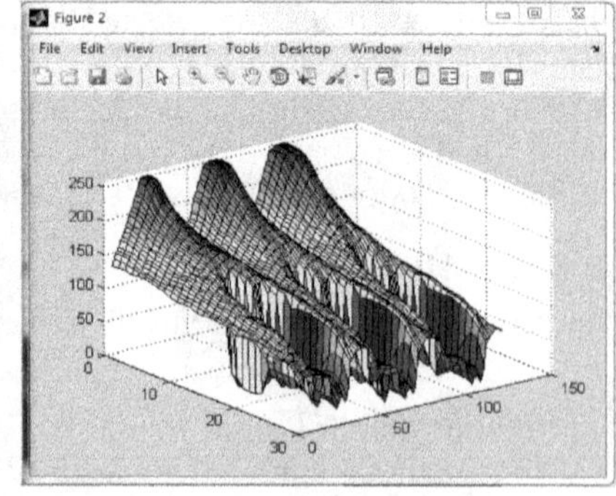

FIGURE 12.10 Background image in graph.

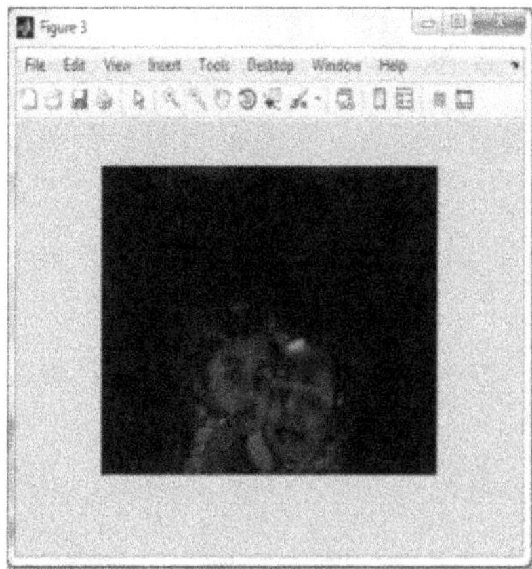

FIGURE 12.11 After background removal.

FIGURE 12.12 RGB to gray scale.

edge points must be well localized as the second condition. In other words, the gap between the detector's edge pixels and the actual edge must be kept to a minimum. The presence of only one response to a single edge is a third criterion. Because the first two were insufficient to totally exclude the possibility of multiple replies to an edge, this was included. The Canny edge detector smooths the image first to remove noise based on these criteria. The picture gradient is then used to highlight places with high spatial derivatives. The program then moves over these zones, suppressing any pixels that are not at maximum brightness (non-maximum suppression). Hysteresis now reduces the gradient array much further. To follow along the remaining pixels that haven't been suppressed, hysteresis is used.

Hysteresis employs two thresholds, with the first being set to zero if the magnitude falls below it (made a non-edge). It is designated as an edge if the magnitude exceeds the high threshold. If the magnitude falls between the two thresholds, it is set to zero unless a path exists between this pixel and a pixel with a gradient greater than T2.

Step 1: A set of steps must be followed in order to implement the Canny edge detector algorithm. Before attempting to locate and detect any edges, the first step is to filter out any noise in the original image. The Gaussian filter is used primarily in the Canny algorithm because it can be computed using a simple mask. Standard convolution methods can be used to accomplish Gaussian smoothing after a sufficient mask has been calculated. A convolution mask is usually a fraction of the size of the actual image. As a result, the mask is slid across the image, one square of pixels at a time, modifying it. The detector's sensitivity to noise decreases as the breadth of the Gaussian mask increases. As the Gaussian width is raised, the localization inaccuracy in the identified edges likewise increases slightly.

Step 2: After smoothing the image and removing the noise, the gradient of the image is used to determine the edge strength. On a picture, the Sobel operator provides a 2-D spatial gradient measurement. The magnitude of the approximate absolute gradient (edge strength) at each position can then be determined. The Sobel operator employs a pair of 3 × 3 convolution masks, one of which estimates the gradient in the x-direction (columns) and the other the gradient in the y-direction (rows). They are as in Figure 12.4.

-1	0	+1
-2	0	+2
-1	0	+1

Gx

+1	+2	+1
0	0	0
-1	-2	-1

Gy

The magnitude, or edge strength, of the gradient is then approximated using the formula:

$$|G| = |Gx| + |Gy|$$

Step 3: The gradient in the x and y directions is used to determine the edge's direction. When sum X equals zero, however, an error is generated. As a result, any time this occurs, a limitation must be specified in the code. When the x-direction gradient is zero, the edge direction must be 90 degrees or 0 degrees, depending on the value of the y-direction gradient. The edge direction will be 0 degrees if Gy is set to zero. Otherwise, the edge will be at a 90-degree angle. The following is the formula for determining the edge direction:

$$\text{Theta} = \text{inv}\tan(Gy/Gx)$$

Step 4: Once the edge direction is known, the next step is to relate the edge direction to a direction that can be traced in an image. So, if the pixels of a 5 × 5 image are aligned as follows:

```
x x x x x
x x x x x
x x a x x
x x x x x
x x x x x
```

Then, it can be seen by looking at pixel "a" that there are only four possible directions when describing the surrounding pixels: 0 degrees (in the horizontal direction), 45 degrees (along the positive diagonal), 90 degrees (in the vertical direction), or

135 degrees (along the diagonal that is negative). So, depending on which of these four directions the edge orientation is closest to, it must now be resolved into one of these four directions (e.g. if the orientation angle is found to be 3 degrees, make it 0 degrees). Consider dividing a semicircle into five distinct areas. Any edge direction that falls within the first range (0 to 22.5 and 157.5 to 180 degrees) is therefore set to 0 degrees. Any edge direction that falls between 22.5 and 67.5 degrees in the second range is set to 45 degrees. Any edge direction that falls between 67.5 and 112.5 degrees in the next range is set to 90 degrees. Finally, every edge direction that falls between 112.5 and 157.5 degrees is set to 135 degrees.

Step 5: Now that the edge directions have been determined, non-maximum suppression must be applied. Non-maximum suppression is used to trace along the edge in the edge direction and suppress any pixel value that is not regarded as being an edge (makes it equal to 0). In the output image, this will result in a thin line.

Step 6: Finally, to eliminate streaking, hysteresis [12] is employed. The operator output varying above and below the threshold causes streaking, which causes an edge contour to break up. When a single threshold, T1, is applied to an image and an edge has an average strength equal to T1, noise will cause the edge to dip below the threshold in some cases. It will also extend above the threshold, giving the appearance of a dashed line. Hysteresis uses two thresholds, a high and a low, to avoid this. Any pixel in the image with a value larger than T1 is considered an edge pixel and is instantly designated as such. Then, any pixels with a value larger than T2 that are related to this edge pixel are likewise selected as edge pixels. If you want to follow an edge, you'll need a T2 gradient to begin, but you won't stop until you reach a T1 gradient.

Segmentation with threshold value 0.15 using different edge detectors is shown in Figures 12.13–12.18.

Segmentation with threshold value 0.15 using Canny edge detector is shown on multiple images

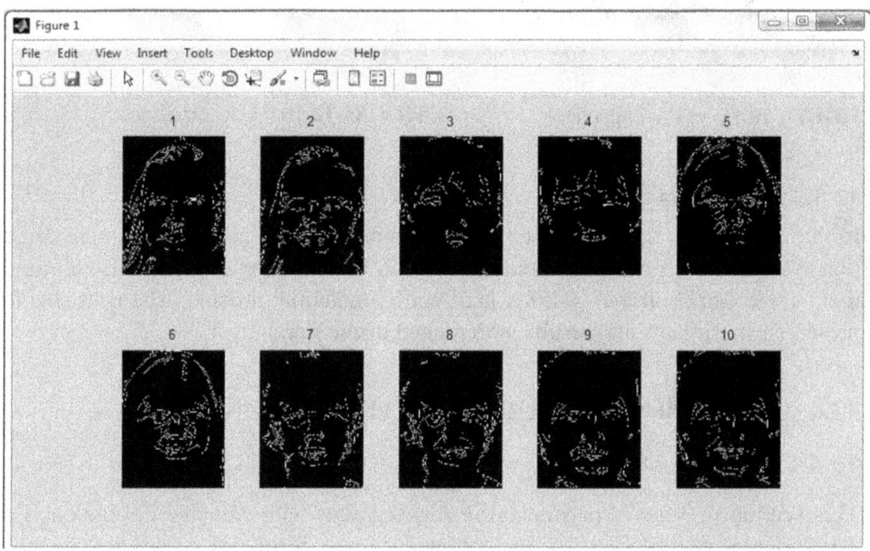

FIGURE 12.13 Sobel detector.

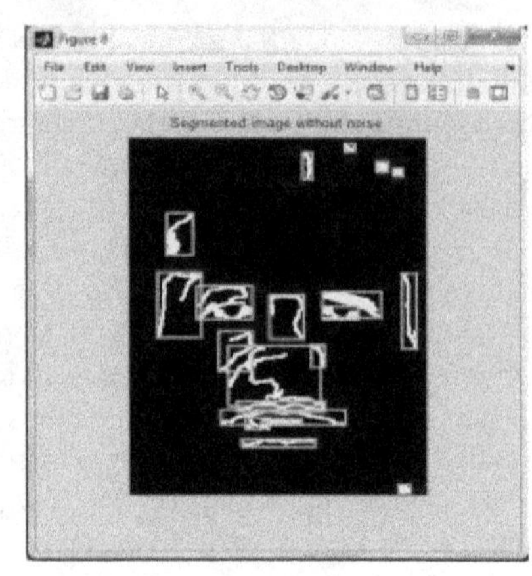

FIGURE 12.14 Canny detector.

FIGURE 12.15 Prewitt detector.

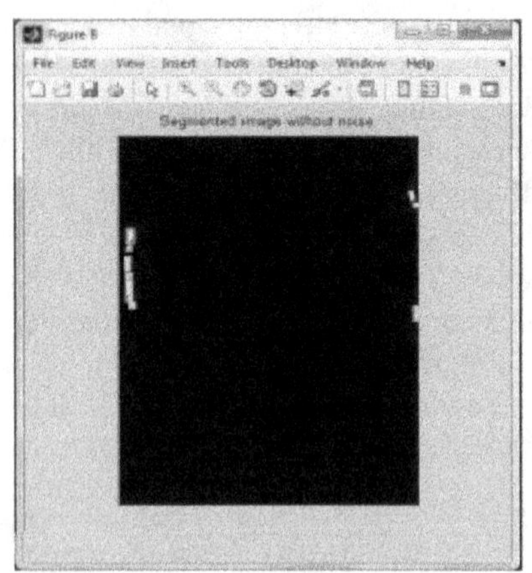

FIGURE 12.16 LoG detector

12.3.3.5 Summary

In this section, we have discussed different modules of the proposed human detection system. This technique is mainly focused on detecting objects, noise elimination, shape detection and linking, and finally, matching module. The next chapter presents experiments and results with related discussion.

12.4 EXPERIMENTS, RESULTS, AND DISCUSSION

12.4.1 Introduction

This section presents experimental results together with valuable discussion. The experimental results focus on the following areas: results of conversion method,

FIGURE 12.17 Zerocross detector.

FIGURE 12.18 Roberts detector.

results of filtering method, results of boundary detection method, and finally, results of human detection method. Section 12.4.2 describes the experiment setup. Section 12.4.3 presents the results of human detection.

12.4.2 Experiment Setup

In this chapter, we use a core2 duo 2.8GHz PC with 1 GB RAM. We also use application software to subtract the background of an image. This image will be input to the system for further processing.

The proposed method is implemented using Microsoft Visual C++(6.0), OpenCV library function, OpenGL programming language, and Adobe Photoshop CS2.

12.4.3 Experimental Results of Proposed System

This proposed system can detect a human from a white background still image.

12.5 CONCLUSION AND FUTURE WORK

This chapter describes human detection from still images using contour/boundary-based matching. The system is tested in different positions for detecting humans. From the experimental results, we conclude that the performance of the system is satisfactory.

12.5.1 Contribution

This chapter has made great contributions in the field of image analysis and object recognition. In this project, we have presented an algorithm for detecting humans [1].

The proposed system runs with satisfactory success rates. The contributions of our work can be summarized as follows:

- A novel method for detecting humans from still images
- A novel method for shape/contour detection from still images
- Satisfactory performance in detecting humans using contour/shape matching
- Average accuracy and precision of human detection method in the system 93.05% and 95.72%, respectively.

12.5.2 Limitations and Future Work

The major limitation of the proposed method is that the system is not dynamic. It can detect humans only from still images. Another limitation is that the system cannot detect whether multiple humans are present. There are situations, such as when a human is immediately behind another human, when this method cannot detect the two humans separately. Another limitation of this method is that the system cannot subtract the background automatically. This has to be done manually for the system input images to detect humans. In future, we will make this method more robust against the various limitations and detect multiple humans.

12.5.3 Concluding Remarks

The ultimate goal of this chapter was to establish an efficient, robust, and user-friendly system to detect humans. Our achievement from this research is satisfactory. We aspire to do more research in the same field.

REFERENCES

1. Harris, C., Stephens, M.: A combined corner and edge detector. Alvey Vision Conference, Bangalore (1998) 147–151.
2. Mikolajczyk, K., Schmid, C.: An affine invariant interest point detector. 7th European Conference on Computer Vision, Volume 1 (2002) 128–142.
3. Lindeberg, T.: Feature detection with automatic scale selection. *International Journal of Computer Vision* 30 (1998) 79–116.
4. Lowe, D.G.: Local feature view clustering for 3D object recognition. Conference on Computer Vision and Pattern Recognition (2001) 682–688.
5. Kadir, T., Brady, M.: Scale, saliency and image description. International Journal of Computer Vision 45 (2001) 83–105.
6. Cordelia, R.R., Schmid, C., Triggs, B.: Learning to parse pictures of people. European Conference on Computer Vision (2002) 700–714.
7. Dalal, N., Triggs, B.: Histograms of oriented gradients for human detection. IEEE Computer Society Conference on Computer Vision and Pattern Recognition, Volume 1 (2005) 886–893.
8. Martin, D.R., Fowlkes, C.C., Malik, J.: Learning to detect natural image boundaries using local brightness, color, and texture cues. *IEEE Transactions on Pattern Analysis and Machine Intelligence* 26 (2004) 530–549.

9. Sirovitch, L., Kirby, M.: Low-dimensional procedure for the characterization of human faces. *Journal of the Optical Society of America* 2 (1987) 586–591.
10. Zabih, R., Woodfill, J.: Non-parametric local transforms for computing visual correspondence. ECCV '94: Proceedings of the Third European Conference on Computer Vision, Volume II (1994) 151–158. on Pattern Analysis and Machine Intelligence 20 (1998) 415–423
11. Mittal, A., Visvanathan, R.: An intensity-augmented ordinal measure for visual correspondence. CVPR '06: Proceedings of the 2006 IEEE Computer Society Conference on Computer Vision and Pattern Recognition (2006) 849–856.
12. Singh, M., Parameswaran, V., Ramesh, V.: Order consistent change detection via fast statistical significance testing. IEEE Conference on Computer Vision and Pattern Recognition, Volume 2008 (2008)
13. Gupta, R., Mittal, A.: SMD: A locally stable monotonic change invariant feature descriptor. 10th European Conference on Computer Vision (2008) 265–277.
14. Ojala, T., Pietikainen, M., Harwood, D.: A comparative study of texture measures with classification based on featured distributions. *Pattern Recognition* 29 (1996) 51–59.
15. Ojala, T., Pietikäinen, M., Mäenpää, T.: Multiresolution gray-scale and rotation invariant texture classification with local binary patterns. *IEEE Transactions on Pattern Analysis and Machine Intelligence* 24 (2002) 971–987.
16. Heikkilä, M., Pietikäinen, M., Schmid, C.: Description of interest regions with local binary patterns. *Pattern Recognition* 42 (2009) 425–436.
17. Gupta, R., Patil, H., Mittal, A.: Robust order-based methods for feature description. In: IEEE Conference on Computer Vision and Pattern Recognition (2010).
18. Papageorgiou, C., Poggio, T.: A trainable system for object detection. *International Journal of Computer Vision* 38 (2000) 15–33.
19. Mohan, A., Papageorgiou, C., Poggio, T.: Example-based object detection in images by components. *IEEE Transactions on Pattern Analysis and Machine Intelligence* 23 (2001) 349–361.
20. Gavrila, D., Philomin, V.: Real-time object detection for smart vehicles. Proceedings of the 7th International Conference on Computer Vision, Volume 87 (1999) 93–98.
21. Gavrila, D.M., Giebel, J., Munder, S.: Vision-based pedestrian detection: The protector system. IEEE Intelligent Vehicles Symposium (2004) 13–18.
22. Jones, M., Viola, P., Viola, P., Jones, M.J., Snow, D., Snow, D.: Detecting pedestrians using patterns of motion and appearance. International Conference on Computer Vision (2003) 734–741.
23. Schapire, R.E.: The boosting approach to machine learning, an overview. MSRI Workshop on Nonlinear Estimation and Classification (2002).
24. Krystian Mikolajczyk, C.S., Zisserman, A.: Human detection based on a probabilistic assembly of robust part detectors. European Conference on Computer Vision (2004) 69–82.
25. Wang, X., Han, T.X., Yan, S.: An hog-lbp human detector with partial occlusion handling. International Conference on Computer Vision (2009).
26. Lanitis, A., Taylor, C.J., Cootes, T.F.: Towards automatic simulation of ageing effects on face images. *IEEE Transactions on Pattern Analysis and Machine Intelligence* 24(4) (2002) 442–455.
27. Panchal, K.K.: 3D Face Recognition on GAVAB Dataset. *International Journal of Engineering Research & Technology* 2(6) (2013). ISSN: 2278-0181. www.ijert.org.

13 A Novel Cross-Slotted Dual-Band Fractal Microstrip Antenna Design for Internet of Things (IoT) Applications

Ram Krishan

CONTENTS

13.1 INTRODUCTION

The Internet of Things (IoT) framework is utilized on a large scale for the incorporation of modern technologies into communication systems [1–3]. The applications of IoT require the interconnection between the devices and can be used in diverse situations [2, 4]. An antenna that can be easily embedded into IoT devices is required for the actual utilization of the IoT applications. For that purpose, microstrip antennas are widely utilized because of their highlights:lightweight, minimal expense, minimization, straightforward production, and effective integration with radio frequency (RF) devices [5–8]. The ISM (industrial, scientific, and measurement) bands of 2.4, 3.96,and 5 GHz, used for wireless local area networks (WLAN),are suitable for IoT applications [9, 10]. Fractal geometry is used for the fractal design of antennas. Mandelbrot, a French mathematician first used the term "fractal" in 1982 for a shape which is split from the whole shape [11]. The term "fractal" originates from

DOI: 10.1201/9781003221333-13

"fractus", a Latin word that means "broken". A fractal antenna [12–14] is depicted as an antenna that utilizes the features of fractals, like self-matching design [15], to expand the limit (both inside and outside) of the material that is fit for sending or receiving electromagnetic radiations [16, 17]. The fractal shape is commonly a reduced-size copy of different measurements of the whole. The fractal shape helps to achieve scaled downsizing and multiband capability of the antenna [18].

13.2 RELATED WORK

This section will explore the previous work done in the field of fractal microstrip antenna development. A compact size antenna operating on dual-band for Worldwide Interoperability for Microwave Access (WiMAX) and WLAN applications is presented in Xiaolei et al., Krishan, and Krishan and Laxmi [19–21]. This antenna resonates at 2.4 and 3.5 GHz frequency bands. The simulated performance of the antenna is studied on various parameters like radiation pattern, efficiency, and gain. A novel planar microstrip antenna that works on 2.4 and 5.5 GHz frequencies is presented [22]. In Mao et al. [10], a monopole antenna for IoT application is fabricated, which can operate on dual-band frequency. This antenna uses defective ground with an H-shaped structure and operates on WLAN frequency bands. An elliptical antenna having ultra-wideband (UWB) features is used for low-cost measurements is proposed in Nobrega et al. [23]. This antenna operates on a UWB frequency range from 1.0 to 13.5 GHz. In Khanna et al. [24], a gap-coupled fractal antenna is designed to work for broader bandwidth. This antenna has 85.4% bandwidth and works for WLAN and WiMAX applications. In Sawant and Kumar [25], aUWB hexagonal antenna designed for wireless communication with feed constitution of a co-planar waveguide is presented. This fractal antenna feed is utilized to enhance the operation of the antenna on upper frequencies (3.1 to 10.6 GHz) under UWB. Singh et al. [14] designed a compact fractal antenna used for wireless systems at a low cost. Minkowski and Koch's curves are integrated into the antenna structure. This antenna is tested using high-frequency simulation software (HFSS) and works for WLAN, ISM band, Bluetooth applications, etc.

13.3 FRACTAL ANTENNA DESIGN AND MEASUREMENTS

FR4 glass epoxy material is used to make the cross-slotted antenna with substrate dielectric constant $\varepsilon_r = 4.4$ and thickness of 1.6 mm. The basic structure of cross-slotted antenna geometry is taken as length and width dimensions of 40 mm. A square patch with length 27.82 mm, width 35.4 mm, and feed width of 2.6 mm is depicted in Figure 13.1. Table 13.1 presents all dimensions of the antenna.

13.3.1 Different Stages of Antenna Creation

Stage 1: The basic geometry of the antenna is derived by following the equation [5, 26].

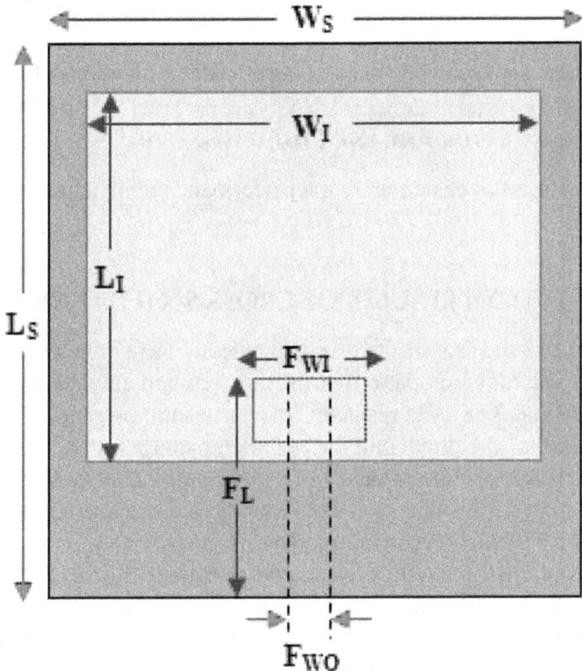

FIGURE 13.1 Antenna measurements.

TABLE 13.1
Antenna Dimensions

Parameter	Value (mm)
W_S	40
L_S	40
W_I	35.4
L_I	27.82
F_{WI}	2.6
F_{WO}	2.6
F_L	16.4

Stage 2: A rectangle of size 35.4 mm × 27.8 mm as width and height, respectively, of the patch for the antenna's first iteration is sliced. The feed of the antenna is taken with dimensions of 2.6 mm × 23.9 mm as width and height, respectively.

Stage 3: Split the square into nine identical squares and afterwards cut a cross (X) from the center with 0.707 mm width and 1.738 mm side-half height.

Stage 4: Again, slice cross (X) with the same dimensions taken in Step 3 from eight squares.

Stage 5: These steps can be repeated to get endless iterations of the antenna.

13.3.2 Parameters for Antenna Characterization

The parameters used to characterize the performance of an antenna arepresented in Table 13.2.

13.4 SIMULATION RESULTS OF CROSS-SLOTTED ANTENNA

The simulations of the cross-slotted fractal antenna were done using HFSS simulation software. The radiation pattern of the cross-slotted antenna in the E-plane and H-plane at 2.49 and 3.98 GHz resonant frequencies are presented in Figure 13.2 (a, b, c, d). The lines of the graph plot the radiation pattern with $\varphi = 90$ degrees and $\varphi = 0$ degrees. The three-dimensional (3-D) peak gain of the antenna is depicted in Figure 13.3(a, b) in a 3-D polar plot. The cross-slotted antenna attains 3.2 and 1.8\dB gain at 2.49 and 3.98 GHz resonant frequencies, respectively.

The result obtained for return loss of the antenna confirms the operation of an antenna on 2.49 and 3.98 GHz frequencies. A return loss value less than –10 dB validates its application for practical use. A –29 and –20 dB return loss was achieved by the antenna at resonant frequencies of 2.49 and 3.98 GHz, respectively. Voltage standing wave ratio (VSWR) depicts the feed matching characteristics of an antenna. The impedance of antenna should be matched with 50 Ohms for all applications. The VSWR value should be lies between 1 and 2 for practical acceptability. The results depicts 1.0104 and 1.2208 VSWR at 2.49 and 3.98 GHz frequencies, respectively.

TABLE 13.2
Parameters for Antenna Characterization

Parameter	Description
Radiation Pattern	The radio wave field strength of an antenna is graphically represented by the radiation pattern.
Antenna Gain (dB)	Antenna gain is the measure of its overall efficiency.
Peak Gain (dB)	This is the maximum gain of the antenna in any direction.
Directivity	Directivity is the gain of an antenna in a particular direction.
Return Loss S11(dB)	Return loss is the ratio of radio waves at the input of antenna that are rejected to those that are accepted. It is determined in decibels (dB).
VSWR	The real voltage made inside a transmission line framework when forward and reflected radio waves are spreading at the same time is VSWR (voltage standing wave ratio).
Bandwidth	The antenna bandwidth is the collection of frequencies over which an antenna can work correctly.
Radiation Efficiency	The antenna efficiency is the power radiated by the antenna divided by signal power fed at the input port.

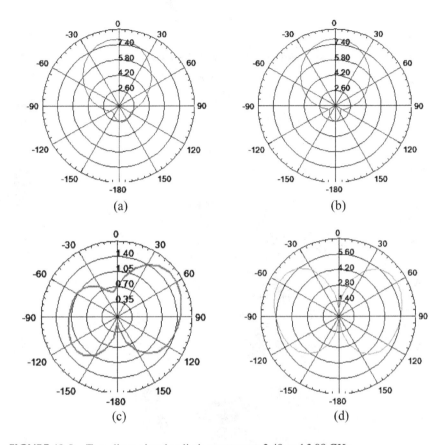

FIGURE 13.2 Two-dimensional radiation pattern at 2.49 and 3.98 GHz.

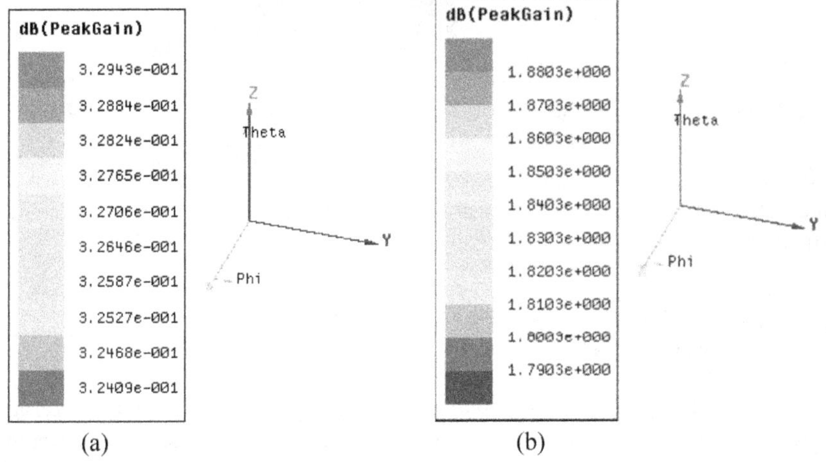

FIGURE 13.3 Peak antenna gain at resonant frequencies 2.49 and 3.98 GHz.

TABLE 13.3
Simulated Results of Cross-Slotted Antenna

	Frequency Band (GHz)	
Parameters	**2.49**	**3.98**
Peak Gain (dB)	3.29	1.88
Return Loss S11(dB)	–29	–20
VSWR	1.0104	1.2208

FIGURE 13.4 Fabricated cross-slotted antenna.

The performance result after simulation of the cross-slotted fractal antenna is presented in Table 13.3. The antenna results at 2.49 and 3.98 GHz resonant frequencies validate its suitability for IoT applications.

13.5 MEASUREMENTS OF FABRICATED CROSS-SLOTTED FRACTAL ANTENNA

The fabricated cross-slotted antenna designed using FR4 glass epoxy material is depicted in Figure 13.4. Vector Network Analyzer (VNA) is used for testing and measuring the experimental results of the antenna. Measured results of the antenna

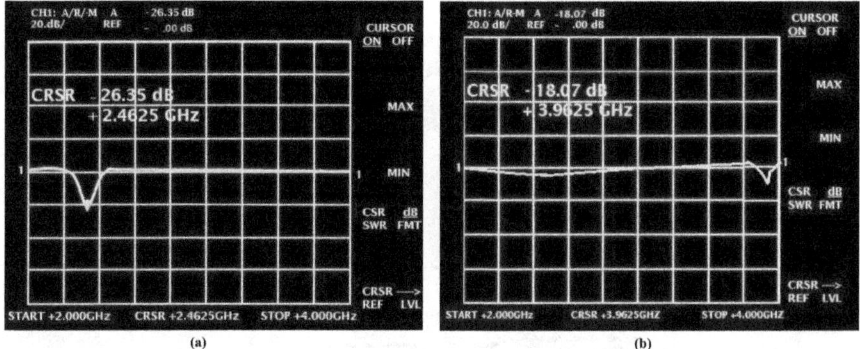

FIGURE 13.5 (a) Return loss versus frequency plot at 2.46 GHz. (b) Return loss versus frequency plot at 3.96 GHz.

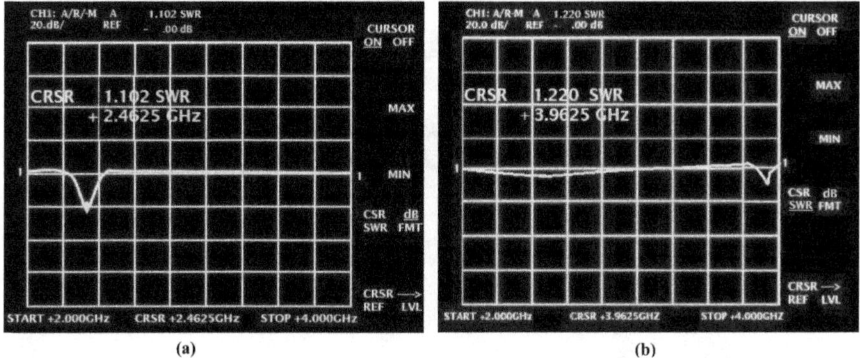

FIGURE 13.6 (a) VSWR versus frequency plot at 2.46 GHz. (b) VSWR versus frequency plot at 3.96 GHz.

are presented as return loss and VSWR parameters of the antenna for 2.46 and 3.96 GHz resonant frequencies.

13.5.1 Return Loss and Voltage Standing Wave Ratio

Figure 13.5 plots the measured return loss result of the fabricated antenna. The result shows that the antenna return loss is plotted as –26 and –18 dB at 2.46 and 3.96 GHz resonant frequency, respectively. Figure 13.6 plots the VSWR result of the fabricated antenna. The result shows that the antenna VSWR is plotted as 1.102 and 1.220 at 2.46 and 3.96 GHz resonant frequency, respectively. Table 13.4 presents the measured return loss and VSWR result of the fabricated cross-slotted antenna. Figure 13.7 shows the simulated and measured return loss results of the cross-slotted antenna.

TABLE 13.4
Measured Results of Cross-Slotted Antenna

	Frequency Band (GHz)	
Parameters	**2.46**	**3.96**
Return Loss S11(dB)	–26	–18
VSWR	1.102	1.220

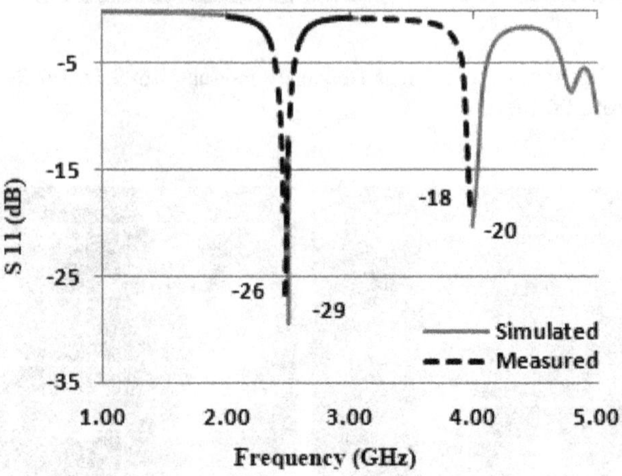

FIGURE 13.7 Antenna return loss simulated and measured.

TABLE 13.5
Comparison of Cross-Slotted Antenna with Other Fractal Antennas

Reference	Size (mm)	Frequency bands (GHz)	Gain
Compact Radiator Monopole Antenna [19]	40 × 35	2.28–2.45	1.35
		3.39–3.62	2.5
Trapezoidal Rings Antenna [27]	36 × 36	2.5–2.69	1.68
		3.3–4.2	3.1
Fractal Loop Antenna [28]	45 × 45	1.81	3.9
Multiband Sierpinski Triangle Antenna [29]	60 × 30	2.25–2.8	2.2
		5.5–6	5.8
Tapered Patch Fractal Antenna [30]	45 × 44	2.85–3.45	3
		4.85–5.50	3.8
Cross-Slotted Antenna	40 × 40	2.41–2.57	3.2
		3.90–4.11	1.8

13.6 CONCLUSION

In this chapter, a fractal microstrip cross-slotted antenna is designed and fabricated for IoT applications. The fractal dimensions of cross slots are used with the substrate of FR4 glass epoxy material to design this antenna. HFSS simulation software and VNAwere used for simulation and measurements of the antenna, respectively. The cross-slotted antenna resonates at 2.41–2.57 and 3.90–4.11 GHz frequency bands. The antenna achieves 3.2 and 1.8 peak gain with 84% radiation efficiency at 2.49 and 3.96 resonate frequencies. Comparative analysis of cross-slotted fractal antenna with other antennas is presented in Table 13.5 to validate the results. It has been observed that simulated and measured results of the cross-slotted antenna confirm its operation for IoT applications.

REFERENCES

1. M.Mustaqim, B. A.Khawaja, H. T.Chattha, K.Shafique, M. J.Zafar, and M.Jamil, "Ultra-wideband antenna for wearable Internet of Things devices and wireless body area network applications," *Int. J. Numer. Model. Electron. Networks, Devices Fields*, vol. 32, no. 6, pp. 1–12, 2019, doi: 10.1002/jnm.2590.
2. T. K.Saha, T. N.Knaus, A.Khosla, and P. K.Sekhar, "A CPW-fed flexible UWB antenna for IoT applications," *Microsyst. Technol.*, vol. 24, 2018, doi: 10.1007/s00542-018-4260-0.
3. S. G.Kirtania, B. A.Younes, A. R.Hossain, T.Karacolak, and P. K.Sekhar, "Cpw-fed flexible ultra-wideband antenna for IoTapplications," *Micromachines*, vol. 12, no. 4,pp. 1–13,2021, doi: 10.3390/mi12040453.
4. Q.Awais, H. T.Chattha, M.Jamil, Y.Jin, F. A.Tahir, and M. U.Rehman, "A novel dual ultra-wideband CPW-fed printed antenna for Internet of Things (IoT) applications," *Wirel. Commun. Mob. Comput.*, vol. 2018, 2018, doi: 10.1155/2018/2179571.
5. C. A.Balanis, *Antenna Theory: Analysis and Design*. Wiley, 2016.
6. A. H.Ramadan, M.Mervat, A.El-Hajj, S.Khoury, M.Al-Husseini, "A reconfigurable u-koch microstrip antenna for wireless applications," *Prog. Electromagn. Res. PIER*, vol. 93, pp. 355–367, 2009.
7. J. J.Huang, F. Q.Shan, and J. Z.She, "A novel multiband and broadband fractal patch antenna," in PIERS 2006: Progress in Electromagnetic Research Symposium, Cambridge, pp. 57–59, 2006, doi: 10.2529/piers051007103829.
8. Y.Liu et al., "Some recent developments of microstrip antenna," *Int. J. Antennas Propag.*, vol. 2012, 2012, doi: 10.1155/2012/428284.
9. Cisco, "WLAN radio frequency design considerations," *Enterp. Mobil.*, 7(3) Des. Guid., pp. 1–36, 2011.
10. Y.Mao, S.Guo, and M.Chen, "Compact dual-band monopole antenna with defected ground plane for Internet of things," *IET Microwaves, Antennas Propag.*, vol. 12, no. 8, pp. 1332–1338, 2018, doi: 10.1049/iet-map.2017.0860.
11. Benoit B.Mandelbrot, *The Fractal Geometry of Nature*. W. H. Freeman and Co., 1982.
12. A. T.Abed, M. J.Abu-AlShaer, and A. M.Jawad, "Fractal antennas for wireless communications," in *Modern Printed-Circuit Antennas*, IntechOpen, 2020, pp. 1–27.
13. B. K.Jeemon, K.Shambavi, and Z. C.Alex, "A multi-fractal antenna for WLAN and WiMAX application," in 2013 IEEE Conference on Information and Communication Technology ICT, pp. 953–956, 2013, doi: 10.1109/CICT.2013.Thuckalay, Tamil Nadu, India.

14. Y.Kumar and S.Singh, "A compact multiband hybrid fractal antenna for multistandard mobile wireless applications," *Wirel. Pers. Commun.*, vol. 84, no. 1, pp. 57–67, 2015, doi: 10.1007/s11277-015-2593-x.
15. K.Song, Y. Z.Yin, B.Chen, S. T.Fan, and F.Gao, "Bandwidth enhancement design of compact UWB step-slot antenna with rotated patch," *Prog. Electromagn. Res. Lett.*, vol. 22, no. March, pp. 39–45, 2011.
16. N.Kushwaha and R.Kumar, "Study of different shape electromagnetic band gap (EBG) structures for single and dual band applications," *J. Microwaves, Optoelectron. Electromagn. Appl.*, vol. 13, no. 1, pp. 16–30, 2014, doi: 10.1590/S2179-10742014000100002.
17. P.Okas, A.Sharma, and R. K.Gangwar, "Circular base loaded modified rectangular monopole radiator for super wideband application," *Microw. Opt. Technol. Lett.*, vol. 59, no. 10, pp. 2421–2428, 2017, doi: 10.1002/mop.30757.
18. R. K.Garg, M. V. D.Nair, S.Singhal, and R.Tomar, "A new type of compact ultra-wideband planar fractal antenna with WLAN band rejection," *Microw. Opt. Technol. Lett.*, vol. 62, no. 7, pp. 2537–2545, 2020, doi: 10.1002/mop.32304.
19. T. I. Y.XiaoleiL.Sun, and S. W.Cheung, "Dual-band monopole antenna with compact radiator for 2.4/3.5 GHz WiMAX applications," *Microw. Opt. Technol. Lett.*, vol. 55, no. 8, pp. 1765–1770, 2013, doi: 10.1002/mop.
20. R.Krishan, "Performance characterization of dual-band microstrip fractal antenna," *Int. J. Inf. Technol. Electr. Eng.*, vol. 8, no. 5, pp. 17–20, 2019.
21. R.Krishan and V.Laxmi, "'X' shape slot-based microstrip fractal antenna for IEEE 802.11 WLAN," in *Advances in Intelligent Systems and Computing*, vol. 553, Springer, 2017, pp. 135–143.
22. M.Ayoub, J.Saxena, and K.Muzaffar, "Design and simulation of a novel dual band microstrip patch antenna with defected ground structure for WLAN/WiMAX," *International Journal of Electronic and Electrical Engineering*, vol. 7, no. 10, pp. 1083–1090, 2014.
23. Clarissa de L.Nobrega, Marcelo R.da Silva, Paulo H. da F.Silva, and A. G.D'Assunção, "Experimental characterization of FSS for WLAN applications with low-cost UWB elliptical," vol. 56, no. 6, pp. 1331–1333, 2014, doi: 10.1002/mop.
24. A.Khanna, D. K.Srivastava, and J. P.Saini, "Engineering science and technology, an international journal bandwidth enhancement of modified square fractal microstrip patch antenna using gap-coupling," *Eng. Sci. Technol. an Int. J.*, vol. 18, pp. 286–293, 2015, doi: 10.1016/j.jestch.2014.12.001.
25. K. K.Sawant and C. R. S.Kumar, "CPW fed hexagonal micro strip fractal antenna for UWB wireless communications," *AEUE: Int. J. Electron. Commun.*, vol. 69, no. 1, pp. 31–38, 2015, doi: 10.1016/j.aeue.2014.07.022.
26. A. I. R.Garg, P.Bhartia, and InderBahl, *Microstrip Antenna Design Handbook*. Artech House Publishers, Boston, MA, London, 2001.
27. S. C.Puri, S.Das, and M.Gopal Tiary, "An UWB trapezoidal rings fractal monopole antenna with dual-notch characteristics," *Int. J. RF Microw. Comput. Eng.*, vol. 29, no. 8, pp. 1–10, 2019, doi: 10.1002/mmce.21777.
28. M.Zeng, A. S.Andrenko, X.Liu, Z.Li, and H. Z.Tan, "A compact fractal loop rectenna for RF energy harvesting," *IEEE Antennas Wirel. Propag. Lett.*, vol. 16, pp. 2424–2427, 2017, doi: 10.1109/LAWP.2017.2722460.
29. T.Benyetho, J.Zbitou, L.El Abdellaoui, H.Bennis, and A.Tribak, "A new fractal multiband antenna for wireless power transmission applications," *Act. Passiv. Electron. Components*, vol. 2018, pp. 1–10, 2018, doi: 10.1155/2018/2084747.
30. A.FerdowsB.Zarrabi, A.M. Shire, Maryam Rahimi, and N. P. Gandji, "Ultra-wideband tapered patch antenna with fractal slots for dual notch application,"*Microw. Opt. Technol. Lett.*, vol. 56, no. 6, pp. 1344–1348, 2014, doi: 10.1002/mop.

14 Examination of Vegetation Health and Its Relation with Normalized Difference Built-Up Index

A Study on Rajarhat Block of North 24 Parganas District of West Bengal, India

Asutosh Goswami, Suhel Sen, and Prodip Kumar Chakraborty

CONTENTS

DOI: 10.1201/9781003221333-14

14.1 INTRODUCTION

A settlement can be defined as a spatial pattern characterized by physical infrastructure made by human beings for living in a given local context [1], and the environmental quality is determined by the shape and spatial configuration of buildings [2]. Built-up lands dominate urban areas, which are characterized by the conversion of the natural environment into artificial impervious built-up lands, and these conversions have significant impact upon the urban environment, hydrological system, and local climate, of which urban heat island phenomena deserve a special mention [3]. So, for the correct planning of human society for security and safety, the identification of settlement features is considered one of the key information sets [1].

The process of urbanisation is considered one of the major backbones of development. The process is becoming accelerated with the increasing growth rate of the population and the consequent rise in their demands and requirements. Because of the rapid rate of urbanisation, land cover in urban areas is changing more drastically over a short period of time than elsewhere [4].

However, during urbanisation, the issue of vegetation health remains neglected. The population increase over the years has played a pivotal role in the expansion of the built-up areas, which also increases the emission and trapping of carbon dioxide, resulting in an increase in air temperature [5]. Besides, the diversity, health, and ecosystem resilience are disturbed by human activities [6].

While quality of life in an urban environment, particularly in a densely built environment, is crucially controlled by the openness to near and distant views, natural light, etc. [2], unscientific urban expansion not only brings about large-scale land use and land cover changes but also affects vegetation quantity as well as quality. High population density and exclusion of the natural environment by man-made structures are considered to be other contributing factors leading to temperature rise [5]. Carbon emission is considered as one of the environmental problems which are associated with energy consumption [7].

For the purpose of construction, natural vegetation is cleared off, which brings about alteration and loss of biodiversity. This loss of biodiversity can occur at the molecular, individual, and community level and also the ecosystem level [6]. Through the process of urbanisation, many grasslands, vegetated areas, or pastoral lands are concretised, which not only brings about a substantial decrease in the quantity of vegetation cover but also greatly affects the quality of the vegetation cover present within the area. A difference can be established between the vegetation index and the built-up index. The vegetation index exploits physical and chemical characteristics to produce typical radiometric information, whereas the built-up index exploits stable structural features of the built-up area [1].

Rajarhat block of North 24 Parganas district in the state of West Bengal is one such area that has experienced a rapid rate of urbanisation in the last few years accompanied by a decline in vegetation health. In the past decades, many vegetation indices were computed to identify canopy temperatures and their characteristics with the aim of minimising the effects of external factors like soil background reflectance [8]. The vegetation health status of in-situ species was assessed using a remote sensing approach by Lausch et al. (2018) [6]. Bokusheva et al. in 2016 [9] analysed the effectiveness of two vegetation health indices to determine wheat yield losses in grain-producing farms in Kazakhstan. Bento et al. in 2018 [10] assessed the relative contribution of normalised difference vegetation index (NDVI) and land surface temperature (LST) to characterising vegetation health in the Euro-Mediterranean region.

Geographic information system (GIS) and remote sensing technology act as an indispensable tool for the analysis of vegetation health through the computation of NDVI, while the scenario of built-up areas can be analysed by using the normalised difference built-up index (NDBI). The relation between these two indices will help to analyse the effect of urbanisation on vegetation health. For monitoring the vegetation activity in a large area, satellite data was found to be more effective compared with the traditional method of field-based studies [8], as accurate data on the size, shape, and spatial pattern of urban built-up areas is required for the study of urban spatial expansion and the resultant urban heat island phenomena [3].

It is very difficult to describe urban design and architecture, as they are found to be very complex [2]. Satellite remote sensing technology plays a vital role in monitoring the built-up land dynamics, and satellite imageries have been in use to discriminate built-up lands from non-built-up lands for the last few decades [11]. The changes in spatial pattern of urban areas are ideally monitored and detected from remotely sensed images, as they are up to date and are capable of providing a panoramic view [4].

A new remote sensing method for the assessment of vegetation stress from Advanced Very-High-Resolution Radiometer (AVHRR)-based vegetation health indices was employed for the assessment of drought in 2000 and 2001. It provides an early warning system for drought compared with the ground-based and NDVI-based techniques, as the former is marked by three channel algorithms and the latter by two channel algorithms [12].

A number of studies have been conducted by researchers [13–33] to identify the vegetation health and the complex nature of the built-up status in urban environs for justifying the interrelationships that exist among vegetation morphology, built-up status, and urban heat island (UHI) phenomena. Such studies used mainly remote sensing data collected from different sources on various time scales. Guha et al. (2018) [34] studied LST along with indices based on vegetation health and urban built-up status for the Italian cities Florence and Naples. Another significant approach was made by Macarof and Statescu (2017) [35] in this regard to compare the indications of surface UHI effect made by NDBI and NDVI and concluded by identifying the feasibility of NDBI for the study of surface UHI. Kaplan et al. (2018) [36] analysed UHI phenomena in Skopje, Macedonia using satellite imageries. Rahman et al. (2009) [37] investigated the application of AVHRR-based vegetation indices to characterize the impact of weather conditions on aus rice yield in Bangladesh. Kogan et al. in 2004 [38] contributed to the derivation of pasture biomass in Mongolia from AVHRR-based vegetation health indices.

This chapter deals with the changes in LST because the assessment of temperature condition affects the change in vegetation health status. This method has been well accepted and validated in all the major agricultural countries of the world [12]. An attempt has been made in the chapter to analyse the vegetation health of Rajarhat block from 1999 to 2019 by using NDVI values, and also, an interrelationship with the NDBI values has been drawn to identify the relationship between vegetation health and urbanisation.

14.2 MATERIALS AND METHODS

Rajarhat community development block is one of the most important community development (CD) blocks of North 24 Parganas in West Bengal. It is located at the latitude of 22°37′ N and longitude of 88°30′16″E. The block is comprised of 30 villages and 9 census towns according to census 2011. The district headquarters of Barasat is at a distance of 23 km. This area has been experiencing a high rate of urbanisation in the last few years and is expected to turn out to be one of the major urban centres in the future (Figure 14.1).

In order to accomplish the task, Landsat Enhanced Thematic Mapper (ETM^+) and Landsat Thematic Mapper images of the study area have been downloaded from the

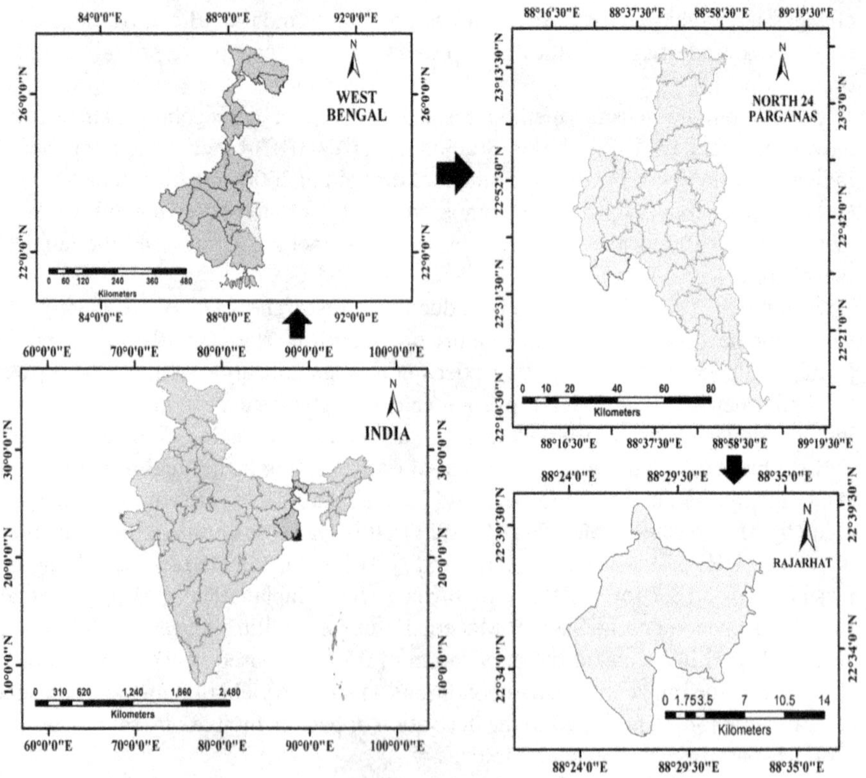

FIGURE 14.1 Location map of the study area.

United States Geological Survey (USGS) Earth Explorer website (https://earthexplorer.usgs.gov) for the three years of 1999, 2009, and 2019. Landsat ETM$^+$ has eight bands, while Landsat TM images have seven bands. Sensor features among Landsat TM and ETM+ are the same from band 1 to band 5 except band 6. In ETM$^+$, band 6 has two sub-bands, namely, high and low. The panchromatic band (band 8) is available only in ETM$^+$. Details of the sensor characteristics are given in Table 14.1.

Satellite images from the dates shown in Table 14.2 were downloaded to fulfil the task.

For the accomplishment of the present study, NDVI and NDBI maps are prepared from the Landsat images, and the relationship is drawn between the two indices. Software like ERDAS Imagine 2015 and ArcGIS 10.3 has been used for the fulfilment of the task.

14.2.1 Normalised Difference Vegetation Index (NDVI)

NDVI is used to monitor the vegetation by using red and near infrared reflectance. The formula for calculating NDVI is as follows:

$$\mathbf{NDVI} = (\mathbf{Near\ Infrared} - \mathbf{RED}) / (\mathbf{Near\ Infrared} + \mathbf{RED})$$

NDVI values range from −1 to +1. Values ranging from 0.3 to 0.6 are considered to be stressed vegetation, while values ranging from 0.6 to 0.8 are considered to be healthy vegetation. For the accomplishment of the present work, NDVI maps have further been classified to generate land use and land cover types. Values below 0, i.e. all the negative values, have been considered as water bodies and wetlands. Values ranging between 0 and 0.15 have been taken as built-up and barren lands. Highly stressed vegetation has been identified by values ranging from 0.15 to 0.30. Finally, the moderate and less stressed vegetation has been indicated by the values ranging from 0.30 to 0.45 and above 0.45, respectively. NDVI maps have been prepared for the Rajarhat block for the years 1999, 2009, and 2019, and finally, the land use maps are prepared on the basis of NDVI values.

14.2.2 Normalized Difference Built-Up Index (NDBI)

NDBI is used for the extraction of built-up features, and its values range from −1 to +1. NDBI is calculated using the following formula:

$$\mathbf{NDBI} = (\mathrm{SWIR} - \mathrm{NIR}) / (\mathrm{SWIR} + \mathrm{NI}\mathbf{R})$$

Where: SWIR denotes short wave infrared and NIR indicates near infrared.

14.3 RESULTS AND DISCUSSION

14.3.1 NDVI and NDBI Scenario of 1999

The NDVI scenario for the year 1999 shows that the highest and lowest NDVI values are found to be 0.57 and −0.2, respectively (Figure 14.2). After classifying the

TABLE 14.1
Sensor Characteristics of Landsat Series

Band Feature	Band 1	Band 2	Band 3	Band 4	Band 5	Band 7	Band 8	Band 6	
	TM/ETM+						ETM+	TM/ETM+	
Name	VISIBLE			NIR	MIR	SIR	PANCHROMATIC	TIR	
	B	G	R					TM	ETM+
Spectral Resolution (μm)	0.45–0.52	0.52–0.60	0.63–0.69	0.76–0.90	1.55–1.75	2.09–2.35	0.52–0.90	10.40–12.50	
Spatial Resolution (m × m)	30 × 30	30 × 30	30 × 30	30 × 30	30 × 30	30 × 30	15 × 15	120 × 120	60 × 60

TABLE 14.2
Details of Satellite Images Downloaded

Acquisition Date	Sensor and Satellite	Reference System/path/row
1999/05/31	Landsat 5 TM	UTM-45N/138/44
2009/05/10	Landsat 5 TM	UTM-45N/138/44
2019/04/28	Landsat 7 ETM+	UTM-45N/138/44

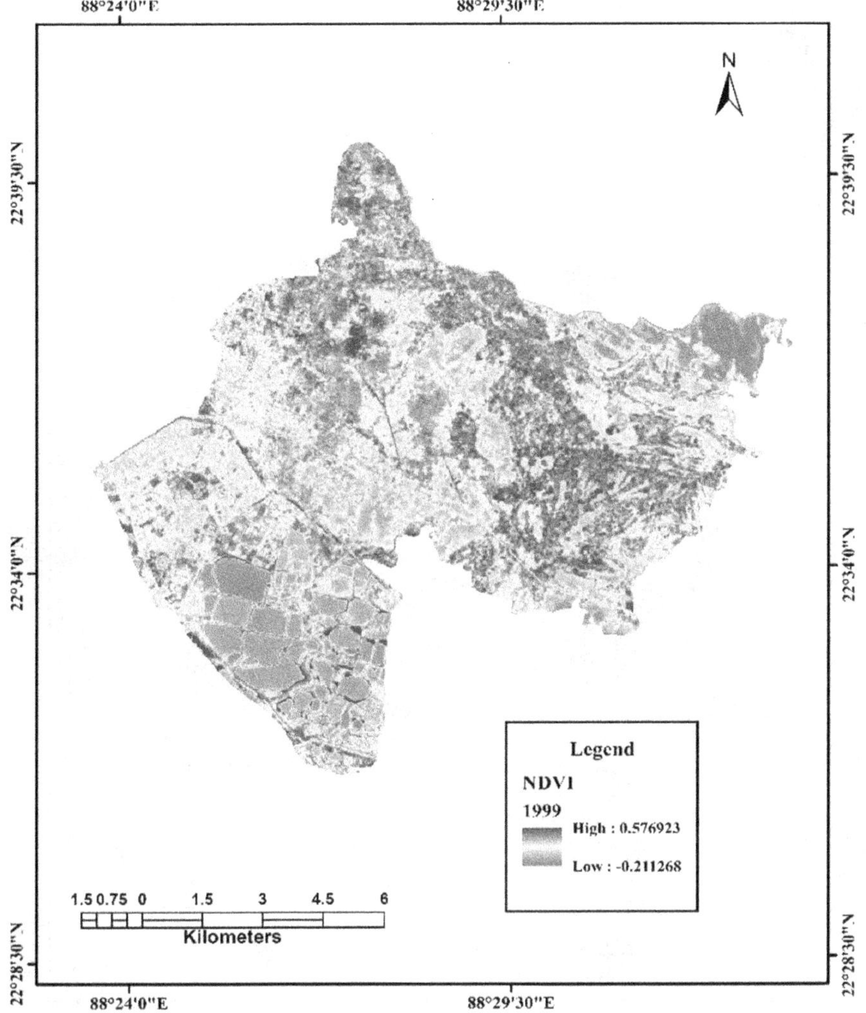

FIGURE 14.2 NDVI map of the study area in 1999.

NDVI map, it is observed that out of the total area of 131 sq km, about 15.24 sq km is covered by water bodies and wetlands, and exhibits NDVI values of less than 0. About 32.52 sq km area is covered by built-up and barren lands with the NDVI values ranging between 0 and 0.15. The area dominated by highly stressed vegetation is calculated as 49.71 sq km with NDVI values ranging between 0.16 and 0.30, while that of moderately stressed vegetation is 32.83 sq km with NDVI values from 0.31 to 0.45 (Figure 14.3). About 1.09 sq km area is dominated by less stressed vegetation with NDVI values above 0.45.

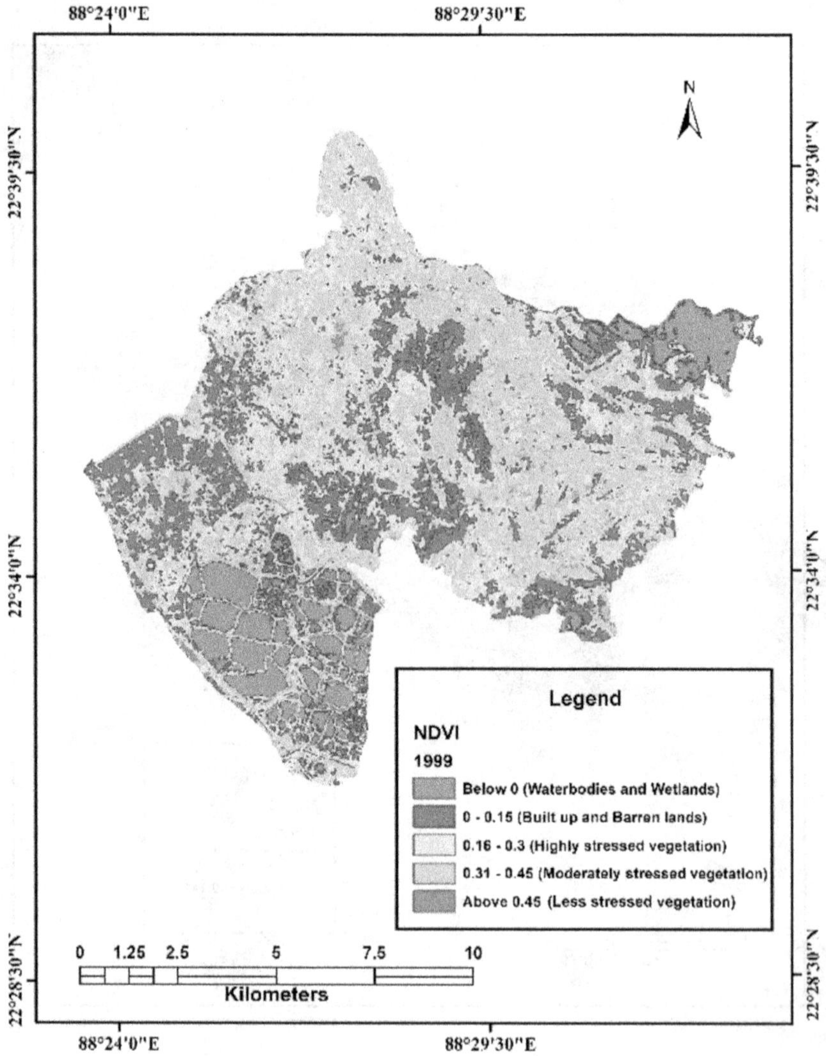

FIGURE 14.3 NDVI classified map of the study area in 1999.

Hence, the situation is found to be alarming, as most of the land area was under the dominance of highly stressed vegetation cover in 1999. Contrary to this, the built-up scenario of 1999 shows that the NDBI values are found to concentrate between −0.39 and 0.40 (Figure 14.4).

14.3.2 NDVI and NDBI Scenario of 2009

The NDVI scenario of the year 2009 shows that the highest NDVI value is calculated as 0.49, while the lowest value is identified as −0.13 (Figure 14.5). On classification

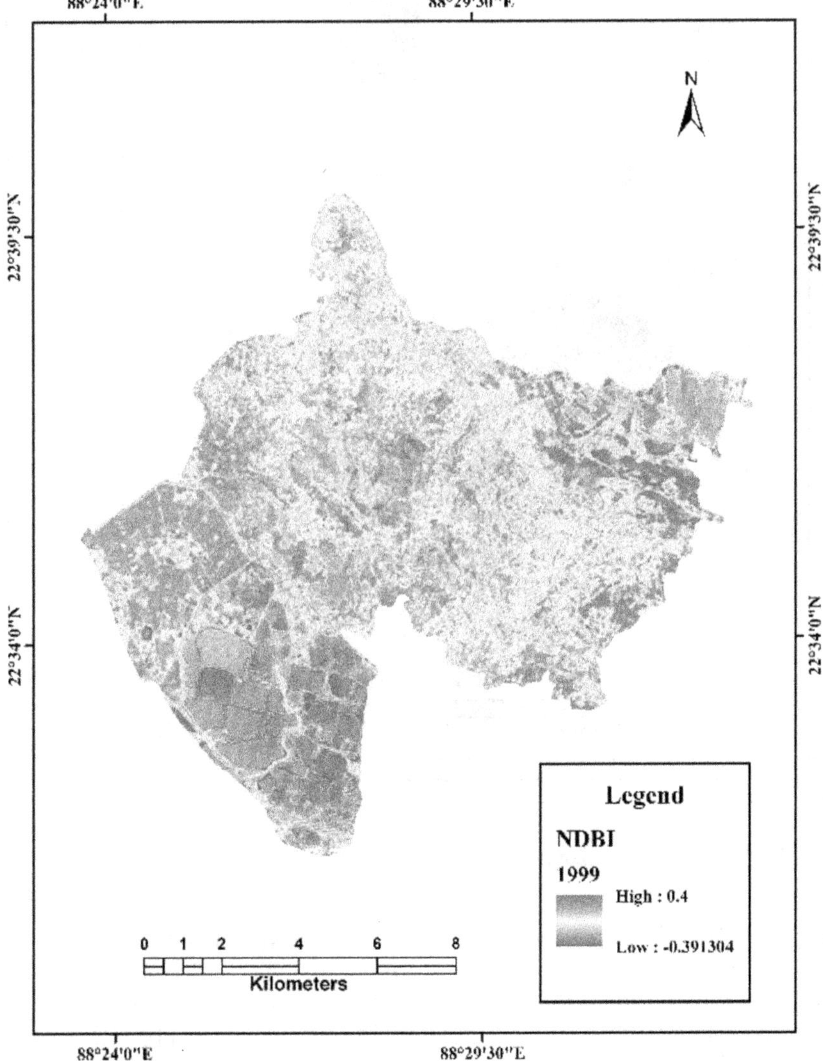

FIGURE 14.4 NDBI map of the study area in 1999.

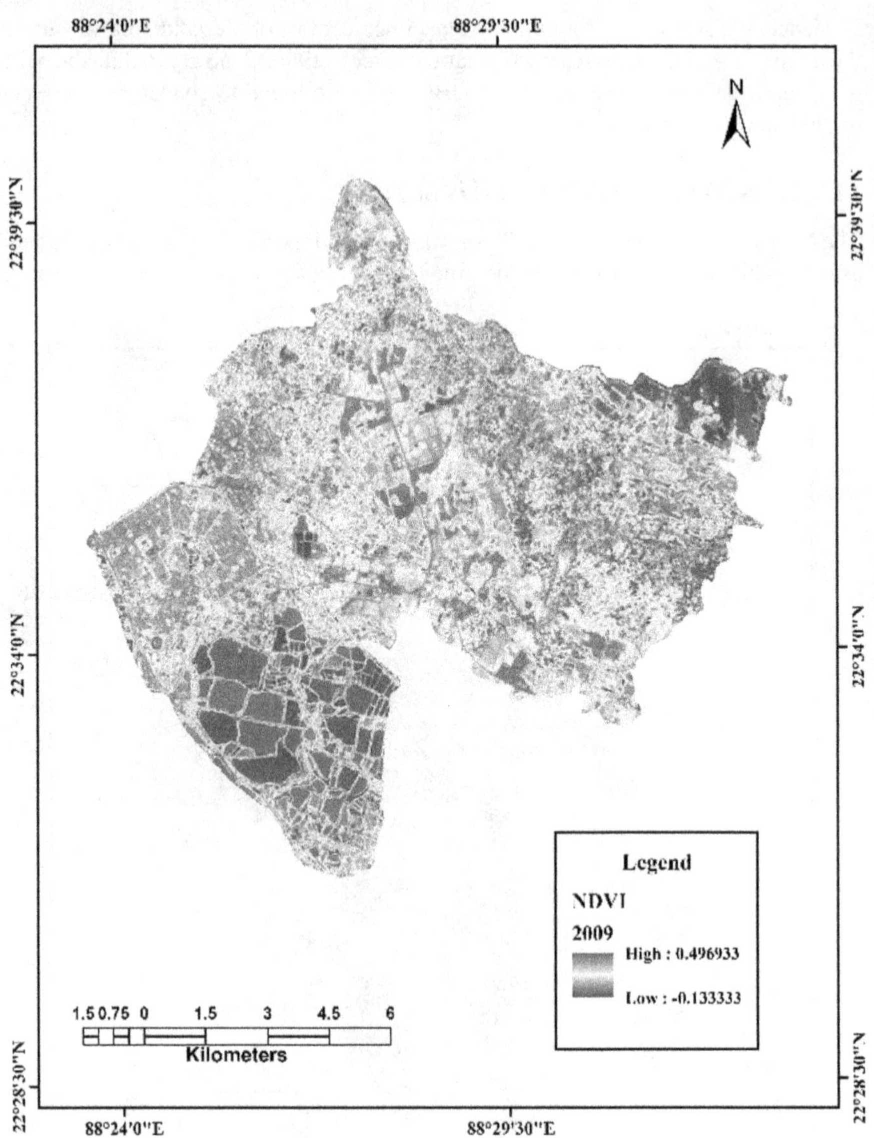

FIGURE 14.5 NDVI map of the study area in 2009.

of the NDVI map, it is seen that out of the total area of 131 sq km, about 18.41 sq km is covered by water bodies and wetlands, and exhibits NDVI values of less than 0 (Figure 14.6). Furthermore, about 47.21 sq km area is covered by built-up and barren lands with NDVI values from 0 to 0.15. The area dominated by highly stressed vegetation is detected as 56.78 sq km, with NDVI values ranging from 0.16 to 0.30, while that of moderately stressed vegetation is 8.95 sq km, for which the NDVI

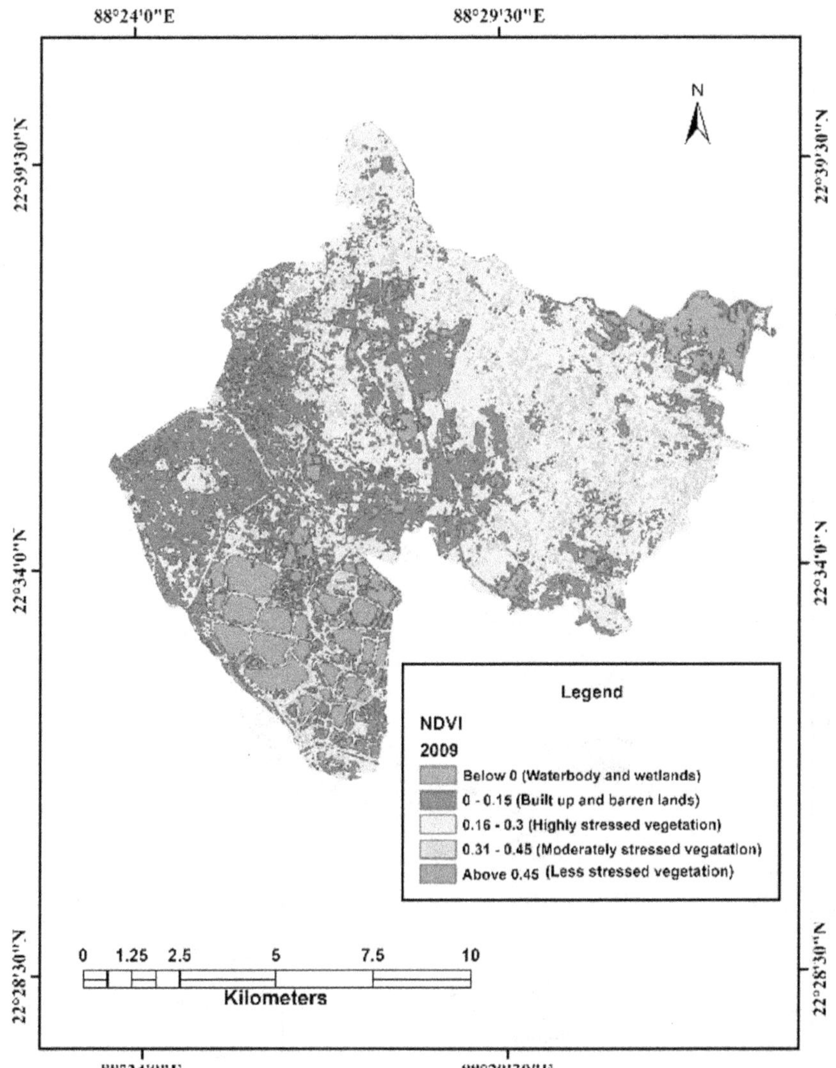

FIGURE 14.6 NDVI classified map of the study area in 2009.

values range between 0.31 and 0.45 (Figure 14.6). Hence, it can be stated that in 2009, also, most of the land area is dominated by highly stressed vegetation cover. The corresponding NDBI values of 2009 range from 0.39 to −0.26 (Figure 14.7).

14.3.3 NDVI and NDBI Scenario of 2019

The NDVI scenario of the year 2019 is presented in Figure 14.8, which shows that the NDVI values of that particular year range from a maximum of 0.45 to a minimum of

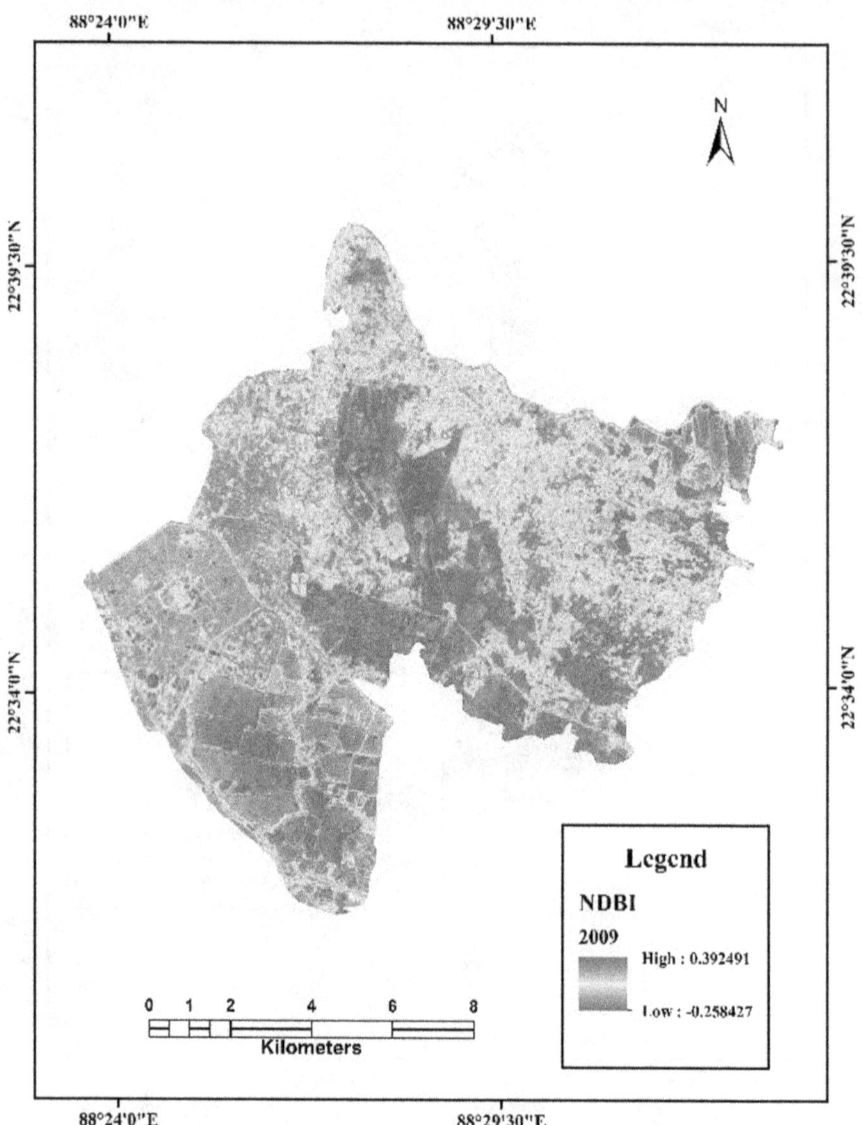

FIGURE 14.7 NDBI map of the study area in 2009.

−0.2. The classification of the NDVI map reveals that 9.29 sq km area is covered by water bodies and wetlands; for which the NDVI values are less than 0. About 63.64 sq km area is under the dominance of built-up and barren lands, with NDVI values from 0 to 0.15. The area dominated by highly stressed vegetation is calculated as 50.78 sq km, with NDVI values ranging between 0.16 and 0.30, while that of moderately stressed vegetation is 7.67 sq km, with NDVI values ranging from 0.31 to

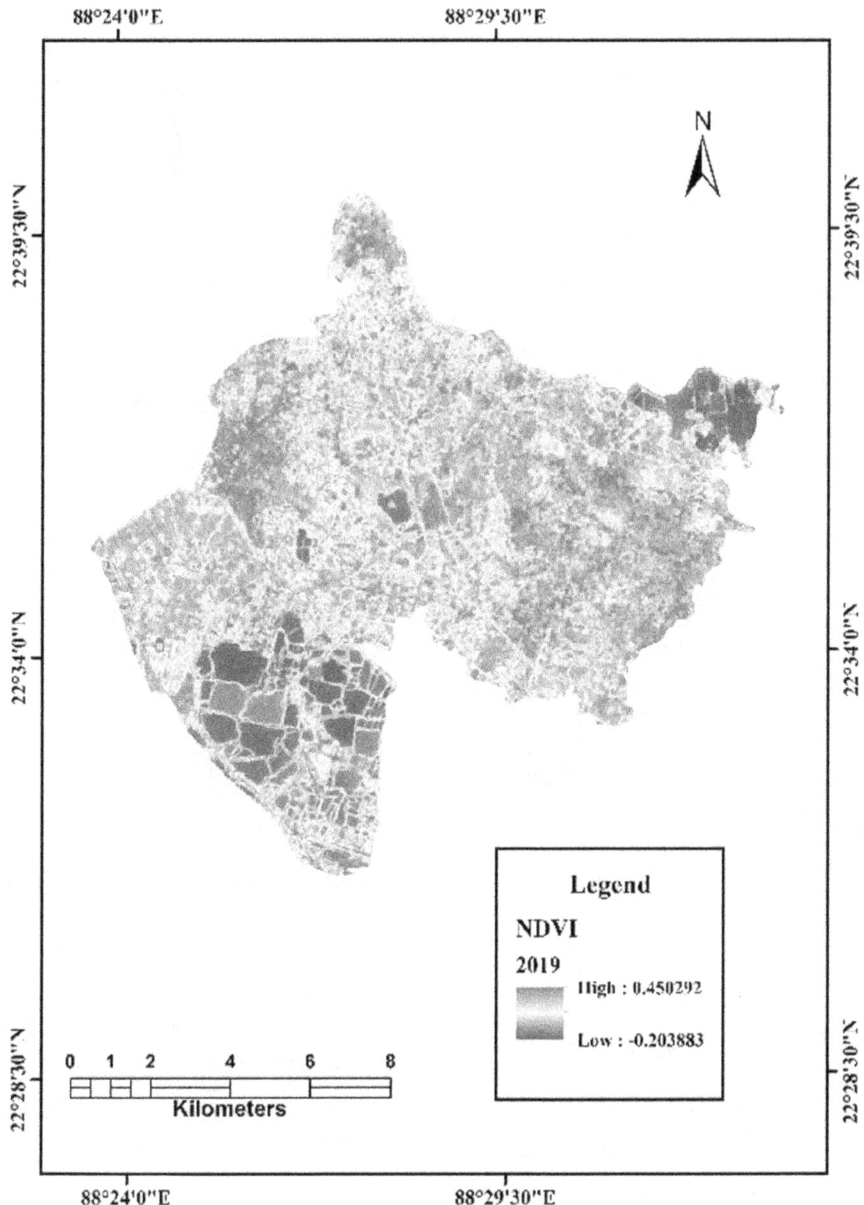

FIGURE 14.8 NDVI map of the study area in 2019.

0.45. Most of the land area was dominated by built-up and barren land (Figure 14.9). The entire vegetation in the block is stressed with an absence of healthy vegetation. The NDBI map for the year 2019 shows that the highest and lowest NDBI values are found to be 0.2 and −0.48, respectively (Figure 14.10).

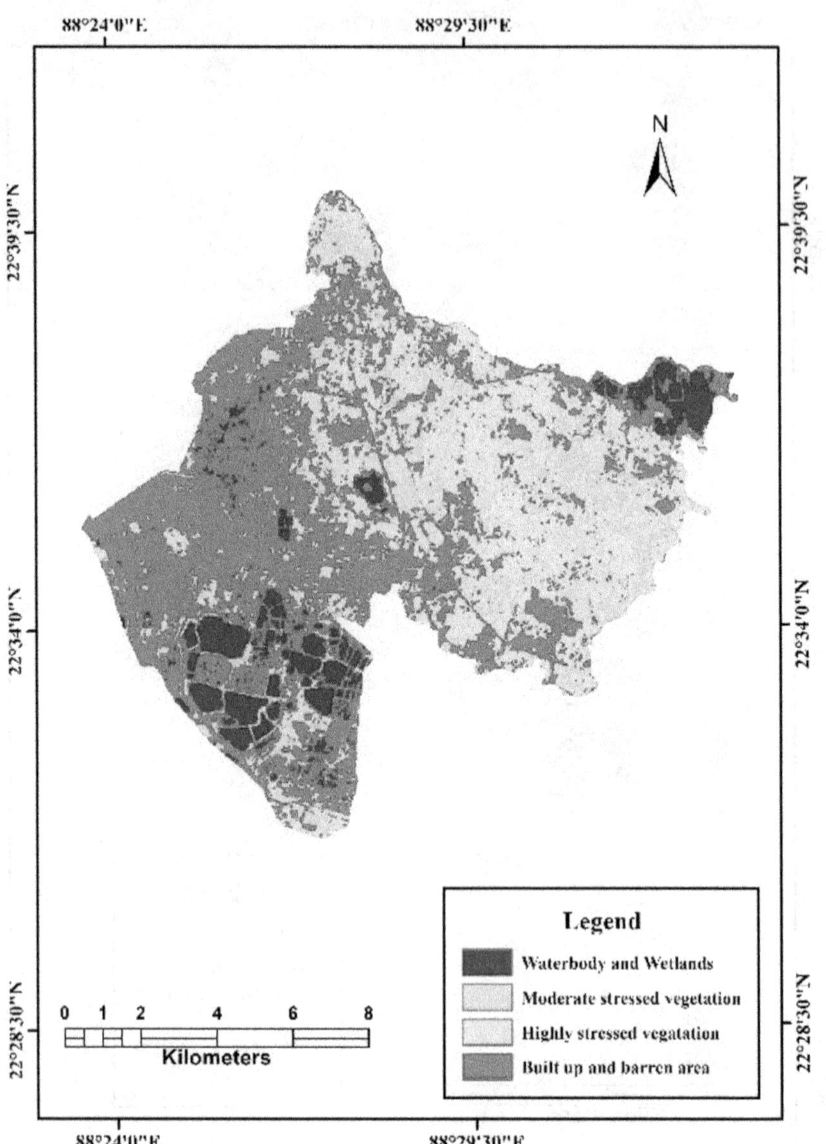

FIGURE 14.9 NDVI classified map of the study area in 2019.

14.3.4 Temporal Change of Land Use Classified on the Basis of NDVI Values

The NDVI maps have been classified to generate the land use maps. It can be seen that the area under water body and wetlands has decreased from 18.41 sq km in 2009 to 9.29 sq km in 2019 due to the rapid rate of urbanisation. Broadly speaking, this

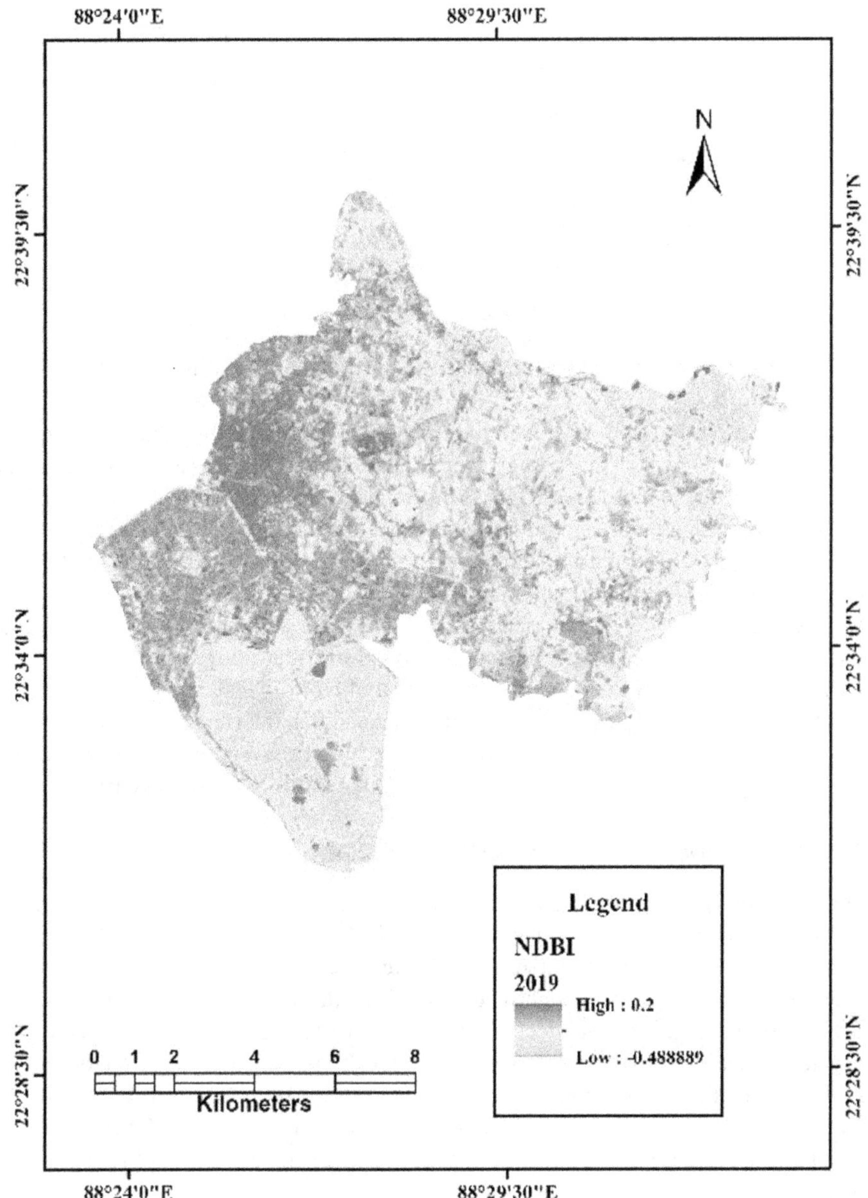

FIGURE 14.10 NDBI map of the study area in 2019.

is mainly due to large-scale filling up of water bodies and wetlands, and their subsequent conversion into built-up areas. The area dominated by built-up and barren lands exhibits a rising trend with areal coverage of 32.52 sq km in 1999, 47.21 sq km in 2009, and rising to 63.64 sq km in 2019. The area under highly stressed vegetation

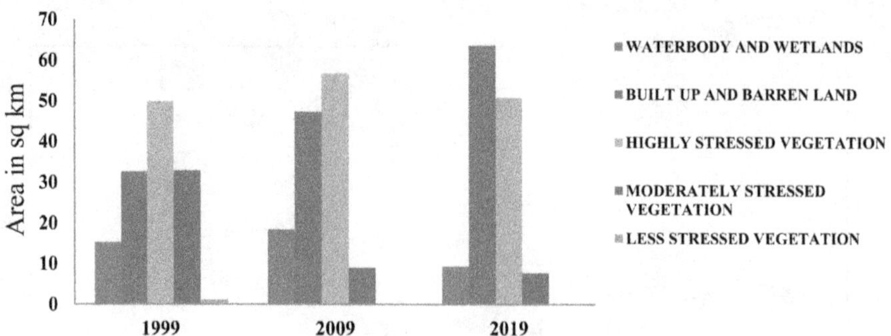

FIGURE 14.11 Temporal change of land use on the basis of NDVI classification.

also shows a rising trend, as it is found to be 49.71 sq km in 1999 and 56.78 sq km in 2009 (Figure 14.11). Surprisingly, the area under highly stressed vegetation has declined to 50.78 sq km in 2019. This is due to large-scale conversion of those highly stressed vegetation areas into built-up and barren lands. The area under moderately stressed vegetation exhibits a rapidly declining trend from 32.83 sq km in 1999 to 8.95 sq km in 2009 and 7.67 sq km in 2019. Less stressed vegetation cover has been detected only in 1999, with an aerial coverage of 1.09 sq km in 1999. In 2009 and 2019, no trace of less stressed vegetation is found over the study area.

This discussion clarifies that the rising trend of built-up and barren lands in the 3 years taken for the present study is mainly due to human interference; there has been large-scale filling up of water bodies and wetlands and conversion of vegetated areas (both highly stressed and moderately stressed) into built-up areas and barren lands.

14.3.5 Temporal Analysis of NDVI and NDBI

Temporal analysis of the NDVI map of the study area reveals that there has been a continuous decline in NDVI value. Since high NDVI value indicates healthy vegetation status, decline in NDVI value will indicate a decline in the quality of vegetation. The highest NDVI value is identified as 0.57 for the year 1999, which declines to 0.49 in 2009 and gets further declined to 0.45 in 2019 indicating a gradual decline in vegetation health status with the passage of time. Temporal analysis of NDBI values indicates that these have remained more or less identical in 1999 and 2009 with the values of 0.40 and 0.39, respectively (Figure 14.12). However, the NDBI value has dropped significantly to 0.20 in 2019 due to the large-scale practice of roof top gardening and rainwater harvesting system.

14.3.6 Analysing the Relationship between the NDVI and NDBI of the Study Area

On drawing the relationship between NDVI and NDBI, it is mentioned that the areas exhibiting high NDVI values record low NDBI values. In the western part of the map, the Salt Lake area is dominated by built-up, barren lands and stressed

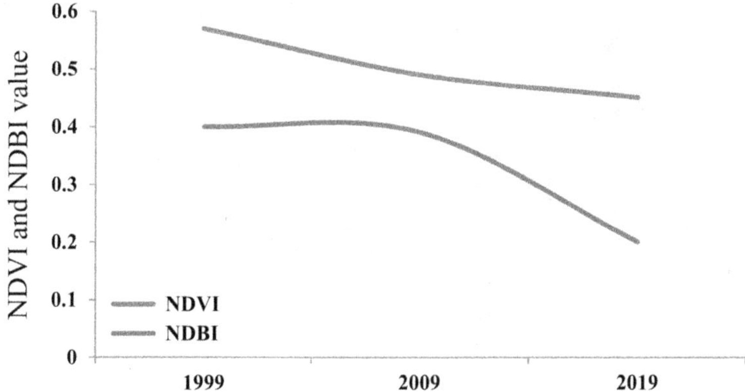

FIGURE 14.12 Temporal analysis of NDVI and NDBI of the study area.

vegetation (Figure 14.3). Consequently, it exhibits moderate to high NDBI values (Figure 14.4). The areas of highly and moderately stressed vegetation located in the study area show medium to high NDBI values. Only a small pocket of least stressed vegetation located at the central part of the map shows a higher NDVI and a lower NDBI value.

It has already been detected from the maps that areas with high NDVI values correspond to low NDBI values, indicating inverse correlation between these two indices. The correlations between NDVI and NDBI values of the study area are found to be strong, with computed R^2 values of 0.911, 0.897, and 0.905 in 1999, 2009, and 2019, respectively (Figure 14.13).

14.4 CONCLUSION

Urbanisation is a continuous process, which cannot be stopped. With further increase in population, the study area will experience a higher rate of urbanisation. However, the issue of vegetation health within the city area has to be kept in mind before further planning and undertaking urban developmental projects. Vegetation within the urban area plays a vital role in maintaining air circulation and a comfortable environment within the urban area. Since Rajarhat is a developing urban centre and is expected to cater for the needs of a huge population in upcoming years, planners should opt for the strategies of sustainable urbanisation with the aim of preserving the urban environment and undertaking the process of urbanisation.

The present study deals with the temporal change of NDVI and NDBI values and their interrelationship. The study area has experienced a continuous decline in the area dominated by water bodies and wetlands, which has a direct impact on the vegetation health. On the contrary, the built-up and barren lands show a rising trend in terms of their aerial coverage. The NDVI values of the study area are found to be decreasing, signifying a gradual decline in vegetation health. Fortunately, in the year 2019, the NDBI value has dropped considerably due to the practice of roof top

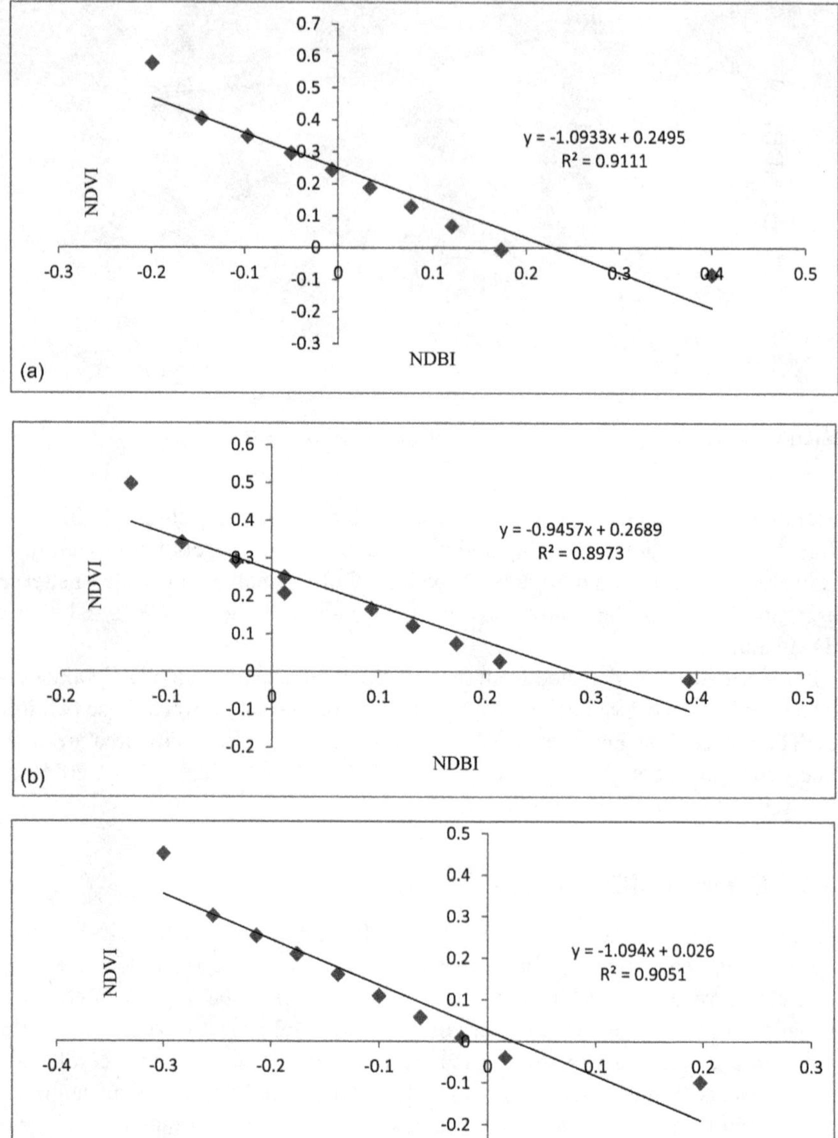

FIGURE 14.13 Correlation between NDVI and NDBI in (a) 1999, (b) 2009, and (c) 2019.

gardening and rainwater harvesting in recent times, which is also a reflection of the increasing awareness of people about nature. Under such circumstances, strategies of sustainable urbanisation have to be adopted for the restoration of the degraded vegetation health.

West Bengal Government and Housing Infrastructure Development Corporation Limited (HIDCO) has already planned strategies for creating a green space within the urban area and is taking proper care of the tree species. Eco Park of New Town in Rajarhat is a manifestation of such an initiative. Public awareness can act as an indispensable tool in this issue, where each and every person living in the area has to realise the importance of vegetation health. Roof top gardening can be considered as one of the strategies to undertake this policy. It has to be kept in mind that healthy vegetation with sustainable urbanisation will ensure a bright and secure living environment for future generations.

ACKNOWLEDGEMENT

The authors are grateful to Agricultural Meteorology Division of the State Agriculture Department, Government of West Bengal and India Meteorological Department, Alipur. The authors would also like to offer their deepest sense of gratitude to Dr Ashis Kr. Paul (Professor, Dept. of Geography, Vidyasagar University, Midnapore) and Dr Swadesh Mishra (former Agro-Meteorologist, Govt. of West Bengal) for giving valuable suggestions, guidance, and supervision during their entire study.

REFERENCES

1. M. Pesaresi, A. Gerhardinger, and F. Kayitakire, "A robust built-up area presence index by anisotropic rotation-invariant textural measure", *IEEE Journal of Selected Topics in Applied Earth Observations and Remote Sensing*, vol. 1, no. 3, pp. 180–192, 2008. Accessed on: Jun. 10, 2021. [Online]. Available doi:10.1109/JSTARS.2008.2002869.
2. D. Fisher-Gewirtzman, and I. A. Wagner, "Spatial openness as a practical metric for evaluating built-up environments", *Environment and Planning B: Planning and Design*, vol. 30, pp. 37–49, 2003. Accessed on: Jun. 10, 2021. [Online]. Available doi: 10.1068/b12861.
3. H. Xu, "Extraction of urban built-up land features from Landsat imagery using a thematic-oriented index combination technique", *Photogrammetric Engineering & Remote Sensing*, vol. 73, no. 12, pp. 1381–1391, 2007.
4. Y. Zha, J. Gao, and S. Ni, "Use of normalized difference built-up index in automatically mapping urban areas from TM imagery", *International Journal of Remote Sensing*, vol. 24, no. 3, pp. 583–594, 2003. Accessed on: Jun. 10, 2021. [Online]. Available doi: 10.1080/01431160210144570.
5. M. O. Alabi, "The built up environment and micro-climate variation in Lokoja, Nigeria", *American International Journal of Contemporary Research*, vol. 2, no. 12, pp. 150–158, 2012.
6. A. Lausch, O. Bastian, S. Klotz et al., "Understanding and assessing vegetation health by in situ species and remote-sensing approaches", *Methods in Ecology and Evolution*, vol. 9, pp. 1799–1809, 2018. Accessed on: Jun. 10, 2021. [Online]. Available doi: 10.1111/2041-210X.13025.
7. R. Yao, B. Li, and K. Steemers, "Energy policy and standard for built environment in China", *Renewable Energy*, vol. 30, no. 2005, pp. 1973–1988, 2005. Accessed on: Jun. 10, 2021. [Online]. Available doi: 10.1016/j.renene.2005.01.013
8. F. Pie, C. Wu, X. Liu, X. Li, K. Yang, Y. Zhou, K. Wang, L. Xu, and G. Xia, "Monitoring the vegetation activity in China using vegetation health indices", *Agricultural and*

Forest Meteorology, vol. 248, no. 2018, pp. 215–227, 2018. Accessed on: Jun. 10, 2021. [Online]. Available doi: org/10.1016/j.agrformet.2017.10.001.

9. R. Bokusheva, F. Kogan, I. Vitkovskaya, S. Conradt, and M. Batyrbayeva, "Satellite-based vegetation health indices as a criteria for insuring against drought-related yield losses", *Agricultural and Forest Meteorology*, vol. 220, no. 2016, pp. 200–206, 2016. Accessed on: Jun. 10, 2021. [Online]. Available doi: org/10.1016/j.agrformet.2015.12.066.
10. V. A. Bento, C. M. Gouveia, C. C. DaCamara, and I. F. Trigo, "A climatological assessment of drought impact on vegetation health index", *Agricultural and Forest Meteorology*, vol. 259, no. 2018, pp. 286–295, 2018. Accessed on: Jun. 10, 2021. [Online]. Available doi: org/10.1016/j.agrformet.2018.05.014.
11. H. Xu, "A new index for delineating built-up land features in satellite imagery", *International Journal of Remote Sensing*, vol. 29, no. 14, pp. 4269–4276, 2008. Accessed on: Jun. 10, 2021. [Online]. Available doi: https:// 10.1080/01431160802039957
12. F. Kogan, "World droughts in the new millennium from AVHRR-based vegetation health indices", *EOS, Transactions, American Geophysical Union*, vol. 83, no. 48, pp. 557–562, 2002.
13. C. J. Tucker, "Red and photographic infrared linear combinations for monitoring vegetation", *Remote Sensing of Environment*, vol. 8, pp. 127–150, 1979.
14. L. Liu, and Y. Zhang, "Urban heat island analysis using the Landsat TM data and ASTER data: A case study in Hong Kong", *Remote Sensing*, vol. 3, pp. 1535–1552, 2011.
15. T. Owen, T. Carlson, and R. Gillies, "An assessment of satellite remotely-sensed land cover parameters in quantitatively describing the climatic effect of urbanization", *International Journal of Remote Sensing*, vol. 19, pp. 1663–1681, 1998.
16. K. P. Gallo, and T. W. Owen, "Satellite-based adjustments for the urban heat island temperature bias", *Journal of Applied Meteorology*, vol. 38, pp. 806–813, 1999.
17. Q. Weng, "A remote sensing-GIS evaluation of urban expansion and its impact on surface temperature in Zhujiang Delta, China", *International Journal of Remote Sensing*, vol. 22, no. 10, pp. 1999–2014, 2001.
18. A. J. Arnfield, "Two decades of urban climate research: A review of turbulence, exchanges of energy and water, and the urban heat island", *International Journal of Climatology*, vol. 23, no. 1, pp. 1–26, 2003. Accessed on: Jun. 10, 2021. [Online]. Available doi: org/10.1002/joc.859.
19. J. A. Voogt, and T. R. Oke, "Thermal remote sensing of urban climates," *Remote Sensing of Environment*, vol. 86, no. 3, pp. 370–384, 2003. Accessed on: Jun. 10, 2021. [Online]. Available doi: 10.1016/S0034-4257(03)00079-8.
20. Q. Weng, D. Lu, and J. Schubring, "Estimation of land surface temperature-vegetation abundance relationship for urban heat island studies", *Remote Sensing of Environment*, vol. 89, no. 4, pp. 467–483, 2004.
21. Q. Weng, and S. Yang, "Managing the adverse thermal effects of urban development in a densely populated Chinese city", *Journal of Environmental Management*, vol. 70, no. 2, pp. 145–156, 2004.
22. Y. Zhang, "Land surface temperature retrieval from CBERS-02 IRMSS thermal infrared data and its applications in quantitative analysis of urban heat island effect", *Journal of Remote Sensing*, vol. 10, pp. 789–797, 2006.
23. R. Amiri, Q. Weng, A. Alimohammadi, and S. K. Alavipanah, "Spatial-temporal dynamics of land surface temperature in relation to fractional vegetation cover and land use/cover in the Tabriz urban area, Iran", *Remote Sensing of Environment*, vol. 113, no. 12, pp. 2606–2617, 2009. Accessed on: Jun. 10, 2021. [Online]. Available doi: org/10.1016/j.rse.2009.07.021.

24. Y. Ma, Y. Kuang, and N. Huang, "Coupling urbanization analyses for studying urban thermal environment and its interplay with biophysical parameters based on TM/ETM+ imagery", *International Journal of Applied Earth Observation and Geoinformation*, vol. 12, pp. 110–118, 2010.
25. C. Tomlinson, L. Chapman, J. E. Thornes, and C. J. Bakeret, "Derivation of Birmingham's summer surface urban heat island from MODIS satellite images", *International Journal of Climatology*, vol. 32, pp. 214–224, 2012.
26. Z. Qin, A. Karnieli, and P. Berliner, "A mono-window algorithm for retrieving land surface temperature from Landsat TM data and its application to the Israel– Egypt border region", International Journal of Remote Sensing, vol. 22, no. 18, pp. 3719–3746, 2001.
27. J. A. Sobrino, N. Raissouni, and Z. L. Li, "A comparative study of land surface emissivity retrieval from NOAA data", *Remote Sensing of the Environment*, vol. 75, no. 2, pp. 256–266, 2001.
28. J. A. Sobrino, J. C. Munoz, and L. Paolini, "Land surface temperature retrieval from Landsat TM5", *Remote Sensing of the Environment*, vol. 9, pp. 434–440, 2004.
29. J. Song, S. Du, X. Feng, and L. Guo, "The relationships between landscape compositions and land surface temperature: Quantifying their resolution sensitivity with spatial regression models", *Landscape and Urban Planning*, vol. 123, pp. 145–157, 2014.
30. A. Asgarian, B. J. Amiri, and Y. Sakieh, "Assessing the effect of green cover spatial patterns on urban land surface temperature using landscape metrics approach", *Urban Ecosystems*, vol. 18, no. 1, pp. 209–222, 2015. Accessed on: Jun. 10, 2021. [Online]. Available doi: org/10.1007/s11252-014-0387-7.
31. P. Fu, and Q. Weng, "Consistent land surface temperature data generation from irregularly spaced Landsat imagery", *Remote Sensing of Environment*, vol. 184, pp. 175–187, 2016. Accessed on: Jun. 10, 2021. [Online]. Available doi: org/10.1016/j.rse.2016.06.019.
32. J. Peng, P. Xie, Y. Liu, and J. Ma, "Urban thermal environment dynamics and associated landscape pattern factors: A case study in the Beijing metropolitan region", *Remote Sensing of Environment*, vol. 173, pp. 145–155, 2016.
33. Z. Zhang, G. He, M. Wang, T. Long, G. Wang, X. Zhang, and W. Jiao, "Towards an operational method for land surface temperature retrieval from Landsat 8 data", *Remote Sensing Letters*, vol. 7, no. 3, pp. 279–288, 2016.
34. S. Guha, H. Govil, A. Dey, and N. Gill, "Analytical study of land surface temperature with NDVI and NDBI using Landsat 8 OLI and TIRS data in Florence and Naples city, Italy", *European Journal of Remote Sensing*, vol. 51, no. 1, pp. 667–678, Accessed on: Jun. 10, 2021. [Online]. Available doi: 10.1080/22797254.2018.1474494.
35. P. Macarof, and F. Statescu, "Comparison of NDBI and NDVI as indicators of surface urban heat island effect in landsat 8 imagery: A case study of Iasi", *Present Environment and Sustainable Development*, vol. 11, no. 2, pp. 141–150, 2017. Accessed on: Jun. 10, 2021. [Online]. Available doi: 10.1515/pesd-2017-0032.
36. G. Kaplan, U. Avdan, and Y. Avdan, "Urban heat island analysis using the landsat 8 satellite data: A case study in Skopje, Macedonia", in Proceedings of the 2nd International Electronic Conference on Remote Sensing (Sciforum Electronic Conference Series, Greece), vol. 2, no. 358, pp. 1–5, 2018. Accessed on: Jun. 10, 2021. [Online]. Available doi:10.3390/ecrs-2-05171
37. A. Rahman, L. Roytman, N. Y. Krakauer, M. Nizamuddin, and M. Goldberg, "Use of vegetation health data for estimation of aus rice yield in Bangladesh", *Sensors*, vol. 9, pp. 2968–2975, 2009. Accessed on: Jun. 10, 2021. [Online]. Available doi: 10.3390/s90402968.
38. F. Kogan, R. Stark, A. Gitelson, L. Jargalsaikhan, C. Dugrajav, and S. Tsooj, "Derivation of pasture biomass in Mongolia from AVHRR-based vegetation health indices", *International Journal of Remote Sensing*, vol. 25, no. 14, pp. 2889–2896, 2004. Accessed on: Jun. 10, 2021. [Online]. Available doi: 10.1080/01431160410001697619

15 Image Processing Implementation for Medical Images to Detect and Classify Various Diseases on the Basis of MRI and Ultrasound Images

Rati Goel

CONTENTS

DOI: 10.1201/9781003221333-15

15.1 INTRODUCTION TO MEDICAL IMAGES

Medical image reconstruction is required for deeper understanding of clinical abnormalities including kidney stone diseases, breast cancer, and brain tumor. Imaging has become an essential component in bio-medical and clinical research. Medical imaging is a technique that allows a physician to create a visual representation of interior organs. It is later used for clinical analysis and medical intervention. A wide variety of medical imaging modalities is available to treat and diagnose the disease. This has become almost compulsory to help radiologists by using digital computer images for accurate clinical diagnosis. Medical imaging techniques such as ultrasound, CT (computed tomography), and magnetic resonance imaging (MRI) are widely used for clinical research, effective planning, and diagnosis [1].

To automate disease detection processes, image processing techniques have a key role. The imaging techniques followed by modern computing abilities lead to the development of highly accurate screening systems to judge diseases or any abnormalities. There are various types of imaging modalities used for medical diagnosis of several diseases. Every technique has a specific method or algorithm depending upon the types of disease inside the human body [2]. A ligament injury is best identified by an MRI, whereas a bone injury is best detected by X-rays. Medical professionals select the imaging technique depending on the type of disease and the strengths of individual imaging techniques. No single imaging technique is best for every disease. Each imaging technique has certain disadvantages and advantages.

For our study, we have considered ultrasound and MRI imaging to provide information about kidney stones, brain tumors, and breast cancer. These imaging methods can be used to find the tumor, cyst, cancer, swelling, or appearance of the stone, to measure the size of the renal stone (calculi) and blockage of urine flow, and to detect anomalies [3]. Ultrasound and MRI images have the drawback of poor quality. The main reason behind the poor quality is speckle noise and disturbance, which reduce the image quality. There are many factors in diagnosis, such as image quality and the expertise of the radiologist, etc. Speckles are one major degradation factor in ultrasound and MRI images. Due to degradation or low quality, abnormalities such as calculi are very complex to recognize by visual perception in ultrasound images. When deciding on the imaging modality, the physician needs to focus on risk and potential benefits in order to acquire better potential benefits. Enhancement techniques of image processing can improve the image quality[4], and segmentation techniques of image processing can detect brain tumors, kidney stones, and any other abnormalities. The main target of the proposed method is to detect tumors, breast cancer, and kidney stones accurately using segmentation techniques. This work has introduced one automatic detection method for brain tumors, kidney stones, and cancer to increase the accuracy and yield and decrease the diagnosis time. Each imaging technique has certain disadvantages and advantages.

15.1.1 Computed Tomography (CT)

CT images are produced using X-ray photons with a digital reconstruction technique. The X-ray beam is generated by a tube, passed through the patient, and captured by the detector. This captured beam is reconstructed to produce a three-dimensional image. The CT scanner uses different reconstruction algorithms at different angles to produce an image. The most commonly used clinical applications in CT studies are CT brain, pelvic CT, CT angiography, and cardiac CT [5]. These are used to locate abnormalities in the body, such as tumors, abscesses, abnormal blood vessels, etc.

15.1.2 Ultrasound

Ultrasound imaging provides cross-sectional images of the body, which are constructed using high-frequency sound waves. In ultrasound procedure, there is no radiation exposure, and hence, it is a very safe process with minimum known adverse effects [6]. The sound waves are emitted by the transducer at some frequency, and the returning echoes are captured at frequencies dependent on the tissues through which the waves traverse. The returned sound wave is digitized, which appears as dots or echoes on the screen. These echoes are used in cardiovascular ultrasound to visualize the peripheral vascular structures of the heart, and in abdominal ultrasound to assess the anatomy of the liver and gallbladder.

15.1.3 Magnetic Resonance Imaging (MRI)

This methodology is generally used in radiology and utilizes magnetic radiation to assess detailed internal structures. It is mainly used in medical diagnosis of soft tissues of the body parts. Normally, an MRI scanner consists of two powerful magnets. One causes the body's water molecules, which are normally scattered, to line up in one direction. The alignment of the hydrogen atoms is then altered by turning on and off the second magnetic field. When the magnetic field is turned off, these hydrogen atoms will switch back to their original state [7]. These changes are detected by the scanner to create a detailed cross-sectional image. This image helps to visualize internal structures such as joints, muscles, and other structures in an effective way.

15.1.4 Fluoroscopy

The body structures are visualized in a real time environment using an imaging modality termed fluoroscopy. These real time images are generated by continuously emitting and capturing the X-ray beams on the screen [8]. To differentiate various structures, high-density contrast agents may be used, which help in the assessment of anatomy and the functions of these structures.

15.1.5 Ophthalmic Imaging

In addition to other medical imaging modalities, there exist several ophthalmic imaging methods, such as OCT (optical coherence tomography) and FA (fluorescein angiography), which plays a major role in the diagnosis of several eye-related pathologies. The necessity of injection is a major disadvantage of this method. The

TABLE 15.1
Risks and Benefits of Imaging Techniques

Risks	Benefits
Exposure to radiation	Accurate diagnosis
Cost	Improved management of condition
Stress	Option for effective treatment
Unrequired testing	Online monitoring
	Early detection of problem

risks and benefits associated with imaging techniques need to be evaluated before recommending them. The risks and benefits of imaging techniques are summarized in Table 15.1.

A particular imaging test should be recommended if it is going to accurately diagnose the disease and thereby benefit the patient.

15.2 HUMAN BODY DISEASES DETECTED BY IMAGE PROCESSING TECHNIQUES

15.2.1 Kidney Stone

A kidney stone is a solid piece of material. The urine minerals are responsible for the formation of kidney stones. This may be genetic or may be due to substance concentration [9]. Kidney stone formation, called renal calculus, is production of crystals in the urine. The formation of kidney stones is natural in human beings. In general, kidney stone formation is detected by severe abdominal pain followed by fever. For effective treatment, early kidney stone detection is very important. The existence of kidney stone affects the kidney function. Ultrasound and CT techniques are preferred to detect the presence of kidney stone [10]. Figure 15.1 shows an ultrasound image of a kidney with stone. Ultrasound is simple and a cheaper imaging technique in comparison with CT. Kidney stones may be classified on the basis of their chemical composition or location. Figure 15.1 shows a normal kidney and an abnormal kidney with stone, but both images are low quality.

15.2.2 Breast Cancer

Breast cancer is a form of cancer that originates in the breast. Figure 15.1 illustrates normal and cancerous breast tissue. When cells start to multiply out of control, then cancer begins. Breast cancer cells frequently form a tumor, which could be normally noticed on an X-ray or else felt like a lump [11]. Mostly it is women who suffer breast cancer, although males can also be affected.

It is necessary to comprehend that most breast lumps are not cancerous but benign. Non-cancerous breast tumors, also known as benign, are abnormal growths, though

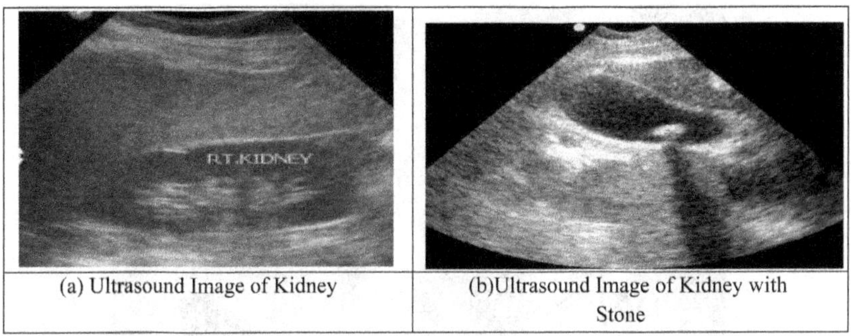

(a) Ultrasound Image of Kidney

(b)Ultrasound Image of Kidney with Stone

FIGURE 15.1 (a) Ultrasound image of kidney. (b) Ultrasound image of kidney with stone.

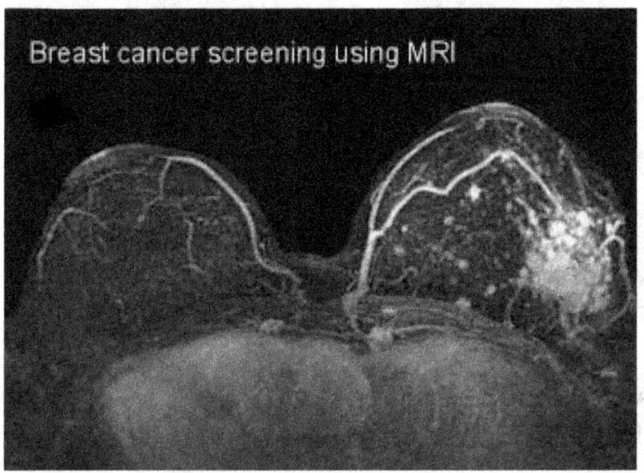

FIGURE 15.2 Breast cancer MRI image.

they don't extend beyond the breast. These cancers aren't life threatening, though a few types of benign breast lumps augment the risk of breast cancer in women [12]. Every breast lump or variation has to be checked by a health care specialist to conclude whether it's benign (non-cancerous) or malignant and additionally, whether it will have an effect on the future risk of cancer. Figure 15.2 shows an MRI image of breast cancer.

15.2.3 Brain Tumor

Brain tumor signifies a malformed mass of tissue wherein cells multiply rapidly and ceaselessly within the brain tissues. The tumor causes, through the affected tissues, effects in the brain and lung. All tumors are related to cancer. MRI techniques play a crucial role as an emergency diagnostic method. Brain tumor segmentation involves separating distinct tumor cells (effective tumor, solid, edema, and necrosis) from normal brain cells (gray matter [GM], white matter [WM], and cerebrospinal fluid

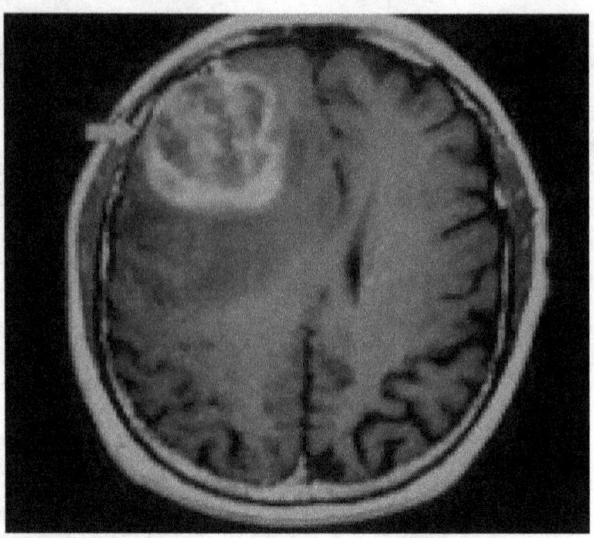

FIGURE 15.3 Brain affected by tumor.

[CSF]). When it comes to brain tumour research into non-natural cells are frequently targets of investigation 13]. A brain tumor is shown in Figure 15.3. The procedure of MRI doesn't involve any pain or radiation and is non-invasive.

15.3 IMAGE PROCESSING TECHNIQUES TO DETECT ABNORMALITIES

Image processing is a widely used methodology in various medical sectors. Image processing involves performing some operations on images to extract some useful information. Image analysis is very helpful in the early detection of various diseases in which the time factor is crucial.

15.3.1 Image Acquisition

The proposed method starts with image acquisition, which is used to take images from the external source of the system. In this method, we acquired the image using the image acquisition toolbox command in MATLAB. In order to read the image in matrix format, we used the abovementioned toolbox [14]. For this work, first, the acquired dicom images are converted into jpeg format to make the image readable for processing.

15.3.2 Image Preprocessing (Conversion RGB to Gray)

In image preprocessing, the original image is obtained from the degraded image. Images are preprocessed and converted to gray images with removal of labels.

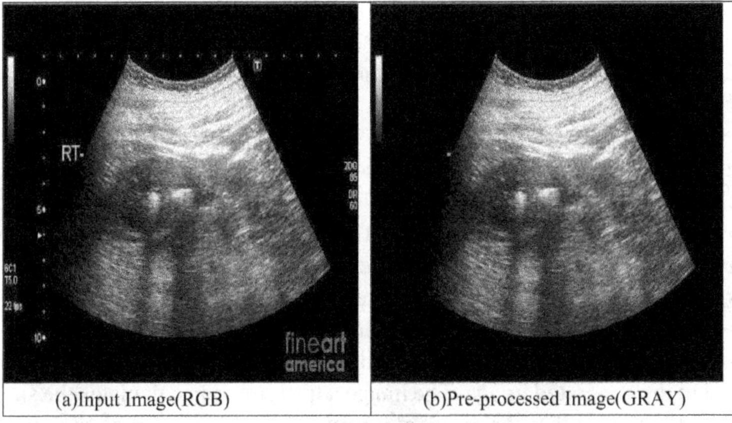

FIGURE 15.4 (a) Input image(RGB). (b) Preprocessed image (gray).

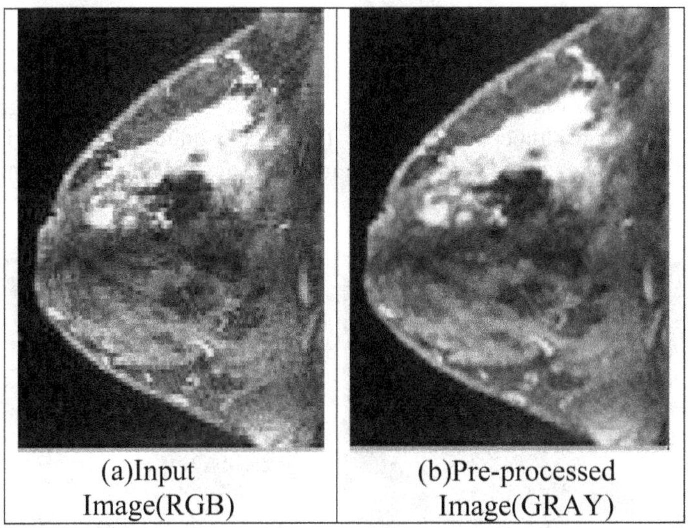

FIGURE 15.5 (a) Input image(RGB). (b) Preprocessed image (gray).

Preprocessing always yields better results for further processing [15]. After preprocessing, the image becomes sharp and clear. This can be easily seen in Figure 15.4a and b and Figure 15.5a and b.

15.3.3 Image Contrast Enhancement by Intensity Adjustment

The primary goal of an image enhancement technique is to process the biomedical image to make it suitable for disease diagnosis by medical professionals. For visual analysis of a medical image, physicians should have a good knowledge of the images of the patient for better diagnosis. Medical images are poorly illuminated, and so many structures are not clearly visible. Many regions/boundaries are vague/fuzzy in nature. So if the quality of the image is improved, processing may become easier [16]. For this reason, medical image enhancement is extremely important. This technique removes the various artifacts and sharpens the features such as contrast, edges, and boundaries. This technique increases the quality of a biomedical image such that meaningful information may be extracted for further processing, as presented in Figure 15.6 and Figure 15.7. The contrast of an image may be decreased or increased as desired by using contrast enhancement techniques [17]. Components of biomedical images may be modified by image enhancement techniques to improve finer details, sharpness, and clarity through interpretation and visual analysis.

This methodology is an image enhancement technique that plots the values adjustment in an image using the image adjust function, with the intensity value scan specified in the generated image. The image adjust function calculates the simulation studies and indicates that the image's contrast is raised, as well as the image's histogram, and sets the adjustment limits automatically. [18]. The intensity adjustment method plots the revised intensity values to the full display range of the datatype.

15.3.4 Median Filter

This is a non-linear filtering technique in image processing. When an image contains a lot of noise, like saltnpepper/impulsive, this filter is quite useful. Filtering can be used as a pre-processing step to improve images before further processing, such as edge detection or any segmentation.

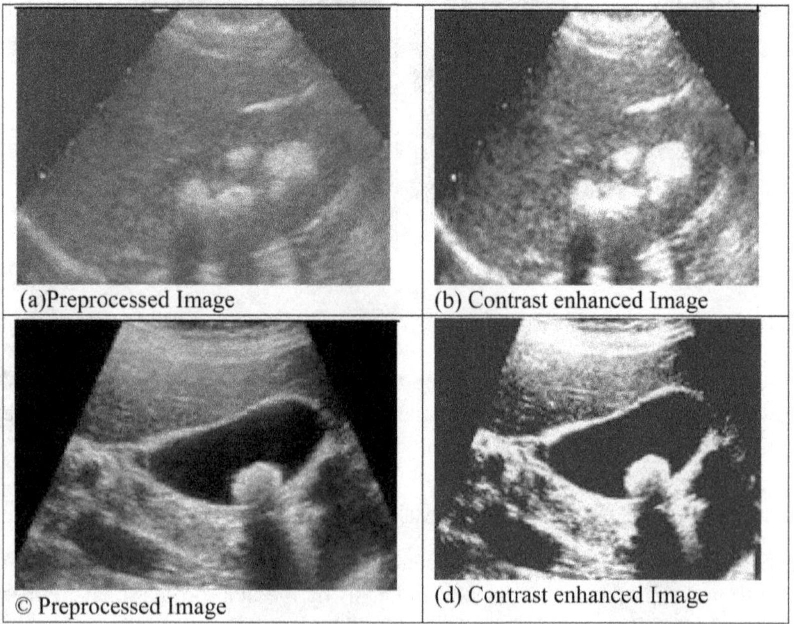

FIGURE 15.6 Contrast enhanced kidney stone images.

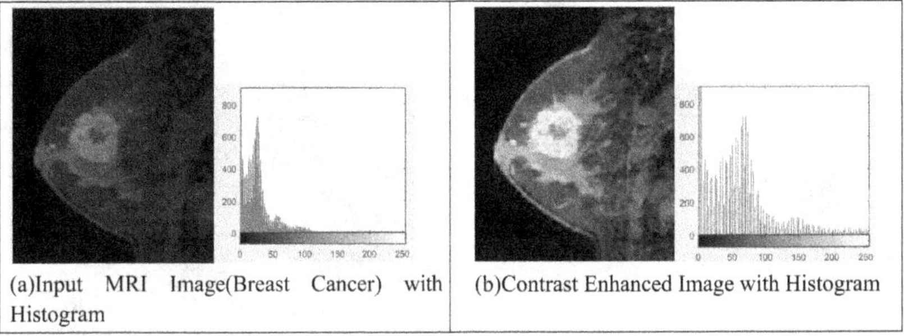

FIGURE 15.7 Breast cancer MRI image. (a) Breast image pre contrast application; (b) breast image post applied contrast. (Tumor is also enhanced.)

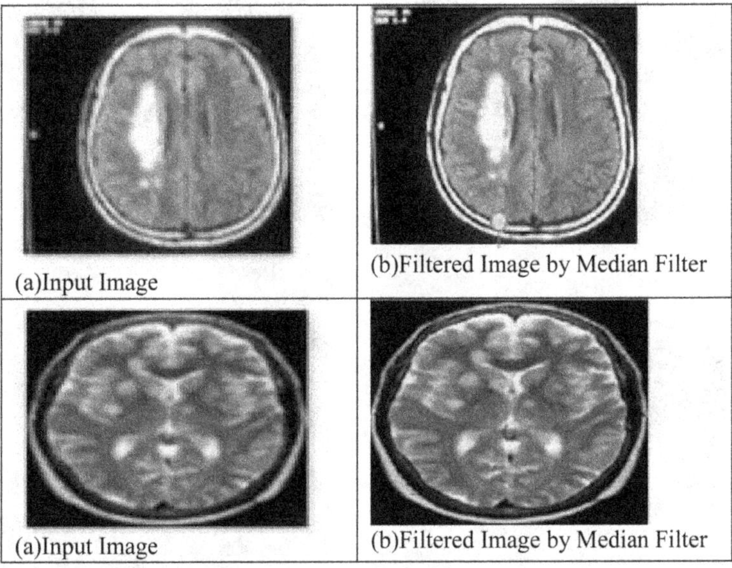

FIGURE 15.8 Brain tumor MRI Images. (a) Input images. (b) Filtered image.

In digital image processing, median filtering is a very widely used methodology. In general, ultrasound images suffer from the drawback of speckle noise. Median filter is particularly effective for impulsive noise and speckle noise [19]. In order to reduce the noise with edge preservation, the median filtering yields a better result. This filtering technique is utilized to give an approximation of the original image without any loss of information. There are many windowing operators, such as the mid filter, median, max and min filter, and mean filter[20]. Median is the best of them.

Preprocessing always yields better results for further processing. After preprocessing, images become sharp and clear. This can be easily seen in Figure 15.8, Figure 15.9, and Figure 15.10.

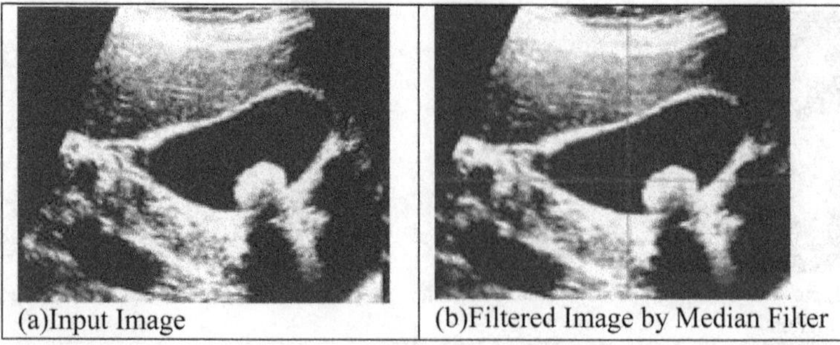

FIGURE 15.9 Kidney stone images. (a) RGB image. (b) Gray filtered image.

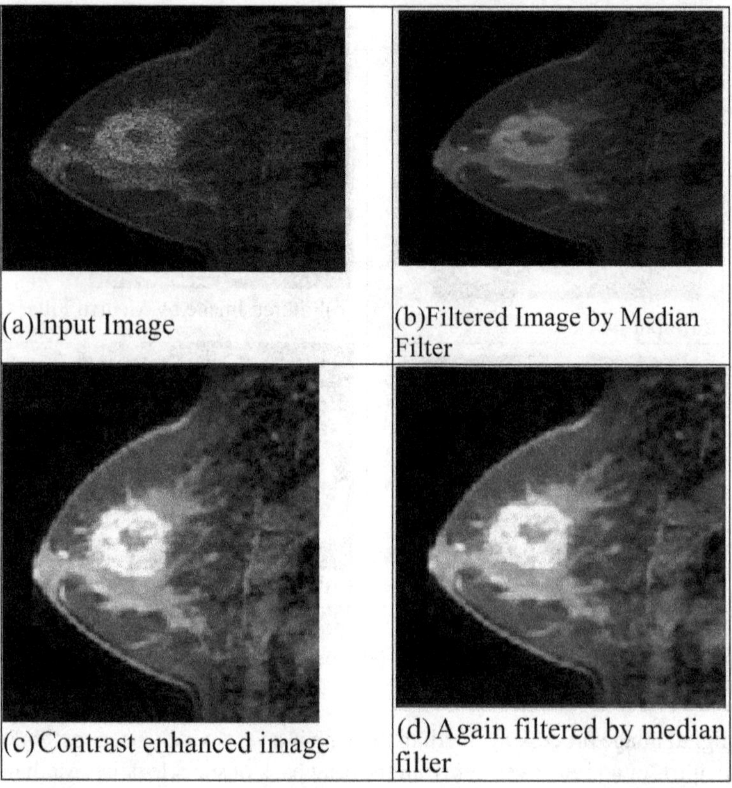

FIGURE 15.10 Breast cancer MRI images. (a) Input image. (b) Filtered image. (c) Contrast enhanced image. (d) Again filtered by median filter (now more clear and sharp).

15.3.5 Segmentation

An essential and challenging procedure of image processing is segmentation. To partition an image having similar properties and features is image segmentation technique. An image can be easily analyzed and simplified meaningfully by the segmentation technique [20]. The most basic and essential part of digital image processing is segmentation. This technique is mainly used for image partitioning into a set of disjoint regions. The values of regions should be homogeneous and uniform attributes based on texture, tone, color, intensity, etc. [21]. Image segmentation has no general theory. Brain tumors and breast cancer are also different as compared with natural tissue, including CSF, GM, and WM. A kidney stone is a different solid part as compared with a normal kidney [22]. To find the stone threshold, segmentation is considered, and clustering segmentation is used to detect tumors and cancer.

Segmentation by morphological operation

Steps for Morphological segmentation are as follows:

i. Read MR brain images as input from database.
ii. Input image is obtained by applying the threshold T, such that

$$T(a,b) = \int_{0,\ \text{otherwise}}^{1,\ g(c,d)\geq T} W$$

where $g(c,d)$ is the intensity of gray scale image.
iii. Apply erosion on the binary image with suitable structuring element, and the resulted image is called an eroded image. A•B = (A ⊕ B) B, where A represents the binary image and B is the structuring element.
iv. Apply dilation, A o B = (A⊕ B) B, where A represents the binary image and B is the structuring element.
v. Create region of interest (ROI) of tumor or cancer region.

15.3.5.1 Clustering Segmentation

Clustering segmentation is used to identify the location of tumors and cancer. Manual segmentation is a time-consuming process. K-means clustering is the most frequently used approach for dividing information into groups or clusters, as shown in Figure 15.11 and Figure 15.12. K-means clustering falls into the category of hard

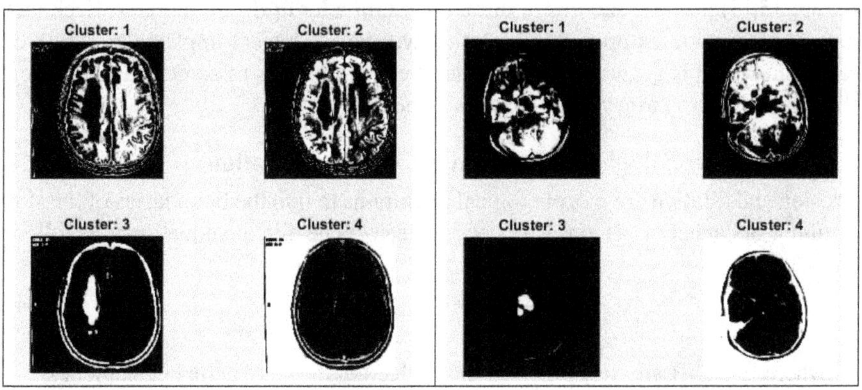

FIGURE 15.11 Segmented brain images by clustering.

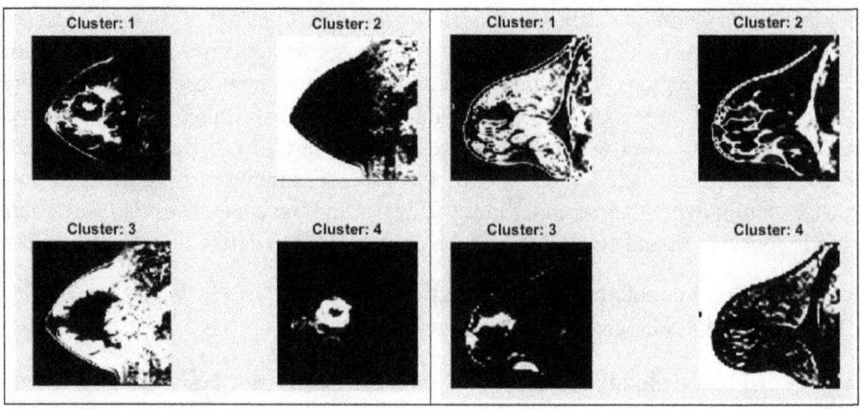

FIGURE 15.12 Segmented breast cancer images by clustering.

clustering [23]. In hard clustering, each pixel is related to one cluster. In this clustering, the membership functions 0 or 1 are used. The zero indicates that the pixel does not belong to any cluster, and 1 indicates that the pixel belongs to a cluster. The K-means clustering technique utilizes hard clustering. In this, each pixel is assigned to a nearby center [24].

15.3.5.2 Threshold Segmentation

Threshold techniques are among the easiest approaches for image segmentation. The pixels of the images are partitioned based on intensity level with help of this method. This method can be defined as single level and multilevel. Objects which are have lighter than background, then it's histogram have two mode, one near to lowest value showing dark and other near to highest point showing lighter [25]. We set the threshold values 15 and 20 manually. Pixels below this threshold become blacker, with the value 0, and pixel values above the threshold become whiter, with the value 1. We get the result in the form of a segmented image. The results seem to be good (see Figure 15.13) because the image did not contain a lot of different parts. Each technique has its own features. Segmentation will be the logical implementation in our research, which is generally used to find the ROI in terms of some characteristic of the images where kidney stone may be suspected.

15.3.5.3 Morphological Operation for Area Localization

Erosion and dilation are morphological operations to find the boundaries of the stone or tumor or cancer or whatever we need to detect. The dilation equation is as follows:

$$AB = \left\{ Z \middle| (B)_Z IA \neq \varnothing \right\} \tag{4.1}$$

where $\varnothing$ and $\hat{B}$ are the empty set and reflection of the structuring element B. The erosion equation is defined as

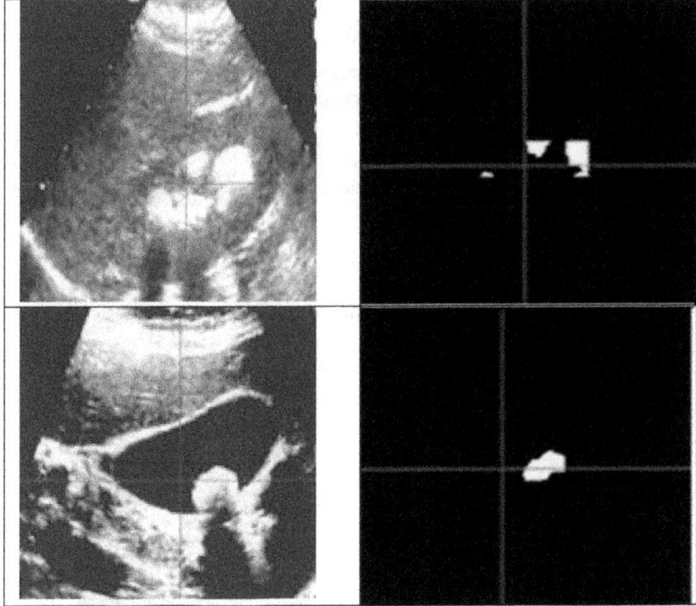

FIGURE 15.13 Segmented kidney stone images by threshold.

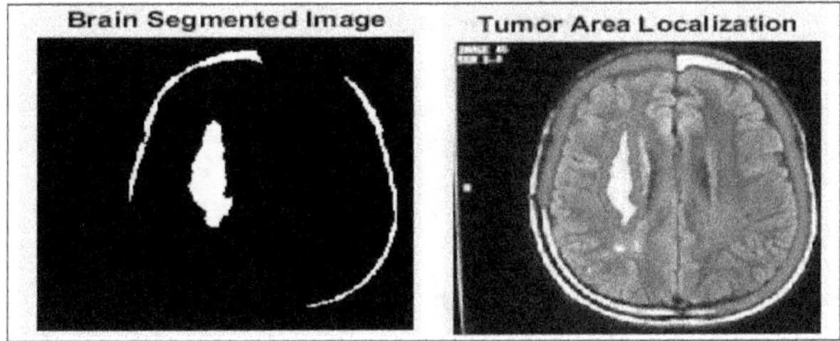

FIGURE 15.14 Segmented and detected boundary of tumor.

$$AB = \left\{Z \middle| \cdot (B)_Z IA^C \neq \varnothing\right\} \quad (4.2)$$

where A^C is the complement of A.

To draw the boundary or find the layout of tumor and cancer in the MRI images, we have applied the erosion technique as shown in Figure 15.14, Figure 15.15, and Figure 15.16.

15.4 CLASSIFICATION BY CONVOLUTION NEURAL NETWORKS

A convolution neural network (CNN) is a deep multilayer feed-forward neural network machine learning algorithm, which resembles the functioning of a human

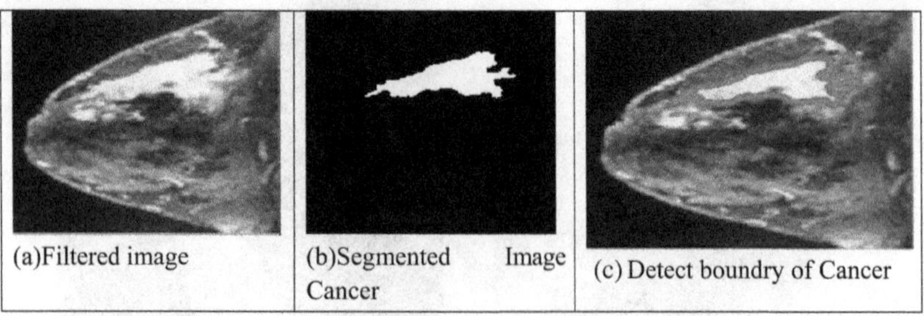

FIGURE 15.15 Detection of breast cancer image.

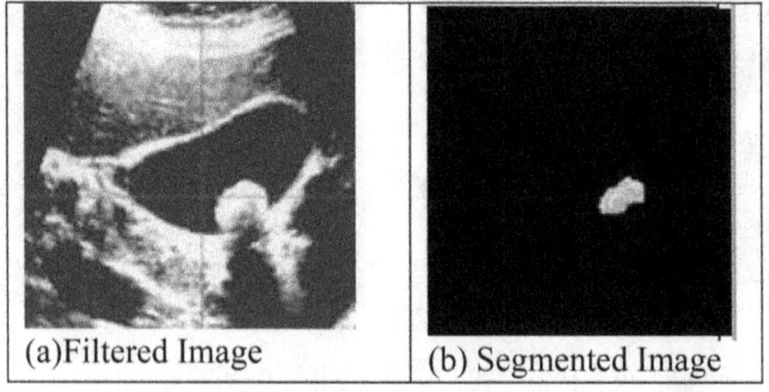

FIGURE 15.16 Detection of kidney stone images.

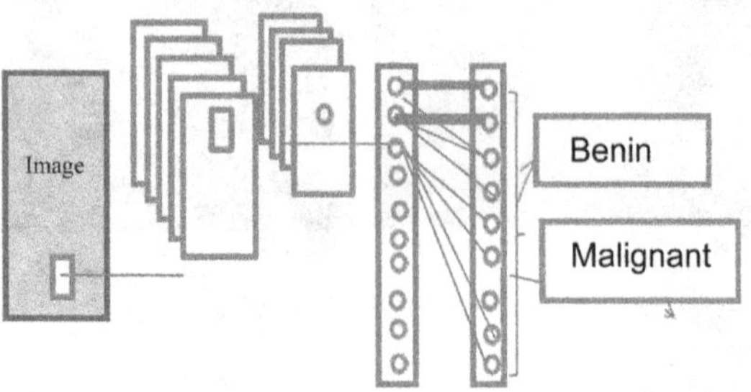

FIGURE 15.17 Architecture of CNN model.

being's visual cortex. The CNN's architecture is typically made up of one input layer, the feature-extraction layer, a classification layer, and lastly, an output layer [26]. The feature-extraction layer is formed of a pooling layer and the convolution layer, whilst the classification layer is usually built with an Multilayer Perceptron (MLP). The pooling layer, the convolution layer, and the fully connected (in the MLP) layer that is found between the input layer and the output layer are non-linear hidden layers. The deeper the layers, the more the complexity of the learned image properties increases [27]. Figure 15.17 represents the detailed architecture of the CNN model with layers along with its parameters.

Following that, we simply apply the wavelet transform to the images, which allows us to calculate the input image features step by step. We use the NN classification to classify the input photographs to the database images. Here, the database images are the MRI/CT images, in that we have some normal images and abnormal images, meaning without cancer and with cancer, respectively [28].

Normally, the classification is used to classify whether the image is normal or abnormal. If a tumor is identified during comparison of the pixels, a message will be displayed that a tumor has been detected after completion of the NN training.

15.5 RESULT ANALYSIS

The images of diseases are accumulated from the Web, and for the development process, the MATLAB environment is used. In this work, 50 images related to brain, 15 images related to kidney, and 5 images related to breast cancer are considered. In the present technique, the statistical analysis is carried out on images. For disease detection, we have used segmentation, which is followed by classification techniques. In the initial step, the dataset of images is collected. When the images are read by MATLAB, all text and related labels inside the image have been removed. Then, images are converted to gray from RGB and resized. Further, a median filter is used to eradicate noise.

For MRI (tumor and cancer), DWT is applied. The CNN classifier is used to categorize the given image as normal or abnormal, this means stages of diseases and clustering of segmentation to find the tumor. Finally, to detect the location of tumors and cancer, operations of clustering segmentation are exploited, as depicted in Figure 15.11 and Figure 15.12. The final segmented image is displayed in Figures 15.19, 15,20, and Figure 15.21, showing it is affected by 30% known as benign, and the other image is showing in Figure 15.20 and Figure 15.22 that it is affected by 50% known as malignant. In Figure 15.18, the brain is in normal condition with no tumor (Table 15.2).

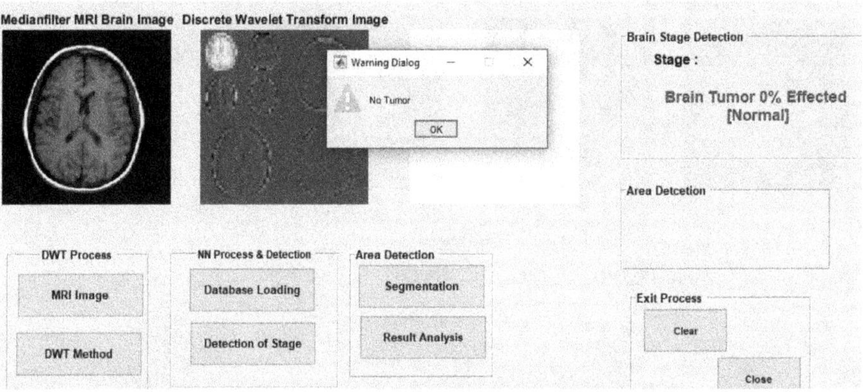

FIGURE 15.18 Detection of normal brain.

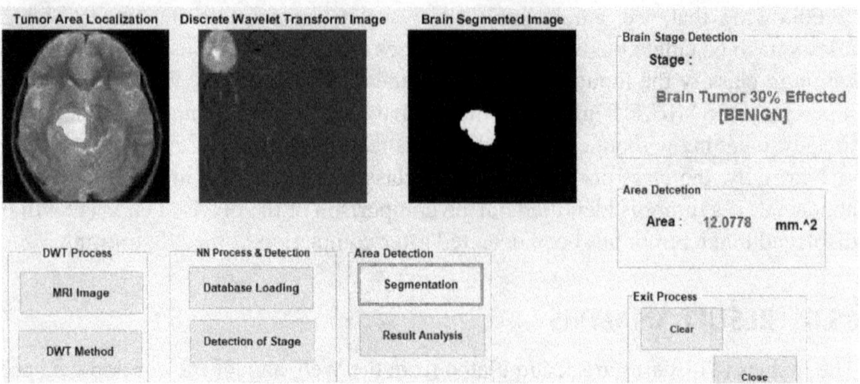

FIGURE 15.19 Detection of brain tumor 30% affected.

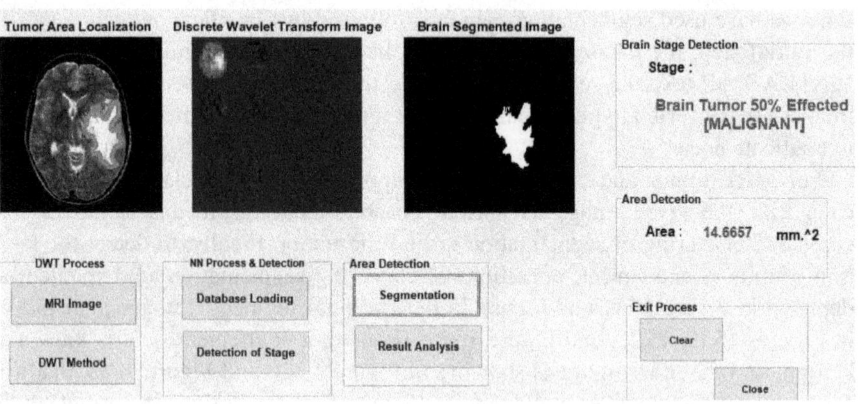

FIGURE 15.20 Detection of brain tumor 50% affected.

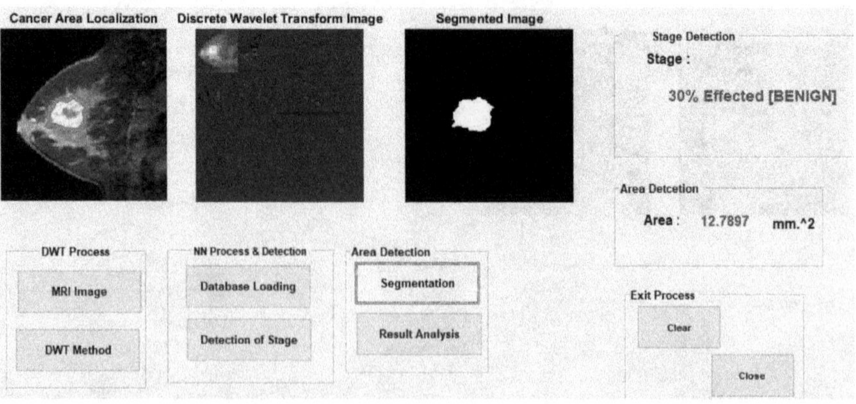

FIGURE 15.21 Detection of cancer 30% affected.

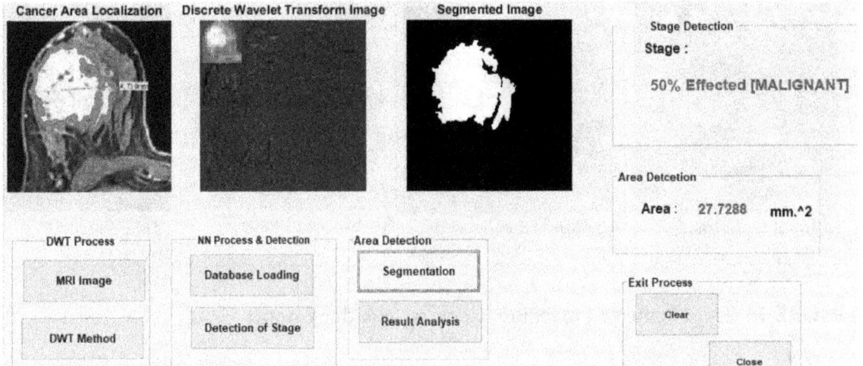

FIGURE 15.22 Detection of cancer 50% affected.

TABLE 15.2
Confusion Matrix for Brain Tumor Detection

Parameters	TP	FN	FP	TN
Value	25	1	1	22
Specificity	95.652	95.652	95.652	95.652
Sensitivity	96.153	96.153	96.153	96.153
Accuracy	95.918	95.918	95.918	95.918

Accuracy: The accuracy of the classification process is based on correct and incorrect results of brain tumor. The accuracy of the classification can be calculated using Equation 15.1:

$$\text{Accuracy(ACC)}\frac{\text{TP}+\text{TN}}{\text{P}+\text{N}}=\frac{\text{TP}+\text{TN}}{\text{TP}+\text{TN}+\text{FP}+\text{FN}} \tag{15.1}$$

We found that by using our methodology, we got above 95% accuracy.

For kidney stone detection, operations of threshold segmentation are exploited. The final segmented image is displayed in Figure 15.17, and the complete procedure to detect kidney stone is displayed in Figure 15.23. For analysis and detection of the sensitivity, accuracy, specificity, and F1 score, the confusion matrix is used at 15, 20 threshold of segmentation, as shown in Table 15.3.

The final system analysis is carried out to judge the accuracy, specificity, and sensitivity by preparing the confusion matrix. The classification system accuracy obtained is approximately 96%. The presented technique can help doctors in early detection for accurate treatment of patients (Figure 15.24).

15.6 CONCLUSION

In real time, MRI and ultrasound images may not be clearly visible. This needs further diagnosis. In the proposed work, accurate kidney stone, brain tumor, and

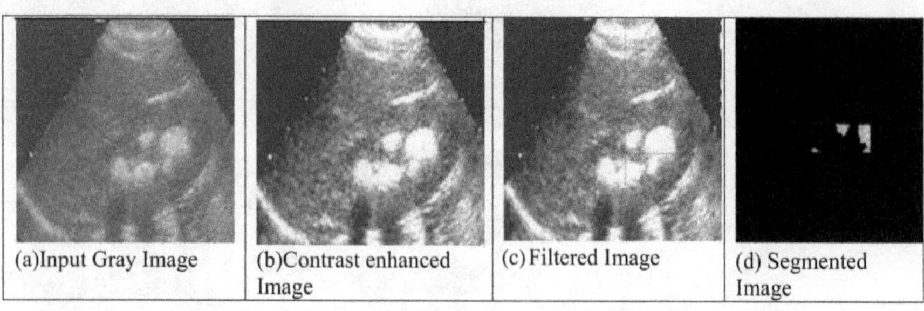

FIGURE 15.23 Complete procedure to detect the kidney stone.

TABLE 15.3
Performance Evaluation by Various Thresholds (Kidney Stone Detection)

1	Threshold(15)	Specificity	.93
		Sensitivity	1
		F1-score	.96
		Accuracy	.96
2	Threshold(20)	Specificity	1
		Sensitivity	1
		F1-score	1
		Accuracy	1

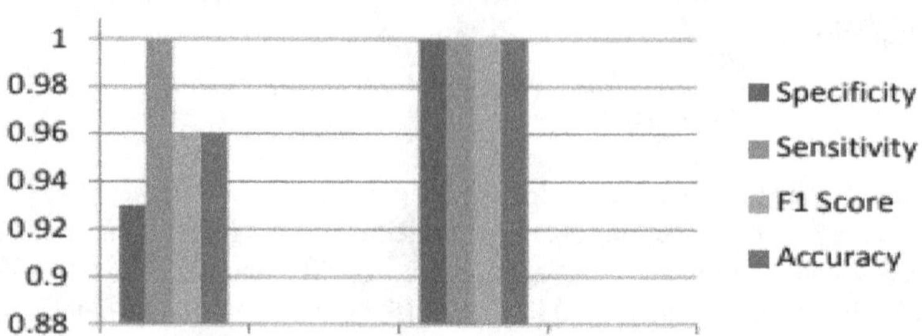

FIGURE 15.24 Performance evaluation by visualization.

breast cancer detection are presented. The presented methodology requires preprocessing, contrast enhancement, and filtration. A CNN model is used to detect the stage, meaning the type of tumor. Afterwards, for detecting the location of disease, clustering and threshold segmentation are used. The analysis is carried out based on accuracy. Thus, the effectiveness of the work has been verified. This technique may be used as an aid for early and accurate diagnosis of diseases by medical professionals for the benefit of students.

REFERENCES

1. P. Rasti and M. Daneshmand, et al., 2016. Medical image illumination enhancement and sharpening by using stationary wavelet transform, in 24th Signal Processing and Communication Application Conference (SIU), Zonguldak, pp. 153–156.
2. B. Mohan Kumar and Prof. Y. S. Thakur, 2017. An introduction to steganography techniques in the field of digital image processing, 7(6).
3. M. S. Alkoffash and Mohammed J. Bawaneh, 2014. A survey of digital image processing techniques in character recognition, *Int J Comput Sci Network Secur* 14(3).
4. T. Rahman and M. S. Uddin, 2013. Speckle noise reduction and segmentation of kidney regions from ultrasound image, in International Conference on Informatics, Electronics & Vision (ICIEV), IEEE.
5. Rahman, Tanzila and Mohammad Shorif Uddin, 2013. Speckle noise reduction and segmentation of kidney regions from ultrasound image, in International Conference on Informatics, Electronics & Vision (ICIEV), IEEE.
6. Adam, Tijjani, U. Hashim and U. S. Sani, 2012. Designing an artificial neural network model for the prediction of kidney problems symptom through patient's metal behavior for pre-clinical medical diagnostic, in International Conference on Biomedical Engineering (ICoBE), IEEE.
7. V. P. G. P. Rathi and S. Palani, April 2011. Detection and characterization of brain tumor using segmentation based on HSOM, wavelet packet feature spaces and ANN, in Proceedings of the 3rd International Conference on Electronics Computer Technology (ICECT'11), vol. 6, pp. 274–277, IEEE, Kanyakumari, India.
8. Demirhan A. and Toru M., n.d. Segmentation of tumor and edema along with healthy tissues of brain using wavelets and neural networks, *IEEE J. Boimed Health Inform* 19(4):1451–1458.
9. Prema T. Akkasaligar, Sunanda Biradar and Veena Kumbar, 2017. *Kidney Stone Detection in Computed Tomography Images*, IEEE.
10. Goel R. and Jain A., 2020. Improved detection of kidney ultrasound using segmentation techniques, in *Advances in Data and Information Science. Lecture Notes in Networks and Systems*, vol. 94. Springer.
11. Campilho, A. Polónia, C. Eloy, E. Castro, G. Aresta, J. Rouco, P. Aguiar and T. Araújo, 2017. Classification of breast cancer histology images using convolutional neural networks, *PLOS ONE* 12(6): e0177544.
12. M. E. Sanders and J. F. Simpson, 2014. *Breast Pathology.* Edited by D. E. Elder. Vol. 6. Demos Medical. ISBN:978-1-936287-68-0.
13. Sawarkare S. et al. n.d. Classification of brain tumor using discrete wavelet transform, principal component analysis and probabilistic neural network, *Int J Res Emerg Sci Technol* 6. E-ISSN:2349-7610.
14. Vaish, Pallavi et al., 2016. Smartphone based automatic abnormality detection of kidney in ultrasound images, in 18th International Conference on e-Health Networking, Applications and Services (Healthcom). IEEE.
15. R. Goel and A. Jain, 2018. The implementation of image enhancement techniques on color n gray scale IMAGEs, in 5th International Conference on Parallel, Distributed and Grid Computing (PDGC), pp. 204–209, doi: 10.1109/PDGC.2018.8745782.
16. K. Tarun and V. Karun, 2010. A theory based on conversion of RGB image to gray image, *Int J Comput Appl* 7(2). (0975 – 8887).
17. P. Janani et al., 2015. Image enhancement techniques: A study, *Indian J Sci Technol* 8(22). September 2015 ISSN: 0974-6846.
18. Jeyalakshmi, T. R. and K. Ramar, 2010. A modified method for speckle noise removal in ultrasound medical images, *Int J Comput Electr Eng* 2(1): 54.

19. Jinhua Yu, Jinglu Tan and Yuanyuan Wang, 2010. Ultrasound speckle reduction by a SUSAN-controlled anisotropic diffusion method, *Trans Pattern Recognit* 43: 3083–3092.
20. M. Arun Kanti et al., 2014. Image processing with sampling and noise filtration in image recognition process, *Int J Eng Inventions* 3(7). e-ISSN: 2278-7461, p-ISSN: 2319-6491
21. Viswanath, Kalannagari and Ramalingam G., 2014. Design and analysis performance of kidney stone detection from ultrasound image by level set segmentation and ANN classification, in International Conference on Advances in Computing, Communications and Informatics (ICACCI, 2014). IEEE.
22. Sneha A. Mane and S. R. Chougule, 2016. Neural network of kidney stone detection, *Int J Sci Res* 5(4). ISSN (Online): 2319-7064.
23. K. Sharma and B. Preet, September 2016. Classification of mammogram images by using CNN classifier, in Proceedings of the 5th International Conference on Advances in Computing, Communications and Informatics, ICACCI 2016, pp. 2743–2749, India.
24. N. Varuna Shree et al., 2018. Identification and classification of brain tumor MRI images with feature extraction using DWT and probabilistic neural network. *Brain Inf* 5: 23–30.

16 Benchmarking of Medical Imaging Technologies

Punit Kumar Singh, Sudhakar Singh, Ashish and Hassan Usaman

CONTENTS

DOI: 10.1201/9781003221333-16

16.1 INTRODUCTION

Biomedical imaging technologies have significantly improved patient safety. Photo-driven therapy with increased precision in disease diagnosis and surgical procedures has reduced the high risk resulting from human error. This chapter offers an outline of emerging imaging technologies and techniques and their ability to unravel medical problems. The first section of the chapter outlines the elementary principles and processes of varied imaging methods currently in use. The second section examines the attributes of picture handling devices and potential improvements of picture directed treatment, extended with a brief spotlight on clinical imaging radiation openness. The writers are optimistic that the chapter will help researchers within the area of medical imaging.

Picture and video handling finds its greatest application in clinical science. Biomedical imaging methods have important roles in both the diagnostic and remedial fields. These strategies have essentially assisted in improving human services to patients. Picture guided treatment has definitely diminished the high risk of human error with improved precision in disease diagnosis and surgery. The historical backdrop of clinical imaging began in the 1890s; it began to be mentioned in the 1980s and has been impressively investigated recently with innovative progress. The basic aim of this section is to offer essential information on the beginnings of various biomedical imaging systems and their progress in recent times. Wilhelm Conrad Rontgen's disclosure of X-ray beams marked the start of clinical imaging. Rontgen discovered electromagnetic waves within the relevant frequency range in 1895 while doing research with a Hittorf–Crookes tube. He named them X-beams after using them to capture a first image. William Coolidge later improved the X-beams with the creation of the Coolidge tube, which allowed a more stunning vision of deeply entrenched living structures and malignancies. Coolidge's usage of a Coolidge tube with tungsten fiber was among the most outstanding uses of X-rays within the world of radiology. Felix Bloch and Edward Purcell discovered nuclear magnetic resonance (NMR) in 1946. This was the initial move toward the development of magnetic resonance imaging (MRI). Raymond Vahan Damadian proposed the first MR body scanner in 1969 with the approach of NMR. Utilizing NMR, he found that tumors could be distinguished from ordinary cells. Godfrey Hounsfield developed a technique for "cutting" an image, utilizing X-radiation at different focuses around the section. He constructed the first computed tomography (CT) picture of human brain utilizing the first model CT scanner. In 1977, the first human MRI body scanner was presented. With the help of math and PC estimations and with inventive augmentation in leading-edge new imaging frameworks were developed. This molded biomedical imaging into an interdisciplinary field, which may help physicists, researchers, mathematicians, pharmacologists, and computational researchers.

16.2 IMAGING TECHNIQUES

All imaging methods must meet the same basic requirements. Emitted light passes through the body/region in question, transmitting or reflecting radiation that is captured by an indicator to create a picture design. Various modalities use different kinds of wave.

16.2.1 Traditional Film Radiography

Before the advent of X-rays, traditional film radiography supplied the primary radiographic picture. In conventional radiography, X-ray film is placed in the center of two screen supports and fluorescent screens. The arrangement is put in the middle of two or three cassettes, as seen in Figure 16.1. CT uses X-rays, whereas MRI and SPECT (Single-photon emission computed tomography) utilize radio frequency waves and gamma rays, respectively.

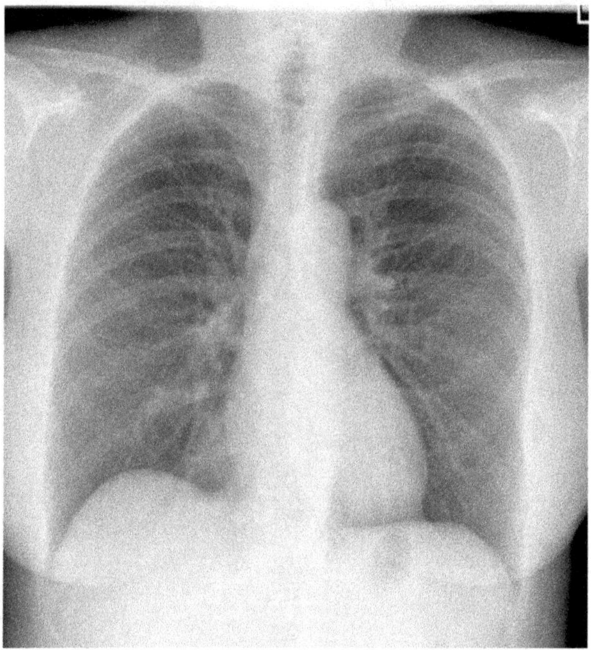

FIGURE 16.1 Chest radiography. (Source:radiopedia.com)

16.2.2 Imaging Radiography

Ordinary film radiography was adjusted with the usage of a reusable phosphor imaging plate and was named computed radiography (CR).. The radiographic shadow is received on the imaging plate, and the plate is placed on a detecting device or a CR viewer. The detecting device includes a helium neon laser and a photomultiplier tube. The laser scans the imaging plate and releases light, which is passed to the photomultiplier tube and, with the assistance of system programming, produces an automatic picture. An example is the image of a thorax taken by CR shown in Figure 16.2. A distinction is made between CR and direct or digital radiography (DR). DR is usually called direct imaging, and CR is understood as indirect imaging. The critical differentiation between CR and DR is that CR features a tape system that works on the imaging plate to produce the image; however, DR receives the unfiltered image. DR is helpful in reducing expense, with dependable capacity and straightforward film preparation. Be that as it may, both CR and DR have points of interest in the explicit portrayal of living systems and explicit calculations to visualize different diseases. In any case, elements like identification productivity, dynamic range, spatial inspection, goals, differentiation, and noise have very little effect on the character of images in either CR or DR.

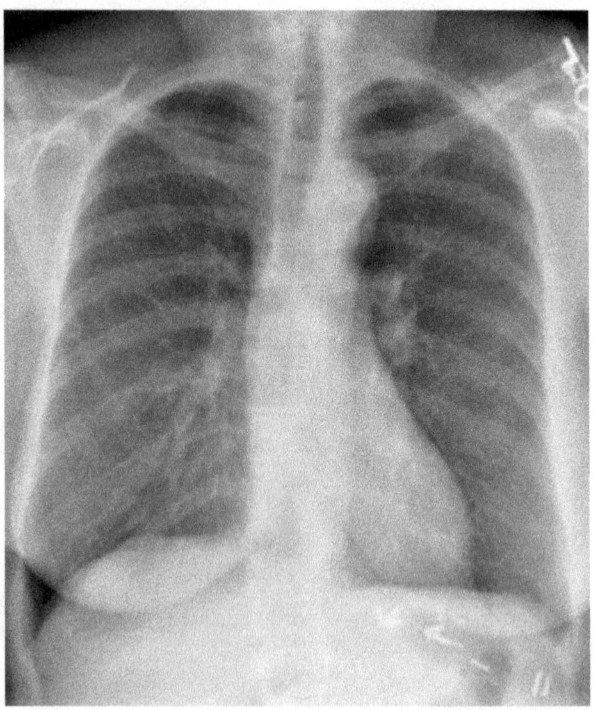

FIGURE 16.2 Computed radiography. (Source:vareximaging.com)

16.2.3 Computed Tomography

Tomography manages the photography of a specimen in segments/cut configuration utilizing any sort of wave. CT is a computer-assisted imaging method that produces cross-sectional images. Using programming, parts of a photograph are overlaid on each other to create a sophisticated three-dimensional (3-D) image. CT is outstandingly useful to assess risk, recognizing irregular tumor development, organization, or recurrence. Figure 16.3 shows the CT picture of a surviving malignant breast tumor. The left breast had been carefully emptied, and the scan was conducted to detect any tumor recurrence after a month. CT is also commonly used to diagnose coronary artery disease (atherosclerosis), vein aneurysms, blood clots, aggravations, and growth, and injuries to the brain, skeletal framework, and internal organs. Figure 16.4 depicts the CT output of liver augmentation, hepatomegaly, as an example. Compared with MRI, CT is a better option for abdominal and pelvic examination.

Advances in CT such as spiral or multicut imaging accomplished the desired resolution of 1 mm and the ability to sweep the entire body in 10 s(Sakas, 2002). Bone or surface pictures are all the more unmistakably visualized in CT. Recently, electron beam CT (EBCT) has been utilized, whereby particular pictures/depictions can be delivered without any moving parts. EBCT has more advantages contrasted with ordinary CT, including more rapid picture acquisition. Differential medium is used to acquire a clear depiction of particular parts/moving items, such as the cerebrum, veins, and so on. CT is usually performed with iodinated contrast media. Depending on the part of the body, the contrast medium might be administered intravenously or orally. If there is a problem with kidney function, intravenous differentiation medium infusion should be avoided, since it may make the kidneys work even harder. It will occasionally result in adverse sensitivity reactions. People adversely

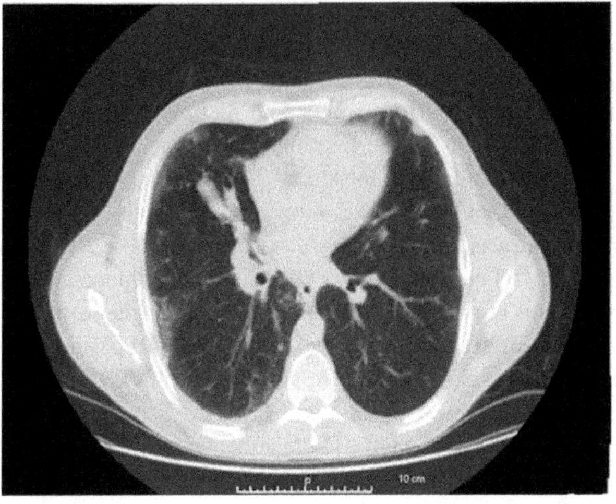

FIGURE 16.3 CT scan of chest images. (Source:cedars-sinai.edu)

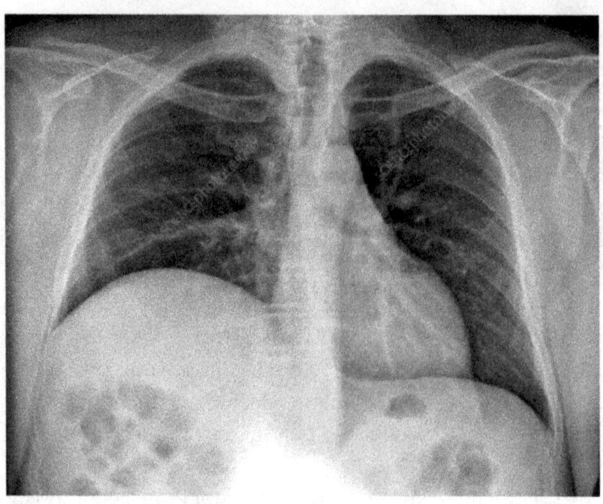

FIGURE 16.4 Enlarged liver (hepatomegaly). (Source:sciencephoto.com)

affected by salts (iodine) are requested to advise the radiologist as an early precautionary measure. Contrast media are excreted in the urine and are for the most part innocuous to patients. In any case, gentle to antagonistic responses are accounted for as often as possible (Jingu et al., 2014).

16.2.4 Magnetic Resonance Imaging (MRI)

X-ray may be a ground-breaking demonstrative procedure for delicate tissues. The X-ray framework generates a solid and uniform magnetic field alongside radiofrequency waves. A reasonably full radiofrequency range is applied to the patient from the scanner. The waves interact with the tissues or any area that holds hydrogen atoms within the body, such as water molecules. The hydrogen atoms are energized and return to their normal state utilizing the energy from the fluctuating magnetic field; this can be detected by the scanner and meticulously interpreted. Therefore, MRI is most appropriate for the perception of delicate tissues, ligaments, and tendons. X-rays are additionally appropriate in the location of certain abnormalities in the cerebrum, as shown in Figure 16.5. Contrast agents like gadolinium are utilized to detect minute contrasts/changes in structures of the body. The many little bit of leeway of utilizing MRI is to shift the complexity of the image. Minute modification of the radio emission recurrence and therefore, the magnetic field can change the differentiation of the image, emphasizing various types of tissue. Another advantage of MRI is that it can develop pictures in any plane (hub/level), which is unfeasible in CT.

There are different sorts of MRI. The use of X-rays to estimate the distribution of water particles inside the body is understood as diffusion MRI. This MRI is crucial within the diagnosis of neurological conditions like multiple sclerosis (MS) and stroke (Bihan et al., 1986). Functional MRI (fMRI) could also be suitable to

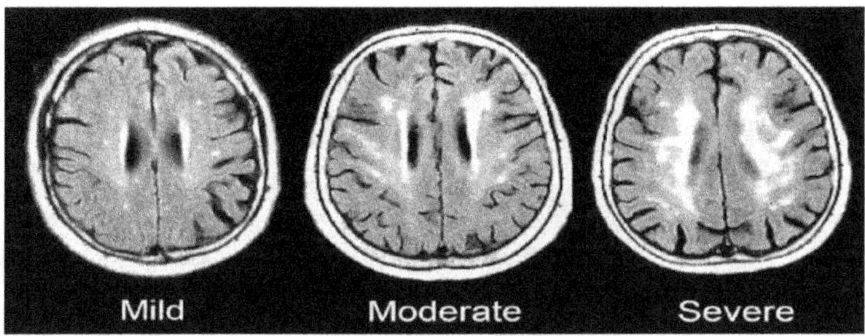

FIGURE 16.5 Image of a cerebrum demonstrating injuries. (Source: betterhealthwhi learning.net)

assess changes in brain activity, which is why neurological clutters are so often used. Another use is continuous MRI, which, as the name implies, screens moving objects in a progressive manner.

16.2.5 Ultrasonography

For sensitive tissues, X-ray might be a game-changing demonstration approach. In combination with radiofrequency waves, the X-ray framework infers a solid and homogenous attractive field. From the scanner, a reasonable amount of complete radiofrequency is administered to the patient. The waves pass through the tissues or any other portion of the body that contains hydrogen particles, such as water atoms. The particle is revitalised and returns to its original state of harmony by drawing energy from a swaying attractive field, which can be detected by the scanner and meticulously handled. For the time being, MRI is the best way to see fragile tissues, ligaments, and tendons. X-rays are also useful in determining the location of some sores in the brain, as seen in Figure 16.5.

Gadolinium and other differentiators are used to recognise minute differences/changes in bodily architecture. The other advantage of using MRI is the ability to change the image's complexity. Minute changes in the radio wave recurrence and therefore the attractive field can alter the image's differentiation, which includes a variety of tissues. Another advantage of MRI is that it can provide images in any plane (hub/level), which is not possible with CT. MRIs come in a variety of shapes and sizes. Diffusion MRI is a type of X-ray that is used to estimate the spread of water particles throughout the body. This MRI is essential in the diagnosis of neurological diseases such as MS and stroke (Bihan et al., 1986). Neurological clutters are often employed because practical MRI (fMRI) might be used to monitor changes in brain activity. (Figures 16.6 and 16.7).

16.2.6 Atomic Medicine

The different imaging modalities covered so far help in visualizing the existing frameworks of the body structures. Medicine or nuclear imaging is that the impression of

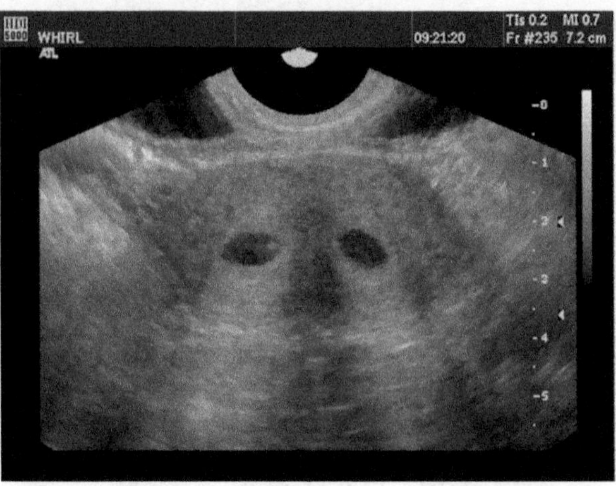

FIGURE 16.6 Transvaginal ultrasonography. (Source: slate.com)

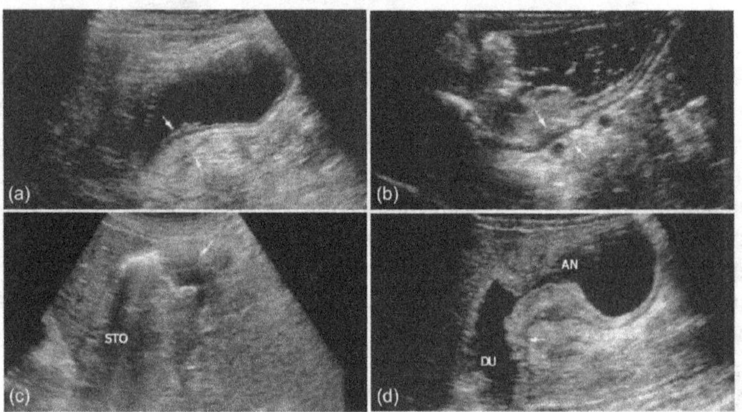

FIGURE 16.7 Transabdominal ultrasonography. (Source: wjgnet.com)

components of the organs. Name ling of radionuclides (radioactive substances/radioisotopes) alongside drug blends in dismemberment components of express organs and distinguishing the second changes at sub-nuclear and cell level is understood as medicine. Scintigraphy and positron release tomography/single-photon outflow figured tomography are the quality nuclear imaging modalities now being utilized.

Because of their extreme affectability, atomic prescriptions are used for early disease detection, such as the early detection of various stages of malignant development, the precise location of sickness, and other deviations from the norm. Patients' reactions to controlled drugs or recurrence of illness, as well as their administration with rapid treatment plans for cutting-edge infections, are effectively conceivable with the assistance of atomic medication. Time utilization may be a key constraint in atomic medicine. The radiotracer might take anything from a couple of hours to days to accumulate within the region under examination. Furthermore, unlike MRI or CT, the aim of the imaging is less obvious.

16.2.7 Scintigraphy

Scintigraphy is a common technique for seeing within the bodily tissues. Patients are given radiopharmaceuticals/radiotracers as an infusion, swallowed, or breathed in while the specific location under examination is exposed. A gamma camera, sometimes called an Anger camera, captures radiation emitted by radiotracers. Two-dimensional pictures are generated with the help of a computer. For most organs, the procedure is referred to as scintigraphy. Cholescintigraphy is a procedure that examines the biliary system (Figure 16.8). Scintimammography, also referred to as atomic medication breast imaging or subatomic breast imaging, can detect the presence of malignant development within the breast.

Radionuclide angiography or radionuclide ventriculography is employed in the location of cardiovascular breakdown because it distinguishes the capacity of the right and left ventricles of the heart. Radioisotope renography is employed to assess the capacity of the kidney. Sestamibi parathyroid scintigraphy is employed in the location of parathyroid adenoma, while bone scintigraphy is employed to find

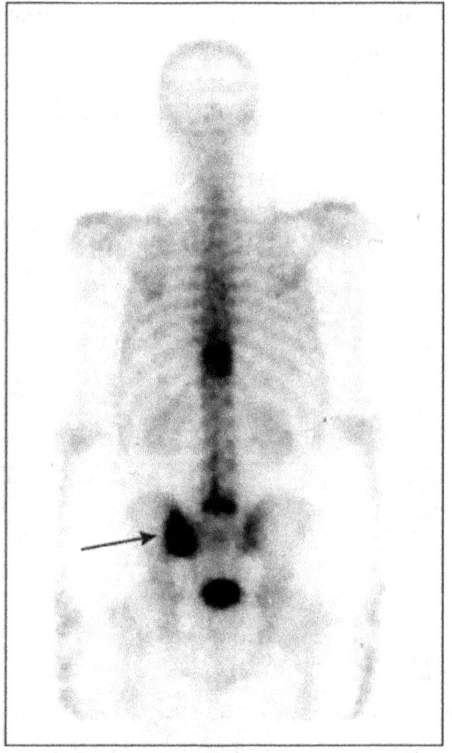

FIGURE 16.8 (Source: sciencedirect.com)

irregularities in bone. When radiolabeled antibodies are infused to locate diseased cells, this is referred to as immunoscintigraphy.

16.2.8 Positron Emission Tomography (PET)

PET is almost like scintigraphy except that it uses a positron-emitting radiotracer. The radiotracer contrast differs depending on whether it is utilized in cancer, neurology, or cardiology. When the positron radiotracer is inserted, it passes through a few millimetres and unites with free electrons to form positronium, which on rot completes a pair of gamma transmits, which is seen and repeated by the PC framework to expire the photographs. The most frequently utilized radioisotopes are short-lived isotopes of carbon-11, oxygen-15, nitrogen-13, and fluorine-18 (Ziegler, 2005). Since short-lived isotopes are utilized, a cyclotron is required for PET. A cyclotron is a kind of molecule smasher. Positron producers (short-lived radioactive isotopes) are generated when stable non-radioactive isotopes are introduced into a cyclotron. PET is reliably utilized within the evaluation of cutoff points like course framework and glucose osmosis. Too as might be expected be the use of PET in peril treatment. Fluorodeoxyglucose (FDG) may be a sugar essential. FDG is labeled with fluorine-18 and intravenously infused into the body (Figure 16.9)

In comparison to normal cells, developing malignant cells will absorb more glucose. As a result, the injected radiotracer will penetrate diseased cells and their aggregates, taking around 40 minutes. Lower-energy-transmitting isotopes, Technetium-99m and Iodine-123, are utilized in SPECT, as against the higher-energy-transmitting isotopes utilized in PET. Because it doesn't need a cyclotron, SPECT is less costly and more widely used. In Table 16.1, the most often used imaging techniques are listed in order along with their focus points and disadvantages.

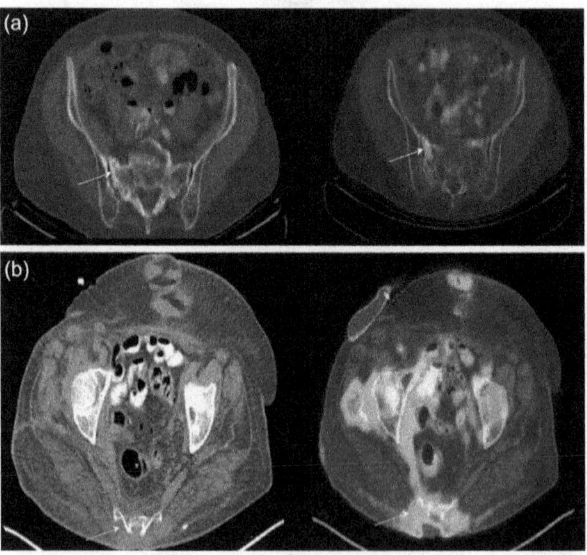

FIGURE 16.9 (Source: wignet.com)

TABLE 16.1
Different Imaging Procedures

	CT	MRI	Ultrasonography	Atomic medicine
Source	**X-rays**	**Radiofrequency waves**	**Ultrasound recurrence**	**Radiotracers**
Time taken for filtering	>30 sec. Relies upon the part under scrutiny.Finished in a short time (10 minutes)	10 minutes to 1 hour, depending on the body part	Relies upon the part of the body: stomach – 20 min, pelvic – 30 min	Relies upon the kind of sweep: two to a few hours after infusion
Radiation amount	2 to 10 mSv	1to100 MHz	1 to 18 MHz	Isotope-dependent contrasts, such as Tc 99m
Advantages	Recognizes tissues varying in physical thickness,financially cost-effective	High sensitivity precision	Noexposure to radiation	High sensitivity precision
Applications	Nervous system science, cardiology, gastroenterology, estimates bone quality	All areas of the body	Heart, liver, kidney, spleen, urinary bladder, pancreas, veins, neonatal brain, thyroid, muscle joints, ligaments, heart, liver, kidney, spleen, urinary bladder, pancreas, veins	Bladder, gut, spleen, adrenal medulla, colon, thyroid, tumor discovery
Hazards/ Precautions	Wellbeing conditions, pregnancy, claustrophobia	Pregnancy, claustrophobia, inadmissible for patients with metal chips/devices pacemakers)	Nil	Not authorized for children under the age of 14.Pregnant women should avoid breastfeeding from the day of delivery and within 24 hours of receiving a gadolinium MRI scan

16.3 OTHER IMAGING TECHNIQUES

Various new imaging strategies are now available or still in development. A few of these systems are mentioned in the following subsections. These systems are frequently less regularly used in the field of clinical imaging because of the absence of some significant elements required for productive imaging, or they are still being developed. Be that as it may, in view of their extraordinary properties, every one of these imaging procedures is increasing in critical significance step by step.

- Impedance of electricity
- Optical coherence tomography (OCT)
- Photoacoustic tomography/thermoacoustic imaging
- Imaging of microwave
- Photoacoustic/thermoacoustic imaging elastography

16.3.1 Electrical Impedance Tomography (EIT)

EIT uses surface cathodes with a low electric current that monitor the developments in tissues during the event of electrical conductivity.. If all else fails, muscle tissues and body liquids, for instance, blood, can be checked in this way. EIT provides two-dimensional pictures, which are obtained from the conductivity of a fairly constant current passed/diffused through the standard models. At higher frequencies, considering dielectric changes, unsurprisingly, the resulting images are obtained as cuts of two-dimensional pictures. Appropriately, this system is where everything is claimed and referred to as Impedance tomography. The imaging procedure is safe and informative when separated from MRI. Because EIT relies on ap-using negligible current without the use of radiations, it is regarded as the safest method with no side effects. EIT is utilized in the screening of chest problems, in lung ventilation inside medical care units, and in cardiovascular, mental, and cervical issues. EIT gauges the electrical conductivity; however, it can't assess conductivity inside normal tissues. Therefore, to overcome these hindrances, MRI is used alongside EIT. The go-to imaging framework is resonance EIT. This combines the usage of both an electrical field and a magnetic field, which provides extremely high-resolution images. The resulting two-dimensional pictures are often converted to three-dimensional pictures by increased calculations. Resonance electrical impedance is intelligently significant and has more expansive applications, when stood aside from EIT (Figure 16.10).

16.3.2 Optical Coherence Tomography (OCT)

OCT is an imaging technology that uses back-reflecting and dissipating long-frequency light beams to enter deeper into the diagnostic area, disperses the light beams, and produces a typical tomogram with greater objectives. The photographs are obtained in the micrometer range with the assistance of low-coherence interferometry. It's normally utilized in diagnosing early location of the retina in ophthalmology. OCT is additionally able to analyze plaques in coronary vessels, in

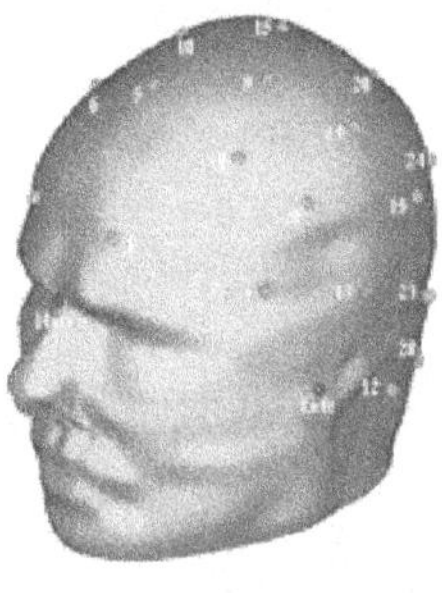
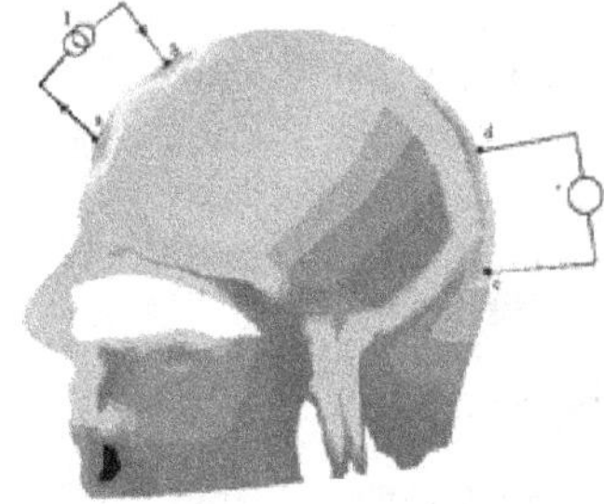

FIGURE 16.10 (Source: blogs.rs.org)

dermatology, and in dentology for root canal treatment. The principal points of interest of OCT include direct imaging with no radiation, usability, and practicality.. OCT is often utilized in conjunction with other analytic imaging methods since it does not emit radiation and is cost-effective. The reconciliation of OCT with endoscopic and catheter units is one such paradigm. The power of OCT to enter the body in a fiberoptic–based approach is the primary use of OCT in these approaches. Due to the method's widespread use, it's being examined in conjunction with other predictive techniques for disease detection and prevention.

16.3.3 Photoacoustic/Thermoacoustic Imaging

Photoacoustic technique depends upon the quality of photoacoustic influence. Photoacoustic imaging utilizes a laser beat that goes through the standard model, holds the essentialness and changes over to warm, accomplishing season of ultrasound waves. These waves are obtained by transducers and are converted to pictures. Thermoacoustic imaging features a relative norm of photoacoustic imaging yet utilizes recurrent radio waves. Photoacoustic imaging is extensively used in the identification of brain wounds, the assessment of haemoglobin assortment, and the screening of chest ailment. Finally, a combination of thermo and photo acoustic imaging is widely used in the conclusion of chest dangerous development. Since it doesn't utilize radiation and won't affect typical cells. there's an increasing tendency to utilize this method in the discovery of tumors.

16.3.4 Microwave Imaging

Microwave imaging may be increasing in popularity as an imaging framework, particularly in the treatment of chest infections. Separately from photoacoustic imaging and ultrasound imaging, microwave imaging has a promising future in chest diseases. Microwave imaging relies on the water content in tissue. Due to angiogenesis, suspect cells for the most part have a high water content, such that they differ from normal cells and are better separated in microwave imaging. A high electrical separation develops between affected cells and ordinary cells in microwave imaging, making this consistently sensitive for confirmation of disease. Microwave imaging is additionally applied in osteoporosis, where it detects the architecture of mineral thickness of bone. Because this imaging strategy is cost-effective, easy to use, and safe, analysts are showing great interest in adapting this procedure for other locations of malignant growth and in the analysis of various infections.

16.3.5 Magnetic Resonance Elastography (MRE)

MRE is a developing system that utilizes the magnetic resonance approach together with elastography. MRE utilizes flexible floods of 10–1000 Hz together with stage differentiate MRI. The adjustments in the wave designs and the estimations of shear modulus are determined from versatile solidness, and from these qualities the outcomes are gotten as pictures. MRE is ordinarily utilized for hepatic fibrosis. The utilization of MRE as a standard imaging methodology is still at its outset.

16.4 REQUIREMENT FOR SEVERAL IMAGING MODALITIES

A solitary imaging structure can give the position of a selected tissue or an organ. As an example, a CT clear can show the closeness of stone or an irregularity of urinary plot, yet frameworks like MRI, PET and ultrasound are utilized for the same clarification. Further, to see kidney supply courses (if there should be an event clearly stenosis) Magnetic reverberation angiogram is employed, anyway to get an annoy, tumor or a pollution CT is employed. Avoiding cystourethrogram is employed to track down the causes of urinary problems in children and abnormal bladder position in women. These techniques are utilized to picture the region inside the urinary system. Despite the existence of several imaging modalities, considering the patient's symptoms, signs, needs, and the quality and goals of the photo, a selected imaging mode is chosen. As an example, sonography provides a report on 53–77% of abnormalities; however, MRI could even more speedily distinguish 91% of dangerous tumors, and PET might be utilized for its high sensitivity in the tumor region. Every procedure has its particular strengths in looking for a specific disease. They also have their own drawbacks; MRI could show the circulatory framework and improvement of neural associations, whereas CT can't. Consequently, optimal methods have been developed to coordinate the expected increases of every imaging strategy and to present the precise diagnosis of the investigated disease. As an example, CT-coupled MRI gives reliable information on both bone and soft tissues close to/around the bone. This assists the specialists in identification of the cause of symptoms in a specific region..

16.5 PICTURE QUALITY, IMAGE PROCESSING, AND VISUALIZATION OF IMAGES

Photographs obtained from different imaging modalities clear the path for specialists to plan an operation or radiotherapy, or to ascertain any risks or different dysfunctions. Picture quality, destinations, and understanding of the outcomes from an image are obviously basic, as minute differences in assessment could change the treatment to be refined for the patient. Careful conclusion and treatment depend, as it were, on the character of the image, because this assumes a big role in understanding biomedical pictures. Consequently, considering the importance of the image, the entire procedure, from getting information/signal from the item to securing the last pictures from the imaging methodology, is examined here.

16.6 PARTS OF IMAGE PROCESSING SYSTEM

The whole procedure, from obtaining a signal from the human body to the final prepared picture as shown on the screen, altogether establishes the fundamental parts of the picture handling framework. Figure 16.11 shows a flow chart of the picture preparation framework.

- *Picture sensors:* A sensor, such as a recorder, camera, or scanner, captures images. A digitizer converts the captured image to a more sophisticated structure (simple-to-computerized converter).
- *Programming handling:* Specialized picture preparing equipment comprises the digitizer and equipment. The fundamental motivation behind the equipment is to accelerate the procedure.
- *Picture show:* From a simple image to a super image can be used as an image processing framework, depending on the necessity.
- *Mass stockpiling:* Images procured from every patient must be retained in a capacity framework. In spite of the fact that the majority of the pictures are

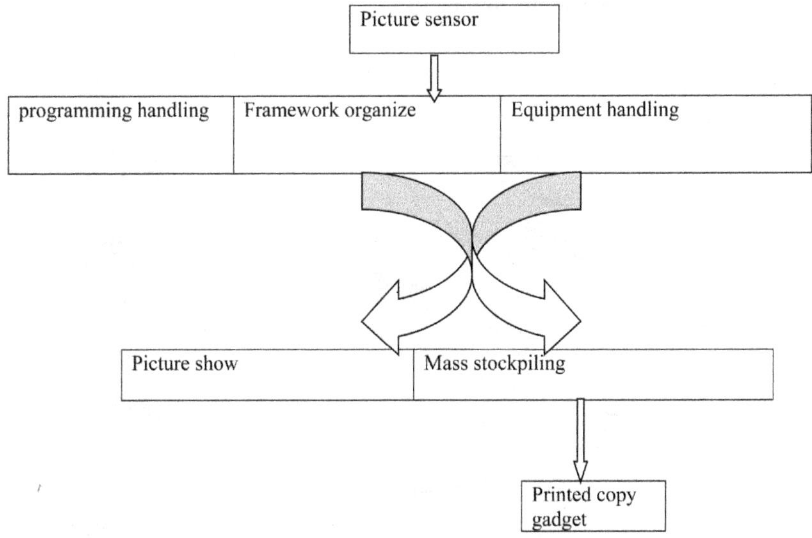

FIGURE 16.11 Segments of picture handling framework.

compressed before storage, enough space is required to store a great many pictures, probably in the thousands. Consequently, satisfactory capacity is required.

- *Printed copy:* Mainly utilized for recording the pictures.
- *Image analysis software:* Image analysis software is commercially available and performs necessary tasks in the preparation framework.
- *Image view:* A device that increases the yield of a picture and its data, such as a screen or a television.

16.6.1 Picture Processing

The picture acquired from the checking equipment may typically have some geometric distortion, obscurity, noise, and so on due to the sensitivity of the equipment. Consequently, once the pictures are recovered from the sensor, they are transferred into the PC. The picture handling programming changes over the first sweep into a more effective picture for the sake of improvement representation. A flowchart of the picture handling steps is depicted in Figure 16.12. performed utilizing framework systems.

1. *Picture obtaining:* Images are gained from camera recordings or from the scanner of the imaging methodology.
2. *Picture reclamation:* Restoration is fundamentally considered to recover the nature of a computerized picture that has been corrupted during advanced picture development,which normally originates from obscure movement or because of instrumental noise or some misfocus of the sensor. Different strategies, for example, backwards channeling or Wiener separating, are utilized to reestablish the first pictures.

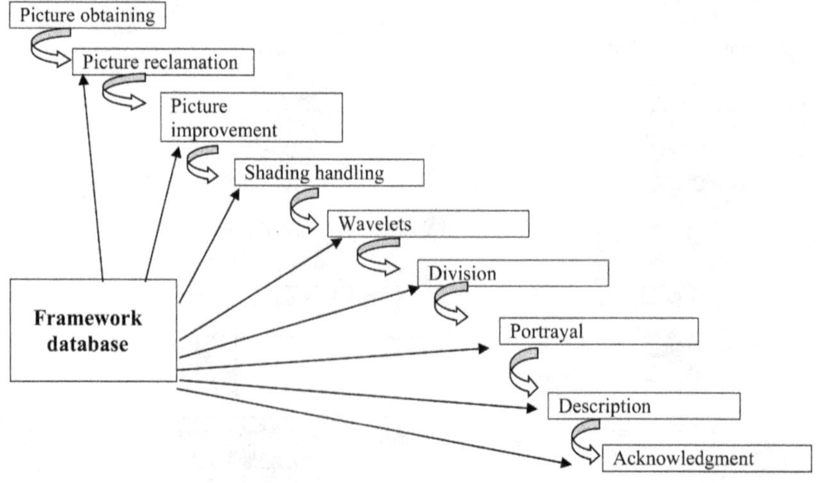

FIGURE 16.12 Steps in picture handling.

16.6.2 Picture Improvement

Enhancement is improving the appearance of the picture. Picture honing, uproar noise reduction, greyscale level management, fading picture brightness, and distinction are all ways of improving the picture's nature. Numerical ideas like Fourier changes and calculations for half toning and zooming are applied to improve the presentation of the picture.

16.6.3 Shading Handling

The term "enhancement" refers to a picture's attractiveness being improved. Picture honing, noise reduction, greyscale management, fading picture brightness, and distinction are some of the tactics used to improve the picture's nature. To increase the picture's presence, mathematical concepts such as Fourier changes and calculations for half toning and zooming are used..

16.6.4 Wavelets

The majority of the images received are of a significant size, necessitating a large amount of additional space. Following that, the images are compressed by restricting their size without compromising the image's quality. Wavelet change is a popular method for compressing images and recovering weak signals from noise.

16.6.5 Division

The partitioning of pictures into different portions or locales to represent the significant territories from the entire picture is known as division. Division calculations, for example, bunching calculations, line and bend identifiers, and areas developing are utilized for this reason. Figure 16.9a represents the first picture and Figure 16.9b shows the fragmented area indicating the site of injury.

16.6.6 Portrayal

After division, the different portions of pixels are spoken to for additional preparing. It might be limit portrayal which center around the outside zones portrayal which center around inward properties.

16.6.7 Description

Describing the locale dependent on the portrayal reasonable for additional handling. This features the quantitative data and the highlights of interest.

16.6.8 Acknowledge

Perceiving solitary characters or checking using estimations. Every one of the means are directed by structure information base which accepts work in planning and association of every module.

16.7 RADIATION EXPOSURE AND RADIATION PROTECTION IN MEDICAL IMAGING

The use of radiation in the clinical sector has been significantly expanded because of advancements in CT, MRI, and other diagnostic methods. In 2006, 62 million CT scans were performed, compared with 20 million in 1995 (Brenner and Hall, 2007), and the number of overlays has increased to two. These radiation sources have the potential to have a significant impact on humans. There are several estimates that may be evaluated for adequate protection of the diagnostic room and the anatomical zone of the patient. This avoids the presence of radiation in different parts of the room and on different parts of the patient's body. Clinicians/radiologists and others in the space use radiation monitors to monitor radiation levels. Improved re-creation calculations and picture handling are generally under research to limit the immediate exposure to radiation from medical devices.

16.8 GENERAL APPLICATIONS OF MEDICAL IMAGING: IMAGING TOWARDS DISEASES

16.8.1 Alzheimer's Disease (AD)

AD can be detected by MRI with statistical methods, which encourages both specialists and patients in treatment and avoidance. Specialists could utilize MRI as a potential biomarker to analyze the phases of illness and its development. This allows the expert to increase or reduce the drug levels and change the combination of prescriptions to be given for each stage of AD. Patients might, of course, see the inner workings of their own bodies, such as wounds or strokes, on MRI images. This necessitates attention to changes in patients' lifestyles, such as smoking, stress, or depression, in order to fully anticipate AD. About 10 percent diminishing change in lifestyle of AD patients. Consequently, the use of MRI to confirm AD could basically benefit the patients.

16.8.2 Malignant Growth

Malignant growth is the chief developing disease,causing a large number of deaths every year. The disease consists of abnormal changes to normal cells, prompting the development of a harmful tumor. These malignant cells can attack different parts of the body, making it a dangerous disease. Liver malignancy is one of the main tumors, causing in excess of six lakhs deaths globally. Tumor/disease cells have a pressurized domain that can dispose of foreign materials,which might be medications or contrast agents. As the contrast agents are quickly inactivated, malignant growth imaging becomes difficult.

16.8.3 Cardiovascular Diseases

SPECT imaging is the most widely recognized method in the determination of cardiac-related issues. SPECT imaging uses a higher level of radiotracer to take pictures. The significant disadvantages of this strategy are the low-quality pictures and more exposure to radiation. Because existing imaging techniques can't penetrate further into tissues to represent three-dimensional plaque creation, it's difficult to completely separatc the three-dimensional architecture of arterialplaques. Wang and his collaborators created a novel imaging technique that allows them to observe the 3-D pattern of plaques.The central idea behind their work is to anticipate the specific engineered safeguards that will be needed when exploring a new area. This framework can enable definite identification of plaque by locating the carbon-hydrogen linkages that shape lipids in vessels with plaque.

16.8.4 Neonatal Abstinence Disorder (NAD)

NAD comprises the side effectsdue to pregnant women drinking alcohol. The disorder is characterized by heart issues, development hindrance, and different variations from the norm in the developmental stages of the baby.. These indications, which are because of the utilization of alcohol, were demonstrated in a model organism,quail. Infusing a known amount of alcohol in developing quail resulted inadverse effects. Imaging the essential living structures and blood stream of the heart using optical soundness tomography revealed amazing results in incubation and blood flow. Another investigation utilizing MRI to foresee the impacts of liquor on NAD uncovered the result of alcohol utilization by pregnant women in a group of children whose mothers hadconsumed alcohol during their fetal development. From the MRI imaging, it was seen that there was a significantreduction, compared with the control group, in the corpus callosum of the brain, which was legitimately identified with mental issues in children. Various difficulties in NAD were emphasized by various specialists from the same radiological society, and the repercussions in the form of neurological derangements were described. Tissue alterations can be seen using diffusion weighted imaging (DWI), which monitors water scattering. Furthermore, proton (hydrogen) magnetic resonance spectroscopy (HMRS) can provide specific subtleties of metabolic alterations to the brain.

16.8.5 Imaging in Drug Development

Medication research and development is a lengthy procedure to which the world contributes billions of dollars. From a large number of medicines evaluated in clinical trials, it takes 15 to 20 years to acquire one Food and Drug Administration (FDA)-approved sedative.

16.8.6 Imaging in Medical Device Manufacturing

Businesses fabricating clinical devices carefully follow the guidelines of the FDA, Nonetheless, mistakes happen during the advancement of clinical devices, for example, heart stents. Manual examination is clearly inadequate. Maybe, imaging methods, for instance, CT, can help in visualizing the three-dimensional construction of each fine space of the clinical device. Further, picture assessment programming, when combined with these methodologies, could give an impression of breaks in construction of the material. High-quality picture examination programming enables the detection of mistakes in the assembly process preceding manual review, saving the time and cost of the assembly procedure.

Imaging procedures have a role not just in reviewing devices during assembly- but also in planning clinical devices for patients. Siemens provides an "image to embed included technology," in which therapeutic equipment are to be designed in a computerised fashion based on the patient's life systems, with item lifecycle executives virtual products. (Siemens, 2015). The development of such customized clinical devices will help in increasingly effective treatment.

Furthermore, upgrading the adjusted movement of clinical imaging for specific patients aids in the faster discovery and development of new drug frameworks. Clinical imaging devices can be installed in washrooms so that changes in urine content can warn patients; for example, pink- or red-tinted urine can indicate hemolysis, which is indirectly related to the progression of cancer. Rather than going to a specialist to perform various tests, such as ultrasound, such clinical devices connected to the patient's home can demonstrate the seriousness of the disease, prompting earlier treatment. This flexible and informative program may be used with any mammography imaging modalities. Regardless, this application should be carefully conducted on iPads in accordance with FDA guidelines.

16.9 CONCLUSION

Early detection of critical illnesses such as malignant development might save a lot of people's lives and their families' happiness. This necessitates breakthroughs and fresh discoveries not just in imaging methods but also in all other therapeutic disciplines. The most important criterion is the proper use of all imaging methods that are beneficial to human civilization.

16.9.1 Future Aspects of Medical Imaging

In spite of the fact that multimodality imaging has brought about better detection, new difficulties have emerged. There is continuous research intohow to conquer the emerging difficulties by (1) incorporating analytic imaging with molecular procedures and (2) developing new strategies further advanced than the current ones. The use of radiogenomic imaging might be one of these occasions. Subatomic profiling of natural tissues is combined with subatomic techniques in radiogenomic imaging. For tumor recognition and related assessments, radiogenomic imaging is now being used. Characteristics are selected as biomarkers that may be determined utilizing radiation oncology. There were 353 chest infection patients tested for their quality enunciation thinking as well as MRI data to isolate the connection between quality verbalization and imaging data. The findings connected 12 imaging aspects to lung infection characteristics and 11 attributes to prostate damage characteristics (Yamamoto et al., 2012). Such an assessment energises the risky development of understanding the sub-nuclear science's secret basic underpinnings..

Techniques have been devised to further increase the adequacy of imaging techniques by further developing contrast agents. F-18 Florbetapir (AmyvidTM) is another tracer approved by the FDA for detecting beta amyloid plaque in the frontal brain of AD patients. This tracer enables PET to distinguish AD at an earlier stage, which is necessary on a regular basis (Abraham, 2013).; Philips overcame biomedical instrument restrictions. Philips released the Vereos PET/CT, a completely updated PET device, in 2013.. The sensitivity increment, volumetric goals, and quantitative precision looked different in comparison to simple structures while using photon counting (Nabeel, 2013).A few processes and estimations are also being produced to further improve the image aims, signal to noise ratio, and acquisition time. Expansion, repetition region, regularization, and learning-based methods are used to create super-objective (SR) frameworks (Thapa et al., 2014).

REFERENCES

Abraham, B. (2013). Imaging's role in the future of Alzheimer's disease. *Electroindustry*, 18(3), 11.

Bihan, D. L., Breton, E., Lallemand, D., Grenier, P., &Cabanis, E. (1986). MR imaging of intravoxel incoherent motions: Application to diffusion and perfusion in neurologic disorders. *Radiology*, 161(2), 401–407. doi:10.1148/radiology.161.2.3763909 PMID:3763909.

Brenner, D. J., &Hall, E. J. (2007). Computed tomography: An increasing source of radiation exposure. *The New England Journal of Medicine*, 357(22), 2277–2284. doi:10.1056/NEJMra072149 PMID:18046031.

Jingu, A., Fukuda, J., Takahashi, A. T., &Tsushima, Y. (2014). Breakthrough reactions of iodinated and gadolinium contrast media after oral steroid premedication protocol. *BMC Medical Imaging*, 14(1), 34. doi:10.1186/1471-2342-14-34 PMID:25287952.

Joshua, E. (2012). Iron oxide nanoparticles for targeted cancer imaging and diagnostics. *Nanomedicine; Nanotechnology, Biology, and Medicine*, 8(3), 275–290. doi:10.1016/j.nano.2011.08.017 PMID:21930108.

Nabeel, U. A. (2013). New innovations in medical imaging technology: Live from RSNA. Retrieved December 18, 2013, from http://in-training.org/new-innovations-in-medical-imaging-technology-live- from-rsna-3961

Sakas, G. (2002). Trends in medical imaging: from 2D to 3D. *Computers & Graphics*, 26(4), 577–587. doi:10.1016/S0097-8493(02)00103-6

Thapa, D., Raahemifar, K., Bobier, W. R., &Lakshminarayanan, V. (2014). Comparison of super resolution algorithms applied to retinal images. *Journal of Biomedical Optics*, 19(5), 056002. doi:10.1117/1. JBO.19.5.056002 PMID:24788371

Wang, T., McElroy, A., Halaney, D., Vela, D., Fung, E., &Hossain, S. et al. (2015). Detection of plaque structure and composition using OCT combined with two-photon luminescence (TPL) imaging. *Lasers in Surgery and Medicine*, 1–10. doi:10.1002/lsm.22366

Yamamoto, S., Maki, D., Korn, R. L., &Kuo, M. D. (2012). Radiogenomic analysis of breast cancer using MRI: A preliminary study to define the landscape. *American Journal of Roentgenology*, 199(3), 654–663. doi:10.2214/AJR.11.7824 PMID:22915408.

Ziegler, S. I. (2005). Positron emission tomography: Principles, technology, and recent developments. *Nuclear Physics. A.*, 752, 679c–687c. doi:10.1016/j.nuclphysa.2005.02.067

17 Application of Image Processing in Plant Leaf Disease Detection

Bhavana Nerkar

CONTENTS

17.1 INTRODUCTION

Image processing for leaf disease detection has been an optimum choice for researchers due to the following advantages [1]:

- It is a contactless system, which ensures that the yield is not tampered with during the evaluation process.
- Processing of the leaf is possible from infinite angles, so any kind of leaf image defect can be evaluated.
- Due to innovations in imaging and classification techniques, the accuracy of these systems will always be improving.

The major components for any leaf-based image classification system are depicted in Figure 17.1. These components include [2–4]:

- Image acquisition devices
- Image pre-processing components
- Segmentation and region of interest evaluation blocks
- Feature extraction and selection blocks
- Training and evaluation blocks
- Post-processing blocks

Each of these blocks is implemented using unique algorithms, which work collectively towards improving the overall accuracy of the system. This is due to the fact

DOI: 10.1201/9781003221333-17

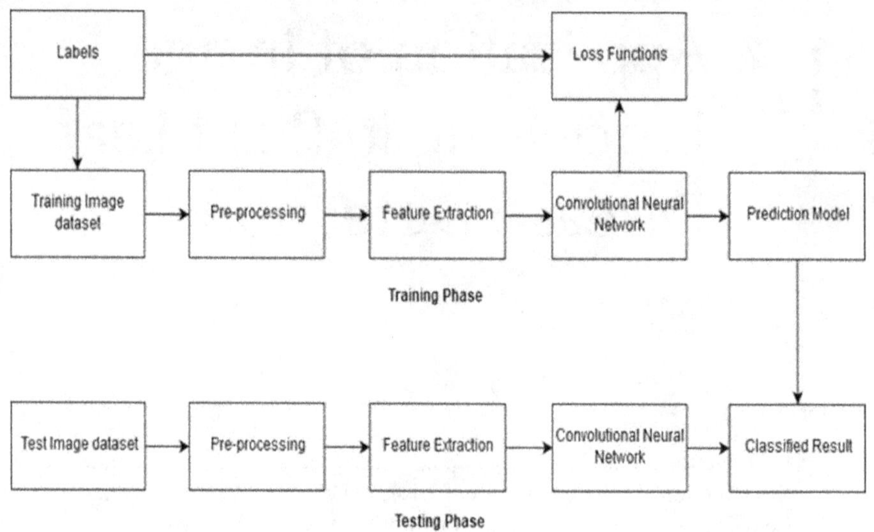

FIGURE 17.1 General leaf image classification system.

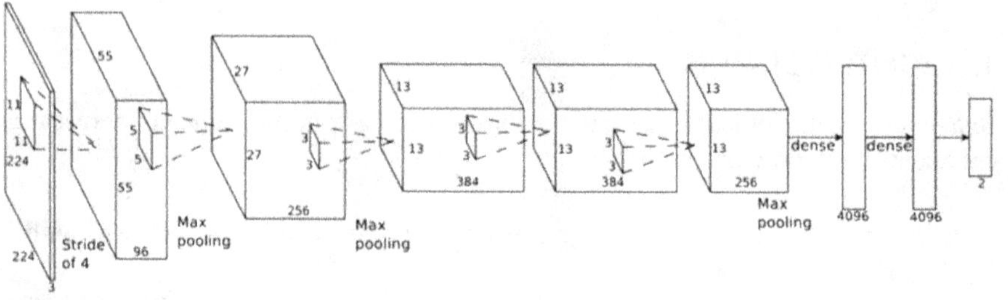

FIGURE 17.2 Architecture of AlexNet CNN.

that image acquisition, pre-processing, and segmentation algorithms have now been standardized for any kind of image processing system. For the case of leaf disease detection systems, acquisition is done using high-end cameras, pre-processing is done using filters such as adaptive median filter, Gaussian filter, and others, while segmentation is done using hierarchical clustering. Most of the work reviewed in this chapter uses these algorithms in one form or another for effective pre-classification [5–7]. Convolutional neural networks (CNNs) have found their way into highly accurate image classification systems over the past decade. Architectures like GoogLenet, AlexNet, VGGNet, and others have been proposed in order to improve the efficiency of classification. Figure 17.2 showcases the architecture for AlexNet CNN.

Using the given architecture, any kind of image can be classified into various classes. The only requirements of this (or any) CNN are [5–9]:

- The dataset must be diverse and should contain a large number of images from different categories.
- These images must be different from each other in terms of angles, colours, depth of capture, and any other possible capturing parameter.
- A high-speed and high-capacity processing unit is needed to train the CNN.

Once the given criterion is fulfilled, then the CNN training can be started. The training of AlexNet CNN follows the given steps [10–12]:

a) All input images are resized to 224 × 224 dimension.
b) Each image is further divided into overlapping strides of 4 × 4.
c) A max-pooling layer [13] is connected to these strides, which converts the image into 96 × 96 blocks, each of size 5 × 5.
d) These blocks are given to another max-pooling layer for dimensionality reduction [14].
e) This layer divides the blocks into blocks of size 3 × 3 and is followed by two more max-pooling layers.
f) These layers are followed by two dense layers to further evaluate image features.
g) Finally, a flat layer is connected to classify the input features into 1 of *N* classes.

Using the given architecture, leaf images can be classified into 1 of *N* classes with very high accuracy. The next section describes various CNN-inspired algorithms for leaf and other agriculture-based classification systems, followed by the discussion about these algorithms in statistical terms for leaf disease classification. This chapter concludes with some interesting observations about the architectures discussed and recommends methods to improve them.

17.2 CONTRIBUTIONS IN THE FIELD OF LEAF DISEASE DETECTION

Exploration in leaf disease recognition utilizing picture handling has moved from general classifiers like k-nearest neighbours (kNN), uphold vector machines (SVM), and so forth to more intricate and computation-intensive calculations like convolutional neural organizations, Q-learning, and so on, like the work done in Mohanty et al. [15], wherein scientists utilized convolutional neural organizations to increase the testing set precision from under 80% to over 98%. They utilized the AlexNet and GoogLeNet-based CNN models to achieve this.. The work proposed by Mohanty et al. [15] is restricted to soybean leaf diseases; however, it can be extended to any sort of leaf. In contrast to Mohanty et al. [15], the work done by Sukhveerkaur [16] utilizes neighbourhood double examples and combines this with SVM for grouping. The researcher assessed bacterial blight, Cercospora leaf spot, powdery mildew, and rust infections in leaves. A preparation set precision of over 98% is claimed, but there is no information on the precision of the test set, which is commonly 8% to 15%, not exactly the preparation set precision for direct classifiers like SVM; therefore, the precision [17] can be considered to be somewhat above 90% (probably). This must be reassessed by scientists so as to check the general framework execution.

The work by Rasel Mia and Roy [18] once again introduces another kind of neural organization, named fluffy ARTMAP neural organization. They utilized a dark level co-event lattice (grey level co-occurrence matrix [GLCM]) and joined it with the proposed neural organization-based classifier so as to get improved precision. Their preparation set precision is over 95%, while the test set precision is around 90%. Returning to CNNs, the work of Saleem and Potgieter [19] takes a shot at maize leaf infections from the PlantVillage dataset. Their design consolidates the chief segment investigation (principal component analysis [PCA]) with the LeNet CNN engineering to get a precision of 97%, which is a tremendous improvement over other organization engineering. This can be utilized as a true norm for handling leaf pictures and can be additionally refined. Another model of CNN that utilizes LeNet is depicted in Priyadharshini and Sivakami [20], wherein scientists utilized comparative handling layers like [20] and got 91% testing set precision for tomato leaf diseases. This supports our choice of CNNs as the organization of decision for our examination. CNN has distinctive organization structures, which must be assessed prior to choosing the one appropriate for our application. For example, the work of Dhingra and Kumar [21] utilizes the ResNet design for recognizing soybean leaf infections. Because of the use of ResNet, the general framework precision improves to 94%. It was tested across 27 diverse CNN models. The proposed ResNet model is discovered to be superior to GoogLeNet and AlexNet models when applied to leaf disease identification.

Easier CNN models that utilize different softmax layers likewise outflank complex non-CNN calculations like neighbourhood twofold examples histograms (local binary pattern histogram [LBPH]) and Haar-WT (wavelet transform) regarding substantial element extraction as observed in Dhingra and Kumar [21], wherein various softmax layers in CNN are utilized. The work suggests that the testing precision of arrangement is over 95%, while strategies like SVM and local binary pattern (LBP)-based techniques can only achieve up to 85%. However, complex models like AlexNet, GoogLeNet, and DenseNet201 have their own particular points of interest for leaf disease discovery. The work in [22] checked this point by applying the AlexNet, GoogLeNet, and DenseNet201 models for apple leaf disease discovery. The calculation with these joined structures can accomplish a precision of 99% on a moderate measured dataset. However, the delay in preparation is enormous; according to our research, building such a mind-boggling model should take at least four days. Such high precision can likewise be accomplished with legitimate element determination and utilizing a basic level layered neural organization. This was done by Boa Sorte et al. [9], wherein measurable credits and nearby double examples were separated and given to a neural organization classifier. This work reduced the preparation time to approximately 60 minutes while getting a high precision of 98%. Less complex elective methodologies of this kind must be utilized so as to improve the framework's productivity, instead of utilizing multi-layer CNNs that burn through numerous computational cycles, and result in a high precision for a respectably estimated dataset. Clever methodologies like the one utilized by Jiang et al. [10] can be considered for additionally improving the framework's productivity. The analysts utilized GoogLeNet Inception structure and Rainbow link for planning their CNN. Utilizing these layers, the general precision of grouping for apple leaf diseases has increased to 97% with an enormous information size of more than 1,000,000 records. Choosing CNNs as the decision of exploration is obvious when the work in [18 is audited. SVM and other direct strategies are restricted by their component handling layer; this is where CNN structures become possibly the most important factor. A comparative audit is done in [19], wherein it is likewise referenced that CNNs are the decision of choice for any sort of leaf disease characterization application. This is additionally checked by [19], wherein an ongoing framework is created utilizing CNN for arrangement of plant leaf diseases, and a precision of over 95% is accomplished. Bio-propelled strategies are another arrangement of calculations, which can additionally improve the precision of CNNs. This can be seen from [20], wherein an audit of different bio-propelled techniques is done so that the best algorithm(s) can be distinguished. It is seen that genetic algorithm is the best-performing calculation for discovering the best highlights from leaf pictures. We have likewise utilized a genetic algorithm-motivated strategy to upgrade the presentation of our CNN. An investigation on the recognition of various leaf diseases utilizing CNN is proposed in [20], which additionally demonstrates that CNN-based classifiers can accomplish over 95% precision for order.

17.3 LEAF DISEASE DETECTION USING CONVOLUTIONAL NEURAL NETWORKS

A deep net architecture inspired by the VGGNet CNN model is discussed in this section. It uses the complex feature extraction and selection capabilities of the VGGNet model and combines them with the efficiency of feature extraction techniques like GLCM, edge maps, and colour maps. The architecture diagram of the proposed deep net classifier can be observed in Figure 17.3. The discussed architecture has the following blocks:

- Image segmentation using hierarchical k-Means clustering
- Feature extraction using GLCM, colour maps, and edge maps
- Concatenation layer to combine these features as pixel levels and append them into the image itself
- VGGNet model for training and evaluation.

The segmentation layer uses hierarchical k-Means clustering, which works using the following steps:

a) Form two clusters from input image using k-Means.
b) Find position and count of green pixels in the image.
c) Remove cluster with lower number of green pixels, and use the one with higher number of green pixels as output segmented image.

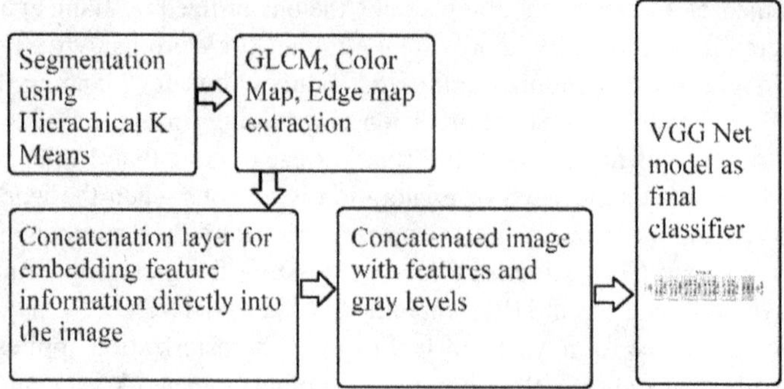

FIGURE 17.3 Overall block diagram.

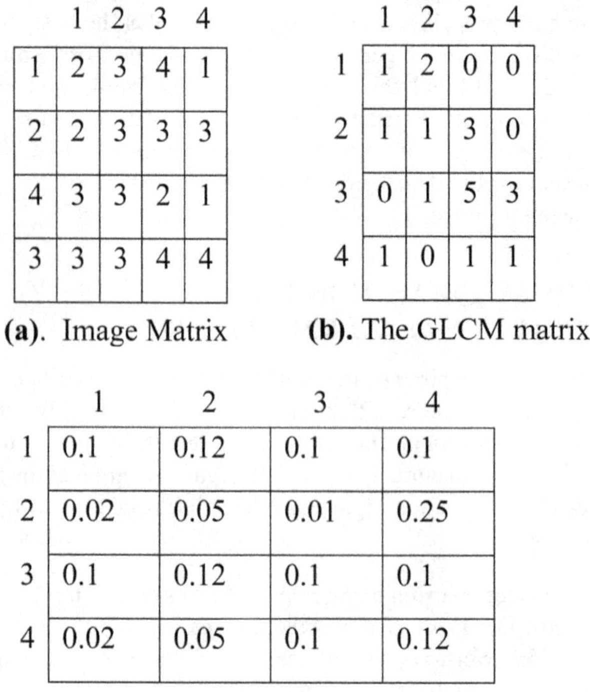

	1	2	3	4
1	2	3	4	1
2	2	3	3	3
4	3	3	2	1
3	3	3	4	4

(a). Image Matrix

	1	2	3	4
1	1	2	0	0
2	1	1	3	0
3	0	1	5	3
4	1	0	1	1

(b). The GLCM matrix

	1	2	3	4
1	0.1	0.12	0.1	0.1
2	0.02	0.05	0.01	0.25
3	0.1	0.12	0.1	0.1
4	0.02	0.05	0.1	0.12

(c). Normalized GLCM matrix

FIGURE 17.4 (a) Image matrix. (b) GLCM matrix. (c) Normalized GLCM matrix.

GLCM [3] is used to evaluate various features like energy, entropy, homogeneity, inertia, correlation, shade value, prominence, and variation, as suggested by Nage and Raut [13].

The input matrix for GLCM is shown in Figure 17.4a, and Figure 17.4b shows the results of GLCM. Finally, Figure 17.4c. showcases the resulting normalized GLCM feature vector array.

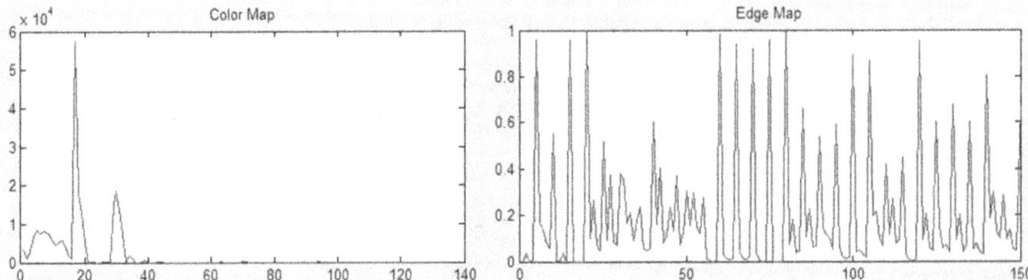

FIGURE 17.5 Colour map and shape map of the image under test.

As well as this, colour maps and edge maps are also estimated and can be observed in Figure 17.5.

All the three features combined together form a perfect descriptor for the leaf image. These features are fully capable of distinguishing between the different types of disease in the image. All these features are appended to the image itself on the band (the band is not relevant here; any band can be used, by keeping the other bands as zeros). Due to this append operation, these features are also considered during the stride-based feature extraction and selection of the VGGNet CNN.

Based on these features, a VGGNet based CNN model is used for final classification as shown in Figure 17.6. The architecture starts with resizing the images into 224 × 224 dimension followed by two convolutional layers of size 3 × 3 (stride). This helps to reduce the size of the image vector to 112 × 112 and is again followed by a series of convolutional layers and max-pooling layers that reduce the size of the image to 14 × 14 × 512. Finally, three fully connected layers are used, which allow the system to classify the leaf diseases into one of the following categories: Apple scab, Apple black decay, Apple sound, Cherry solid, Cherry fine buildup, Corn Cercospora dim spot, Corn normal rust, Corn solid, Corn northern leaf blight, Grape dark decay, Grape esca, Grape sound, Grape leaf curse, Orange haunglongbing, Peach bacterial spot, Peach solid, Pepper chime bacterial spot, Pepper ringer solid, Potato early blight, Potato solid, Potato late blight, Raspberry solid, Soybean sound, Squash fine mould, Strawberry sound, Strawberry leaf burn, Tomato bacterial spot, Tomato early scourge, Tomato solid, Tomato late curse, Tomato leaf mould, Tomato Septoria leaf spot, Tomato two-spotted insect parasite, Tomato target spot, Tomato mosaic infection, and Tomato yellow leaf curl virus. These images were taken from the dataset used in Turkoglu [8] because of its extensive testing. The results are evaluated in the next section.

17.4 RESULTS AND OBSERVATIONS

We compared the performance of the proposed model using custom collected datasets. In order to perform validation, a training set of 70% values, a testing set of 20% values, and a validation set of 10% values are used. Different kinds of images were used for this purpose. The details about these diseases, and the number of images used for each disease, can be seen in Table 17.1,

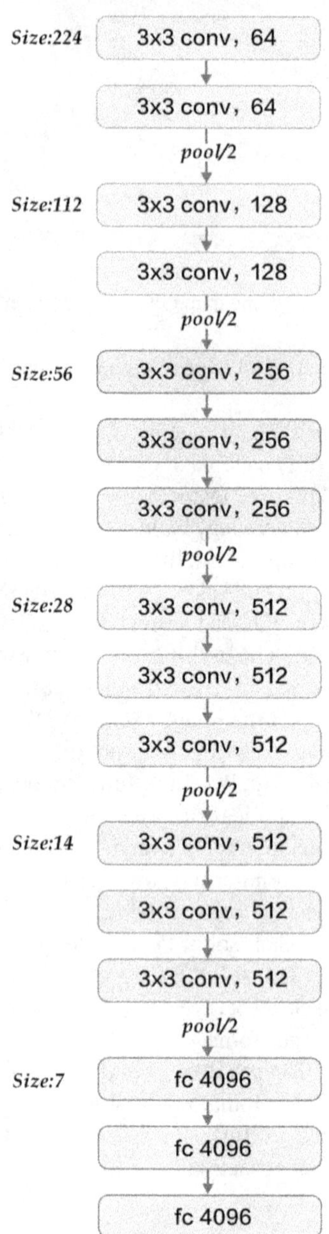

FIGURE 17.6 The VGGNet CNN architecture.

Each of these disease images is divided into the same 7:2:1 ratio for training, testing, and validation. The results on a combined dataset for different algorithms can be seen in Tables 17.2 and 17.3.

TABLE 17.1
Number of Images with Type of Diseases

SN	Disease Name	Information	Sample Image
1	Bacterial blight	It is one of the destructive diseases affecting cotton crops, having blackish colour of veins and spots. Number of images used for evaluation: 976	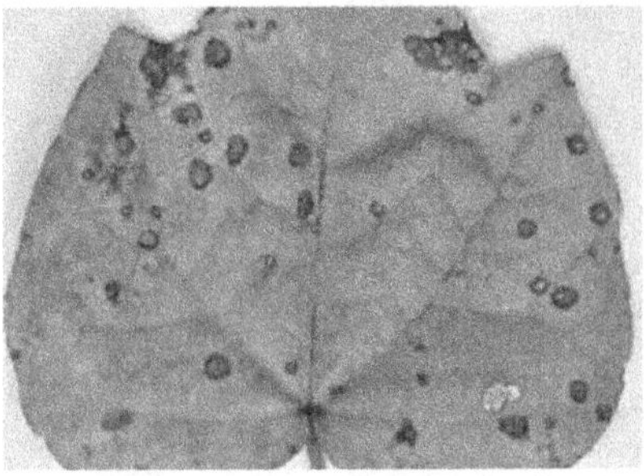
2	Reddening	It has occurred since cotton was introduced in India. Red pigment in the leaves generally. Number of images used for evaluation: 895	
3	Alternaria	Producing numerous brownish leaf spots. Number of images used for evaluation: 1157	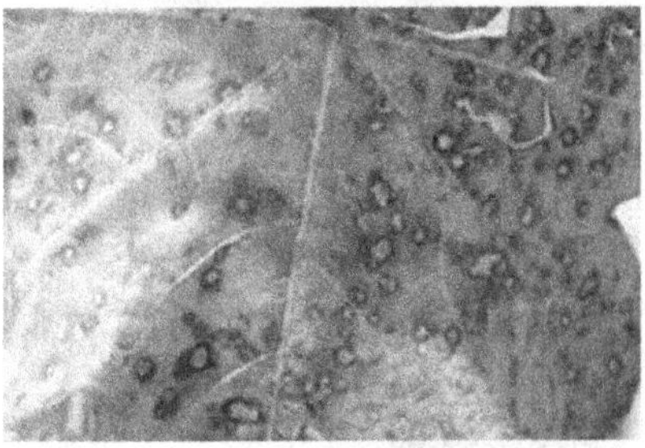
4	Grey mildew	White angular spots occur on both sides of the leaves. Number of images used for evaluation: 946	

TABLE 17.2
Accuracy of Different Classifiers

Img. Tested	Acc. (%) DWT + LBP [2]	Acc. (%) NN with Cpix [3]	Acc. (%) RCNN [4]	Acc. (%) Bio-CNN [7]	Acc. (%) Multi CNN [8]	Acc. (%) Proposed CNN
10	90.00	90.00	100.00	100.00	100.00	100.00
25	95.00	93.00	95.00	100.00	100.00	100.00
40	95.20	96.00	96.80	97.10	100.00	100.00
50	95.60	96.50	97.30	97.60	98.60	100.00
75	95.60	96.70	97.60	97.80	98.70	100.00
100	95.70	96.80	97.80	97.90	98.70	99.30
200	95.70	96.70	97.90	98.10	98.80	98.90
500	95.70	96.80	97.90	98.20	98.80	99.40

TABLE 17.3
Delay versus Number of Images Tested

Img. Tested	Delay (ms) DWT + LBP [2]	Delay (ms) NN with Cpix [3]	Delay (ms) RCNN [4]	Delay (ms) Bio-CNN [7]	Delay (ms) Multi CNN [8]	Delay (ms) Proposed CNN
10	2.36	6.90	8.42	5.52	11.60	1.24
25	2.87	7.60	9.52	6.25	13.12	1.51
40	3.69	7.90	10.54	6.91	14.52	1.94
50	5.98	12.90	17.16	11.26	23.65	3.15
75	5.99	18.20	21.99	14.43	30.31	3.15
100	6.98	35.70	38.80	25.46	53.47	3.67
200	7.59	72.60	72.90	47.84	100.47	3.99
500	8.60	125.20	121.64	79.82	167.63	4.53

The accuracy of the proposed CNN is observed to be greater than all the other models, thereby making it useful for large-scale, real-time deployments (Figure 17.7).

The delay of the proposed CNN is also lower than all the other compared models, making it better and faster [2–4, 7, 8] (Figures 17.8–17.10).

Results indicate that the accuracy of the proposed model is higher when compared with existing models; this accuracy can be further estimated on a per disease basis from Table 17.4, wherein it can be seen that the proposed model has over 98% accuracy in most cases.

The accuracy is averaged over a large set of values and then tabulated. The accuracy indicates massive improvement for each disease type, thereby making the algorithm applicable for a wide variety of applications. Similar values of precision are evaluated for each disease type and tabulated in Table 17.5.

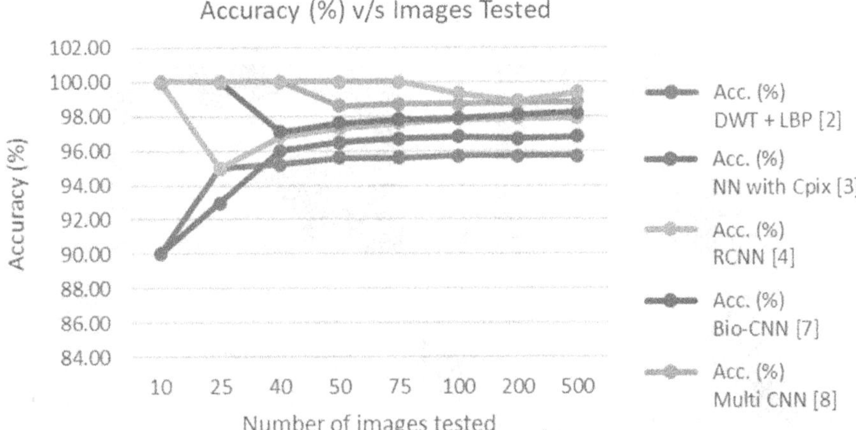

FIGURE 17.7 Accuracy of different algorithms.

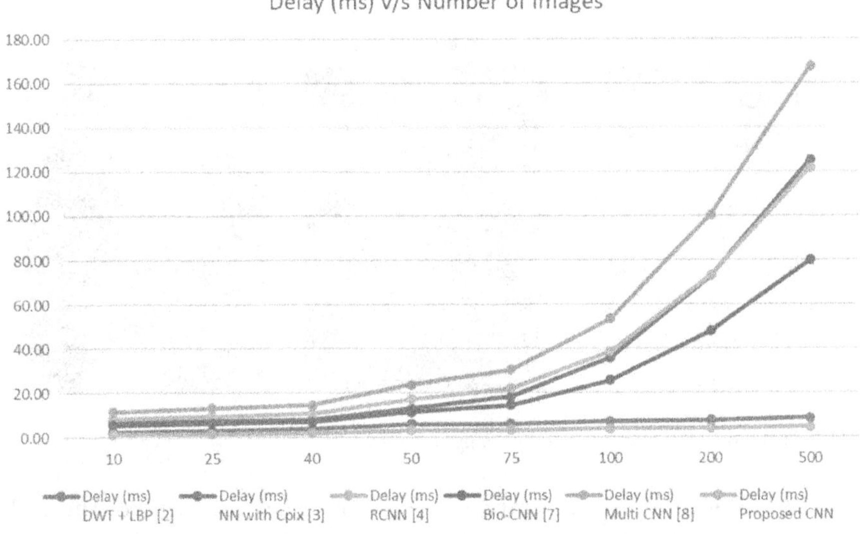

FIGURE 17.8 Delay performance of the proposed CNN.

The precision is averaged over a large set of values and then tabulated. The precision indicates massive improvement for each disease type, thereby making the algorithm applicable for a wide variety of applications. Similar values of recall are evaluated for each disease type and tabulated in Table 17.6.

This recall is averaged over a large set of values and then tabulated. The recall indicates massive improvement for each disease type, thereby making the algorithm applicable for a wide variety of applications. Based on this, values of fMeasure are evaluated for each disease type and tabulated in Table 17.7.

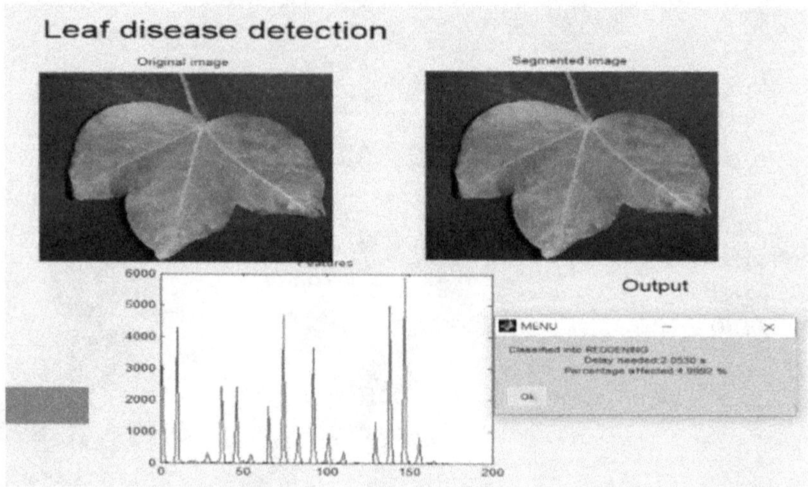

FIGURE 17.9 Disease image.

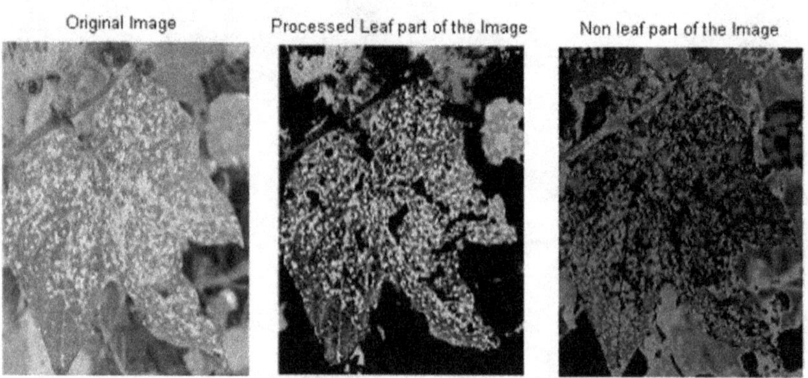

FIGURE 17.10 Disease image processing sequence.

This fMeasure is averaged over a large set of values and then tabulated. The fMeasure indicates massive improvement for each disease type, thereby making the algorithm applicable for a wide variety of applications.

17.5 CONCLUSION

The results indicate that the delay, accuracy, precision, and recall have been optimized by the proposed classifier. In order to showcase the improvement in both delay and accuracy, we can observe Table 17.4, wherein a comparison of mean delay and mean accuracy over different image frames is done, and it is seen that the delay is reduced by more than 40%, while the accuracy is improved by more than 5% relative to the most effective classifiers. These results encourage us to use the proposed

TABLE 17.4
Accuracy on Different Leaf Disease Types

Disease	Accuracy % [3]	Accuracy % [4]	Accuracy % [7]	Accuracy % [8]	Accuracy Proposed %
Apple scab	92.82	94.81	95.94	97.55	99.20
Apple black decay	92.85	92.85	94.67	97.16	99.91
Apple sound	90.96	91.20	95.24	95.63	99.78
Cherry solid	93.54	94.78	95.77	97.94	99.62
Cherry fine buildup	92.57	92.37	96.00	95.93	98.27
Corn Cercospora dim spot	91.20	94.00	93.78	95.57	98.48
Corn normal rust	93.24	91.61	93.79	95.87	98.38
Corn solid	93.26	94.21	94.46	95.51	99.33
Corn northern leaf blight	93.44	91.43	95.87	96.71	99.29
Grape dark decay	92.33	91.14	94.65	96.24	98.70
Grape esca	92.36	94.12	94.50	97.40	98.05
Grape sound	93.03	91.50	94.93	96.00	98.36
Grape leaf curse	90.04	91.55	94.87	97.82	98.42
Orange haunglongbing	90.27	91.39	94.18	95.90	98.94
Peach bacterial spot	91.24	91.09	93.97	96.38	99.00
Peach solid	92.23	93.18	94.69	97.40	98.73
Pepper chime bacterial spot	91.36	91.46	93.21	97.18	99.17
Pepper ringer solid	91.54	92.25	94.30	97.97	98.33
Potato early scourge	92.46	94.46	96.39	97.74	98.86
Potato solid	92.19	91.08	96.22	95.13	99.27
Potato late curse	92.66	93.31	96.14	95.90	99.40
Raspberry solid	93.24	94.27	94.80	95.87	99.26
Soybean sound	92.36	91.60	93.15	97.20	98.44
Squash fine mould	90.92	93.68	96.10	97.07	99.52
Strawberry sound	91.95	93.59	94.92	97.73	98.30
Strawberry leaf burn	90.36	93.05	93.40	97.55	98.98
Tomato bacterial spot	90.31	91.86	93.94	95.34	99.00
Tomato early scourge	92.81	92.65	94.28	96.64	98.30
Tomato solid	91.53	94.34	94.38	96.61	99.40
Tomato late curse	90.66	94.66	96.93	96.63	98.49
Tomato leaf mould	93.58	94.96	96.51	97.59	99.59
Tomato Septoria leaf spot	91.02	94.53	95.06	96.13	98.78
Tomato two-spotted insect parasite	91.02	92.00	95.52	97.11	99.36
Tomato target spot	93.37	92.62	93.15	97.23	99.33
Tomato mosaic infection	92.21	91.45	95.68	95.05	99.16
Tomato yellow leaf curl virus	93.39	92.78	96.17	96.54	99.74

TABLE 17.5
Precision on Different Leaf Disease Types

Disease	Precision [3]	Precision [4]	Precision [7]	Precision [8]	Precision Proposed
Apple scab	85.94	89.12	88.68	91.31	94.09
Apple black decay	87.71	88.73	90.77	92.61	94.85
Apple sound	87.38	87.91	88.65	90.55	93.19
Cherry solid	87.80	87.67	90.12	92.07	94.84
Cherry fine buildup	86.73	88.63	91.05	90.86	93.42
Corn Cercospora dim spot	85.82	87.38	91.76	91.86	94.06
Corn normal rust	85.91	87.84	89.29	92.05	94.25
Corn solid	88.05	88.79	89.19	91.87	94.54
Corn northern leaf blight	86.72	87.51	88.73	90.44	94.51
Grape dark decay	88.94	87.50	91.74	90.31	93.88
Grape esca	87.30	87.65	88.47	91.08	93.87
Grape sound	88.82	90.25	90.51	92.84	93.66
Grape leaf curse	86.73	88.71	91.79	91.97	93.62
Orange haunglongbing	87.18	87.05	89.89	91.68	93.80
Peach bacterial spot	88.98	86.73	90.99	91.70	94.20
Peach solid	87.33	89.12	89.94	92.35	94.79
Pepper chime bacterial spot	85.92	89.64	89.05	91.93	93.26
Pepper ringer solid	88.05	89.81	91.58	90.84	93.58
Potato early scourge	87.80	86.61	92.08	90.79	94.26
Potato solid	87.76	89.71	91.95	92.64	93.39
Potato late curse	86.04	87.58	91.95	91.33	93.33
Raspberry solid	86.58	89.70	90.39	91.00	93.63
Soybean sound	87.89	87.82	90.77	90.58	93.54
Squash fine mould	86.04	88.51	91.89	91.62	94.68
Strawberry sound	86.01	87.20	90.15	92.12	94.46
Strawberry leaf burn	85.67	90.24	89.55	90.74	93.64
Tomato bacterial spot	86.52	87.23	89.08	90.68	94.47
Tomato early scourge	88.73	89.00	89.33	92.50	94.85
Tomato solid	86.26	89.30	90.28	92.33	94.08
Tomato late curse	88.58	86.59	90.77	92.87	94.16
Tomato leaf mould	88.68	89.33	90.02	92.06	94.02
Tomato Septoria leaf spot	86.86	89.86	91.76	92.43	93.55
Tomato two-spotted insect parasite	86.27	87.96	90.46	92.07	93.25
Tomato target spot	87.04	87.35	91.59	92.27	93.71
Tomato mosaic infection	87.49	89.30	91.73	90.82	93.75
Tomato yellow leaf curl virus	88.63	87.91	90.73	92.95	93.24

TABLE 17.6
Recall on Different Leaf Disease Types

Disease	Recall [3]	Recall [4]	Recall [7]	Recall [8]	Recall Proposed
Apple scab	90.17	88.85	93.24	92.67	97.13
Apple black decay	87.79	92.20	91.38	93.05	95.59
Apple sound	89.72	90.35	90.76	94.88	96.30
Cherry solid	88.54	90.18	91.97	93.57	95.95
Cherry fine buildup	90.15	91.45	92.33	92.80	96.92
Corn Cercospora dim spot	88.36	90.27	94.10	92.64	96.75
Corn normal rust	88.56	92.33	92.47	95.15	96.63
Corn solid	89.08	91.07	92.15	95.16	96.48
Corn northern leaf blight	90.75	91.95	93.97	93.44	96.09
Grape dark decay	90.17	91.45	93.32	94.53	96.38
Grape esca	91.02	89.00	91.20	93.72	95.64
Grape sound	88.08	89.48	94.56	94.93	96.62
Grape leaf curse	90.45	91.54	92.80	94.64	96.86
Orange haunglongbing	89.21	92.50	92.59	93.49	96.51
Peach bacterial spot	89.67	89.87	94.05	93.44	95.82
Peach solid	89.96	91.43	93.94	94.12	97.03
Pepper chime bacterial spot	88.69	90.73	92.22	94.38	95.92
Pepper ringer solid	88.00	91.46	93.45	95.03	97.20
Potato early scourge	89.77	90.37	92.98	94.43	97.33
Potato solid	88.35	92.40	90.70	92.89	97.31
Potato late curse	89.26	91.92	92.11	92.92	96.70
Raspberry solid	90.81	91.78	91.40	94.33	96.95
Soybean sound	89.36	91.71	93.82	94.93	96.05
Squash fine mould	87.85	91.78	93.45	93.85	95.90
Strawberry sound	90.58	90.08	92.70	92.78	96.40
Strawberry leaf burn	89.42	89.80	91.78	93.64	96.71
Tomato bacterial spot	91.44	90.90	93.42	94.68	96.31
Tomato early scourge	90.09	89.41	92.93	93.56	96.69
Tomato solid	89.55	91.57	94.50	93.92	96.20
Tomato late curse	89.98	89.40	91.93	93.76	97.38
Tomato leaf mould	91.25	91.41	93.45	93.91	96.79
Tomato Septoria leaf spot	89.07	91.63	93.92	94.07	95.64
Tomato two-spotted insect parasite	88.89	90.34	92.42	93.00	96.79
Tomato target spot	90.15	91.18	91.01	93.90	96.83
Tomato mosaic infection	91.58	90.23	94.38	94.14	97.11
Tomato yellow leaf curl virus	89.79	91.87	90.95	93.11	97.32

TABLE 17.7
fMeasure on Different Leaf Disease Types

Disease	fMeasure [3]	fMeasure [4]	fMeasure [7]	fMeasure [8]	fMeasure Proposed
Apple scab	90.07	89.98	89.75	94.09	95.39
Apple black decay	87.75	89.57	91.88	93.55	94.95
Apple sound	86.82	88.09	92.73	93.66	94.79
Cherry solid	87.34	88.66	89.63	92.10	96.02
Cherry fine buildup	88.37	90.09	89.84	92.67	95.51
Corn Cercospora dim spot	89.24	88.99	92.12	92.37	94.53
Corn normal rust	88.61	87.83	90.13	92.06	95.82
Corn solid	87.73	89.92	93.16	94.20	95.15
Corn northern leaf blight	87.96	91.22	90.21	94.19	94.46
Grape dark decay	87.25	89.65	92.97	92.45	94.45
Grape esca	86.82	89.81	90.05	91.74	94.34
Grape sound	87.88	88.90	91.27	93.01	94.45
Grape leaf curse	87.89	91.24	90.39	93.38	94.52
Orange haunglongbing	89.80	88.84	92.50	94.00	94.80
Peach bacterial spot	89.14	91.37	90.60	92.31	95.97
Peach solid	89.99	90.24	89.50	92.93	95.66
Pepper chime bacterial spot	86.88	90.96	90.23	93.14	94.90
Pepper ringer solid	90.25	89.18	91.45	94.11	94.49
Potato early scourge	89.57	89.18	91.87	93.08	95.23
Potato solid	89.09	87.86	89.60	93.90	95.35
Potato late curse	88.69	88.26	91.70	92.11	95.87
Raspberry solid	88.29	89.47	90.35	92.30	94.41
Soybean sound	88.06	87.84	92.56	92.22	94.66
Squash fine mould	86.79	87.95	93.14	92.11	95.60
Strawberry sound	90.40	87.88	90.25	93.63	95.09
Strawberry leaf burn	88.84	89.76	90.90	92.78	96.22
Tomato Bacterial spot	90.36	89.40	92.15	93.34	94.78
Tomato early scourge	89.73	88.76	90.83	93.55	94.46
Tomato solid	87.25	88.67	92.46	92.94	94.84
Tomato late curse	88.62	87.93	90.88	94.03	95.63
Tomato leaf mould	87.84	88.76	89.84	91.99	95.20
Tomato Septoria leaf spot	88.44	90.52	91.28	93.94	95.85
Tomato two-spotted insect parasite	88.45	87.90	90.04	92.12	95.15
Tomato target spot	88.05	89.78	93.15	93.29	95.69
Tomato mosaic infection	86.74	88.10	90.12	92.19	95.36
Tomato yellow leaf curl virus	87.52	89.99	92.04	93.79	94.97

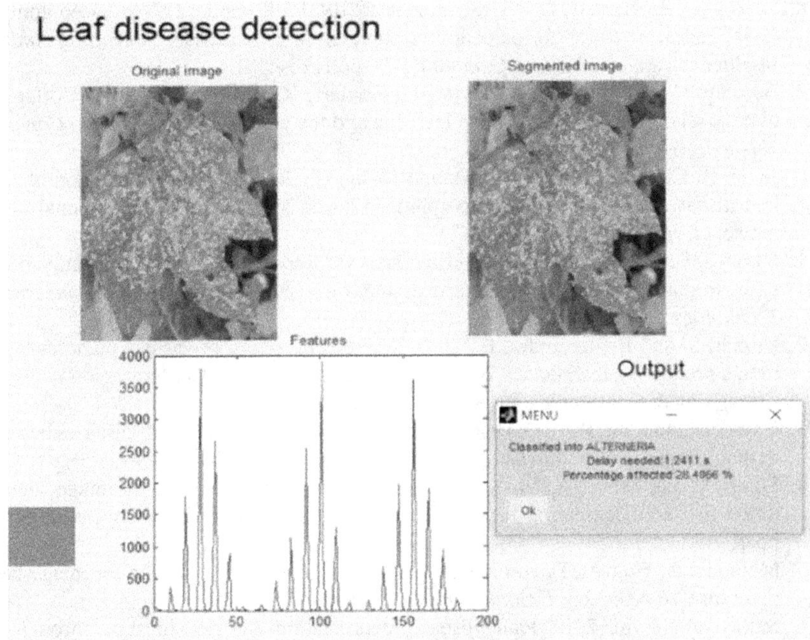

FIGURE 17.11 Alternaria disease image.

system for real-time leaf image disease detection. In future, generative adversarial network (GAN)-based networks can be integrated with the proposed system in order to further improve its performance as shown in Figure 17.11.

REFERENCES

1. Jadhav, S., Udupi, V. and Patil, S., 2020. Identification of plant diseases using convolutional neural networks. *International Journal of Information Technology*, 11, pp. 280–289.
2. Oo, Y. and Htun, N., 2018. Plant leaf disease detection and classification using image processing. *International Journal of Research and Engineering*, 5(9), pp.516–523.
3. Khitthuk, C., 2018. Plant leaf disease diagnosis from color imagery using co-occurrence matrix and artificial intelligence system. *IEECON*, p.4.
4. Priyadharshini, R.A., Arivazhagan, S., Arun, M. and Mirnalini, A., 2019. Maize leaf disease classification using deep convolutional neural networks. *Neural Computing and Applications*, 31(12)pp. 12–24.
5. TM, P., Pranathi, A. and Sai Ashritha, K., 2018. Tomato leaf disease detection using convolutional neural networks. In Proceedings of 2018 Eleventh International Conference on Contemporary Computing (IC3), p.5.
6. Wu, Q., Zhang, K. and Meng, J., 2019. Identification of soybean leaf diseases via deep learning. *Journal of the Institution of Engineers (India): Series A*, 100(4), pp.659–666.
7. Liang, W., Zhang, H., Zhang, G. and Cao, H., 2019. Rice blast disease recognition using a deep convolutional neural network. *Scientific Reports*, 9(1), pp.1–8.

8. Turkoglu, M., Hanbay, D. and Sengur, A., 2019. Multi-model LSTM-based convolutional neural networks for detection of apple diseases and pests. *Journal of Ambient Intelligence and Humanized Computing 23*, pp.1120–1131.
9. Boa Sorte, L., Ferraz, C., Fambrini, F., Goulart, R. and Saito, J., 2019. Coffee leaf disease recognition based on deep learning and texture attributes. *Procedia Computer Science*, 159, pp.135–144.
10. Jiang, P., Chen, Y., Liu, B., He, D. and Liang, C., 2019. Real-time detection of apple leaf diseases using deep learning approach based on improved convolutional neural networks. *IEEE Access*, vol.81, p.7.
11. Saranya, S., Chandra Kiran, N. and Jyotheeswar Reddy, K., 2018. Identification of diseases in plant parts using image processing. *International Journal of Engineering & Technology* 7(2.8), p. 461.
12. Kumar, S. and Raghavendra, B., 2019. Diseases detection of various plant leaf using image processing techniques: A review. In 5th International Conference on Advanced Computing & Communication Systems (ICACCS). p.5.
13. Nage, M. and Raut, P., 2019. Detection and identification of plant leaf diseases based on python. *International Journal of Engineering Research & Technology*, 8(5), p.5.
14. Singh, V. and Misra, A., 2017. Detection of plant leaf diseases using image segmentation and soft computing techniques. *Information Processing in Agriculture*, 4(1), pp.41–49.
15. Mohanty, S., Hughes, D. and Salathé, M., 2016. Using deep learning for image-based plant disease detection. *Frontiers in Plant Science*, 7.
16. Sukhveerkaur, S., 2019. Plant disease identification and classification through leaf images, *Archives of Computational Methods in Engineering* 26.2, 507–530.
17. Mokhtar, U., El Bendary, N., Hassenian, A.E., Emary, E., Mahmoud, M.A., Hefny, H. and Tolba, M.F., 2015. SVM-based detection of tomato leaves diseases. In *Intelligent Systems' 2014* (pp. 641–652). Springer, Cham.
18. Mia, M.R. and Roy, S., 2020. Mango leaf disease recognition using neural network and support vector machine. In *Iran Journal of Computer Science*. Springer, Cham.
19. Saleem, M.H. and Potgieter, J., 2019. Plant disease detection and classification by deep learning, www.mdpi.com/journal/plants, Plants.
20. Priyadharshini, M.K. and Sivakami, R., March, 2019. Sooty mould mango disease identification using deep learning. *International Journal of Innovative Technology and Exploring Engineering*, 8(5S). ISSN: 2278-3075, pp. 250–265.
21. Dhingra, G. and Kumar, V., 2017. *Study of Digital Image Processing Techniques for Leaf Disease Detection and Classification*. Springer, Cham.
22. Ferentinos, K.P., 2018. Deep learning models for plant disease detection and diagnosis. *Computers and Electronics in Agriculture*, 145, pp. 311–318.

18 Monitoring Air Pollution with the Help of Tree Bark and Advanced Technology IoT and AI Techniques at Indore City

Priya Trivedi and Arun Kumar Rana

CONTENTS

18.1 INTRODUCTION

Being constantly exposed to the environment, plants absorb, accumulate, and integrate pollutants impinging on their foliar surfaces. Sometimes, they even show specific responses and thus, can be used as phyto-indicators. In the case of trees, as well as the foliar surface, the bark also absorbs and accumulates air pollutants. Like leaves, bark also provides a large surface area, second only to leaves for the absorption and accumulation of pollutants. Bark is the hard and tough outer covering of tree trunks and branches. Its texture, thickness, and flexibility depend on the type of tree. Although bark serves the same purpose, it looks different from tree to tree to

DOI: 10.1201/9781003221333-18

protect the tree from injury and disease. Some trees have very thick bark that also protects from fire. Others have bad-tasting chemicals in the bark that discourage hungry insects, and some bark is covered with spines or thorns that keep browsing mammals away. At the time of year when deciduous trees and shrubs have no foliage, it is the bark that provides important nutrients and moisture needed for growth. Bark includes all the tissues outside the vascular cambium. However, in popular use, the term "bark" is often used in reference to only the cork, or only to the periderm (cork, cork cambium, and phelloderm) in plants showing secondary growth. The bark provides the patterns seen in trees, adding to the diversity in nature and enhancing human aesthetic pleasure. Bark also provides many other valuable things to human beings. It is the source of the drug quinine, the commonly used salicylic acid (aspirin), and numerous cancer drugs, as well as a wide variety of other benefits.

Bark is also exposed to the environment for reaction with air pollutants, as it provides a very large surface area, second only to leaves. In comparison to leaves, where the surface is mostly smooth, in bark, it is rough. Over the rough surface of bark, the chance of pollutant absorption and accumulation is greater, and this may change bark characteristics, like pH and electrical conductivity, etc. The accumulation of air pollutants in bark is purely a physio-chemical process. The pollutants either passively accumulate on the surface of the bark or become absorbed through ion exchange processes in the outer parts of the dead cork layer. The microclimate was found to have a significant effect on the amount of pollutant deposition on bark. The bark on trees remains for many years, and so it records very precisely all changes occurring in the environment and functions as a sensitive bio-indicator. The first study of air quality using tree bark dates back to the second half of the 20th century, and such studies are constantly being developed today. Reports published almost every year present the potential of using tree bark of different tree species for monitoring atmospheric pollution [1], but we find India to be an exceptional country.

Indore is considered the industrial capital of the state; thus, investigations regarding air pollution were carried out to study the effects of urban air pollution on some morphological, physiological, and biochemical aspects of selected plant species [2]. As far as the effect of air pollution on tree bark is concerned, it remains relatively unstudied except Joshi [3]. As well as this, we know that if we talk about the earth as a whole, ambient air quality is always and closely linked to the ecosystems and climate of different places. In this context, scientists, environmentalists, researchers, and students of many countries, including India, are finding solutions to monitor and reduce air pollution and designing Internet of Things (IoT) and artificial intelligence (AI) techniques for evaluating environment quality in a faster and cheaper way [4–8]. Thus, the present work is undertaken to study the effects of ambient air pollution on physicochemical characteristics of bark, quantitative analysis of sulfate and nitrate, and a review on IoT and AI solutions with reference to air pollution [9–13].

18.2 LITERATURE SURVEY

Extensive work has been carried out for over 100 years on the effects of air pollution on vegetation by researchers in different parts of the world. Most of the work is related to foliar injury and effects on pigments, protein, nucleic acid, amino acids, and enzymes. Some workers have investigated the effects of air pollution on floral behavior [14–16]. A considerable amount of work has been carried out in many countries, like Poland, Nigeria, China, South Africa, Greece, Iran, Turkey, Jordan, Thailand, and many European countries. Bark pH and conductivity have been studied by many researchers [17–19]. Work on heavy metal decomposition has also been carried out by many scientists [20–22], mainly on species such as *Pinus*, *Quercus*, *Fagus*, *Salix* and *Ulmus*, *Acer pseudoplatanus* L., and *Taxus baccata* in different countries. In India, such work is carried out on *Cassia glauca*, *Cassia fistula*, *Bauhinia variegate*, *Ipomoea fistulosa*, *Mangifera indica*, *Azadirachta indica*, and different species of Eucalyptus. Many researchers have recorded sulfate and nitrate accumulation in species *of Pinus*, *Quercus*, and *Ulmus* [23, 24]. A few papers briefly describe solutions to air pollution through IoT and AI techniques [25, 26].

18.3 AIM AND OBJECTIVE

- The main aim of this study was to provide information about the nature, sources, and types of air pollutants in the study area in different seasons.
- The objective was also to investigate the deposition of air pollutants on the bark of two different tree species with reference to different polluted areas in Indore city.
- A review is also provided on AI solutions and IoT techniques to monitor air pollution at low cost in the city.

18.4 STUDY AREA

In Indore, which is considered an educational and medical hub, the air quality has not been safe for many years. As per the reports of the Central Environment and Forest Department (2009), [27] Indore was the fourth most polluted city at the national level. Looking to the critical position of air quality, the government banned the establishment of a few industries here in the years 2010 and 2013. In 2012, the state environmental minister mentioned in the state legislative assembly (Vidhan Sabha) that the air quality of Indore city is lower than the air quality of Delhi, Kolkata, and Chennai [28]. At present, the concentrations of particulate matter (PM) PM_{10} and $PM_{2.5}$ are affecting air quality. The report "State of India's Environment – 2016" by the Center of Science and Environment New Delhi described the air quality of the city as "Critical" in relation to PM_{10} [29, 30]. After the declaration of Indore as "Number one clean city" at the national level in 2017 and 2018, there is now some improvement in air quality [31]. Figure 18.1 shows the trend in annual average concentrations of SO_2, NO_X, and PM_{10} in Indore city in the years 2015 and 2016.

Also, the rapid industrial and commercial development, coupled with the rise in population in the recent past, has contributed to a large-scale increase in traffic in the city. The total number of vehicles registered in Indore by regional transport office (RTO) up to December 17 is 14,93,230. This increasing intensity of traffic has resulted in the manifestation of a number of problems which pose a potential

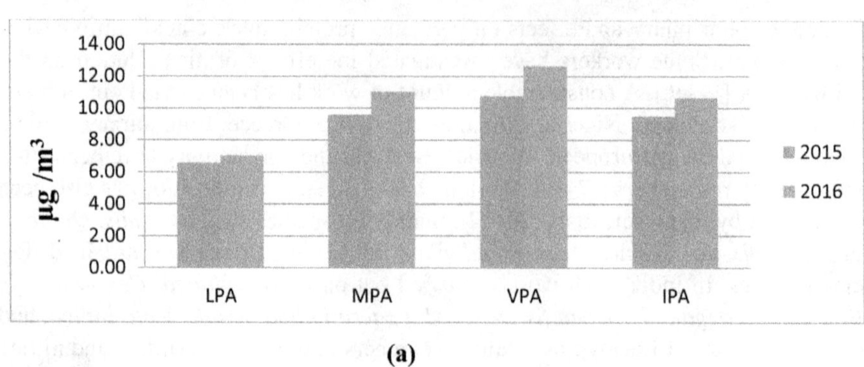

(a)

NOx

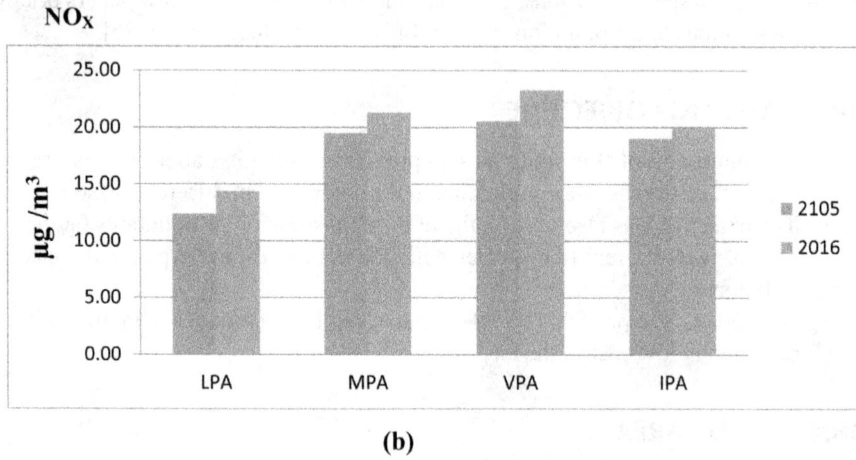

(b)

PM10

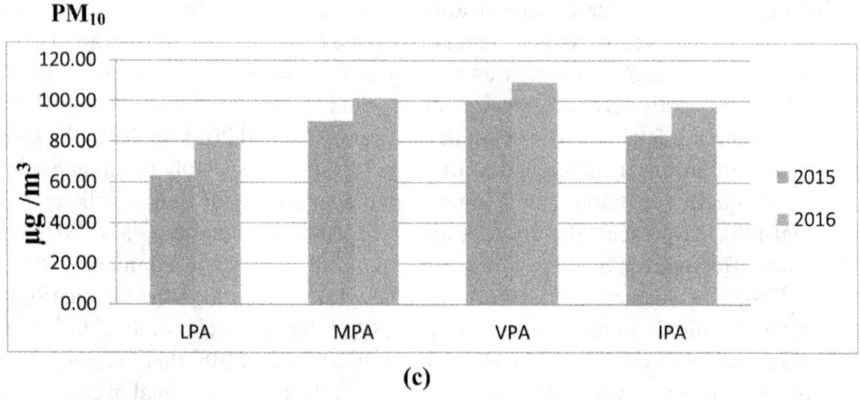

(c)

FIGURE 18.1 Trend in annual average concentrations of (a) SO_2, (b) NO_x, and (c) PM_{10} in Indore city.

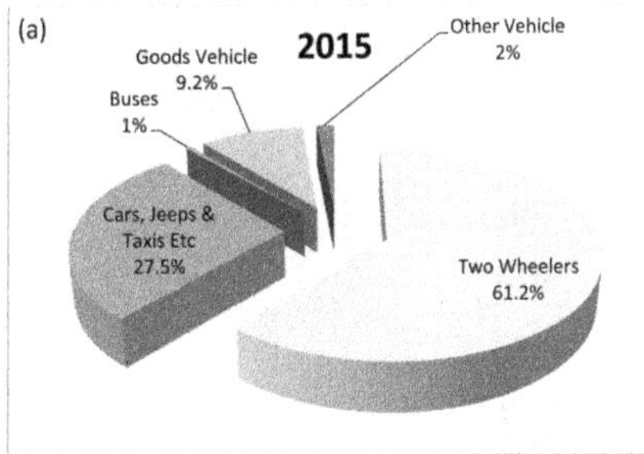

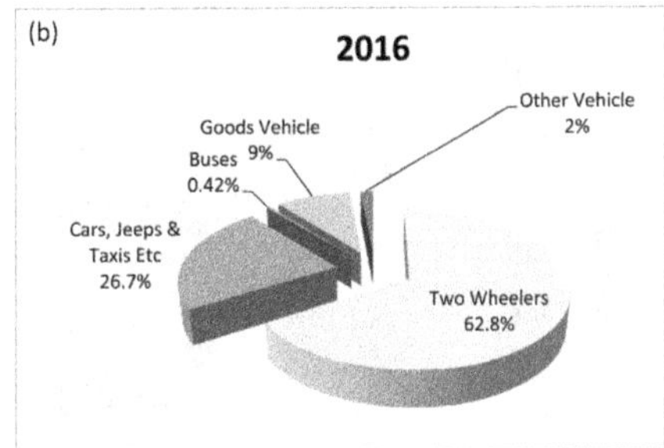

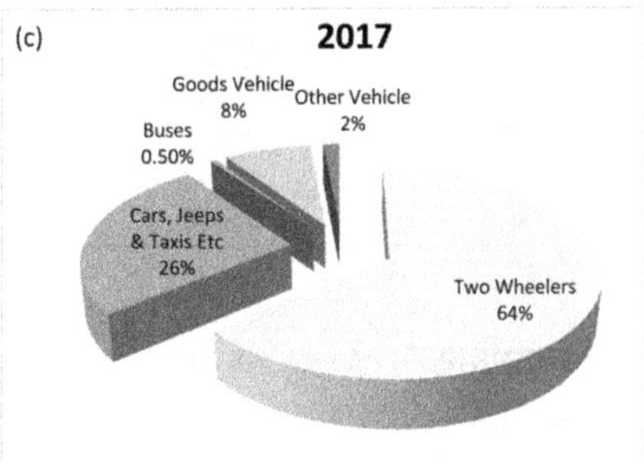

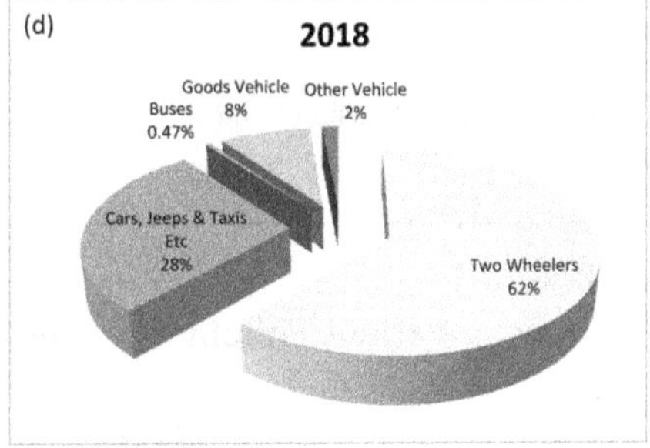

FIGURE 18.2 Percentage distribution of different types of vehicles (from 2015 to March 2018). (a) 2015, (b) 2016, (c) 2017, (d) 2018.

threat to the economic vitality and productive efficiency of the city, as shown in Figure 18.2.

18.5 POLLUTION AREAS

Pollution areas were selected on the basis of the sources and nature of pollutants. Four areas were selected for our research work to evaluate the different types of pollutants and related sources (shown in Table 18.1). The first area was mixed pollution area (MPA), which was located in scheme No.78. The second was vehicular pollution area (VPA); this sampling area is a part of the eastern ring road between Khajarana and Bengali square. The last was industrial pollution area (IPA) in Sanwer Road, the industrial cluster situated on Ujjain road. Data were also collected from Low pollution area (LPA), Ralamandal village. It is located in Indore tehsil of Indore district and is situated 10 km from Indore in the north-east direction. It was taken as a reference area for comparison.

18.6 EXPERIMENTAL TREES

We have chosen *Acacia nilotica* L. (Babul) and *Azadirachta indica* A. Juss. (Neem) tree because they both have medicinal uses as well as being the dominant species used for avenues in Indore city and are also commonly found on the road side. Tree bark of mature plant species of similar height and trunk diameter were selected

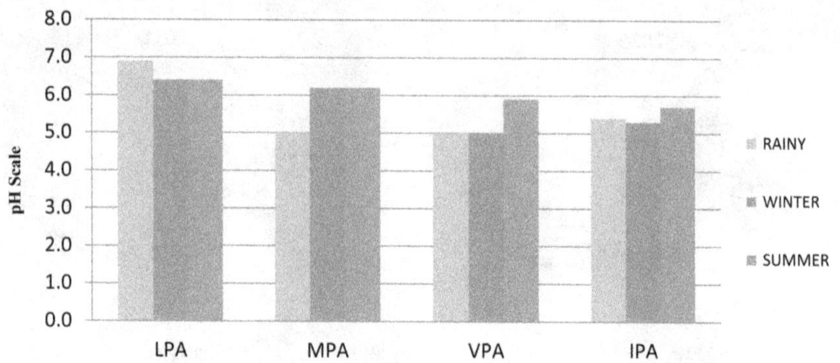

FIGURE 18.3 Seasonal changes in bark extract pH of *Acacia nilotica.*

TABLE 18.1
Sources of Various Pollutants at Different Sampling Areas of Indore City

Area	Pollutants	Sources
MPA	SO_2	diesel vehicles, etc.
	NO_x	Burning of fuel in DG set and vehicles
	PM_{10}	Vehicles, road dust, transportation and construction activities, etc.
VPA	SO_2	Diesel vehicles
	NO_x	Burning of fuel in DG set and diesel vehicles
	PM_{10}	Vehicles, road dust, transportation, construction activities, etc.
IPA	SO_2	Coal based boiler, DG set, diesel vehicles, etc.
	NO_x	Burning of fuel in industrial boilers, DG set, and diesel vehicles
	PM_{10}	DG sets, coal based boilers, diesel vehicles, road dust, transportation and construction activities, etc.
LPA	SO_2	Diesel vehicles, etc.
	NO_x	Burning of fuel, vehicles, etc.
	PM_{10}	Vehicles, road dust, etc.

along the road sides. All four areas were chosen because they represent different conditions of land use, source and nature of pollutants, and traffic patterns.

18.7 MATERIAL

Bark is the main material used for the present study. Bark samples from experimental tree species were collected in triplicate in all three seasons, i.e. rainy (August), winter (December), and summer (April), during 2015 and 2016. About 2 to 5 mm thick chips of bark were removed with a sharp knife from all the directions around the tree at a height of 5–6 feet above the ground level and placed in a zipper poly bag. For uniformity in the samples, trees of the same height, canopy cover, and main

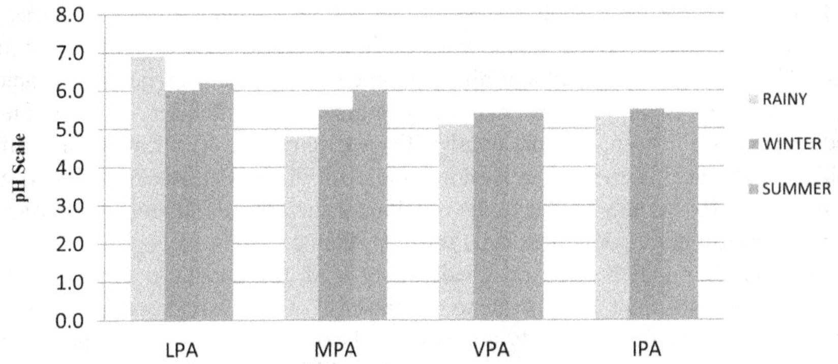

FIGURE 18.4 Seasonal changes in bark extract pH of *Azadirachta indica.*

trunk diameter were considered at all sampling areas. Bark samples of same tree species were also collected from LPA, which serves as a reference for comparison. The samples were brought to the laboratory for further analysis.

18.8 METHODS

Standard methods were used for studying different parameters. The pH and conductivity of the bark were determined by the method given by Grodzinska, 1971 using an EI-112 digital pH meter and an EI-611 digital conductivity meter, respectively [17]. For the estimation of sulfate–sulfur, the method given by Patterson (1958) was adopted [32]. The nitrate content in tree bark was determined using the rapid colorimetric method given by Cataldo et al. (1975) [33].

18.9 OBSERVATION

While observing the physicochemical characteristics of bark, it was found that in *Acacia nilotica* L., the bark was dark brown and rough with deep vertical furrows, blackish grey to brown fissured. On studying its medicinal properties, it was noticed that the bark contains a large quantity of tannin, which is a powerful astringent; its decoction is largely used as a gargle and mouth wash in cancerous and syphilitic infections, while the juice of the bark mixed with milk is dropped into the eye as a conjunctivitis cure. The bark is also used extensively for colds, biliousness, burning sensation, bronchitis, cough, dysentery, piles, and urinary discharge. The bark of *Azadirachta indica* A. Juss was dark brown, sometimes nearly black, thick and rough, with deep, narrow, vertical fissures. Its bark is used for the cure of burning sensation, skin disease, fever, itching, urinary disorders, leprosy, blood complaints, leukoderma, earache, and all wounds.

pH values were recorded from 4.7 to 7.2. The maximum reduction was observed in VPA in both *Acacia nilotica* and *Azadirachta indica.* The reduction in bark pH as observed in the present study is due to the accumulation of gaseous pollutants like

SO_2 and NO_x. The more acidic nature of bark samples from VPA clearly indicates the predominance of acidic gases, which is in agreement with the SO_2 concentration (11.70 μg/m^3) in the ambient air of polluted areas. The presence of the acidic pollutants reduces the bark pH. According to researchers, this trend may be due to industrial gas emission, predominantly SO_2, which causes acidification of rainfall with sulfuric acid. Lower values (more acidic) in plants like *Acacia nilotica* and *Azadirachta indica* may be due to the rough bark surface, which more readily accumulates atmospheric pollutants, including PM, than a smooth surface.

It is said that bark's electrical conductivity is an indicator of concentrations of soluble pollutants in air, mainly due to the presence of sulfate and other compounds. Moreover, electrical conductivity is a numerical expression of the ability of an aqueous solution to carry an electric current. This ability depends on the presence of ions, their total concentration, mobility, valency, and relative concentration, and on the temperature at the time of measurement. Electrical conductivity is a far more sensitive indicator of air pollution than bark pH. It changes even at small emissions of SO_2, which normally do not produce changes in bark pH. However, the maximum % increase of electrical conductivity, as shown in Figures 18.5 and 18.6, was found in the summer season in both the tree species in different pollution areas.

However, an increasing trend in sulfate accumulation was indicated in Figures 18.7 and 18.8 in polluted areas, MPA, VPA and IPA, over LPA, as observed in both the plant species. The rainy season showed higher accumulation in comparison to the winter and summer seasons. Among the three polluted areas, IPA showed higher accumulation than VPA and MPA. In polluted areas, the plants are reported to have higher sulfate content, though in leaves, but this agrees with our findings.

In general, nitrate accumulation was higher in the summer season in all pollution areas for both tree species (shown in Figures 18.9 and 18.10). The maximum % increase was found to be in *Acacia nilotica* at VPA in the winter season. NO_x with

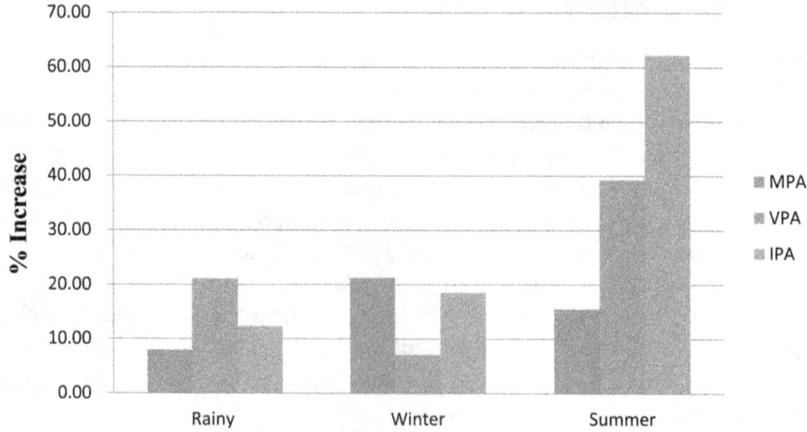

FIGURE 18.5 Seasonal % increase over LPA in bark extract electrical conductivity of *Acacia nilotica*.

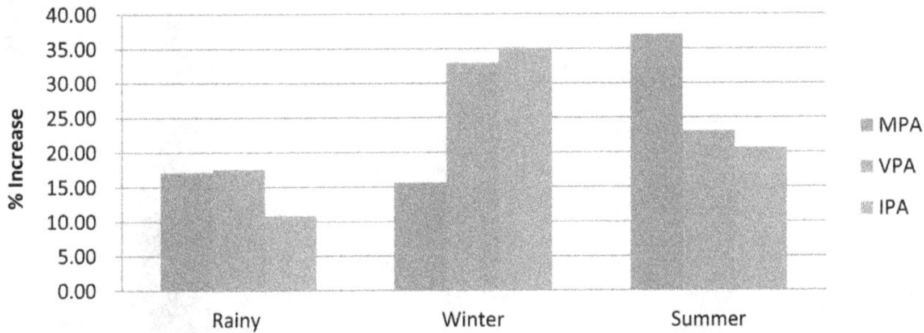

FIGURE 18.6 Seasonal % increase over LPA in bark extract electrical conductivity of *Azadirachta indica.*

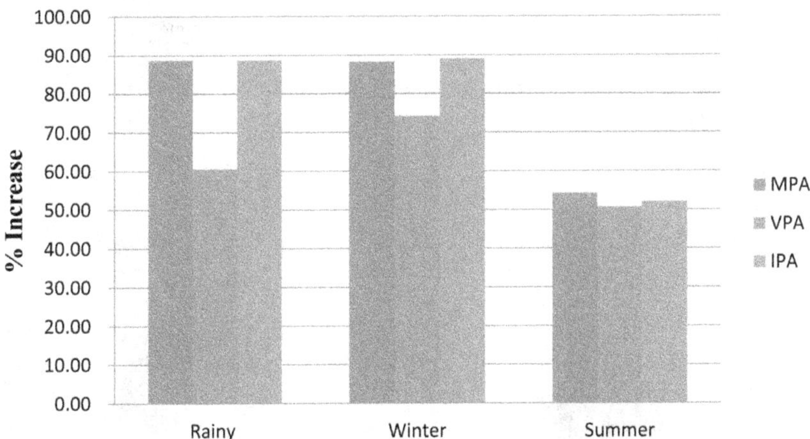

FIGURE 18.7 Seasonal % increase in sulphate accumulation over LPA in bark extract of *Acacia nilotica.*

SO_2 and suspended particulate matter (SPM) when absorbed by tree bark may cause an increase in accumulation of both nitrate and sulfate. However, from an air pollution standpoint, the natural scavenging processes are probably not rapid enough to materially affect hour-to-hour concentration of NO_x in urban areas.

A positive correlation was also seen between total sulfate accumulation with respect to SO_2 concentration and total nitrate accumulation with respect to NO_x in extracts of tree bark of both *Acacia nilotica and Azadirachta indica*, respectively presented in Figure 18.11 and Figure 18.12.

18.10 RESULTS AND DISCUSSION

On the basis of bark extract pH and electrical conductivity, which were found to be higher in different pollution areas in comparison to the reference area, it can be inferred that a higher amount of soluble gases such as SO_2 and NOx along with

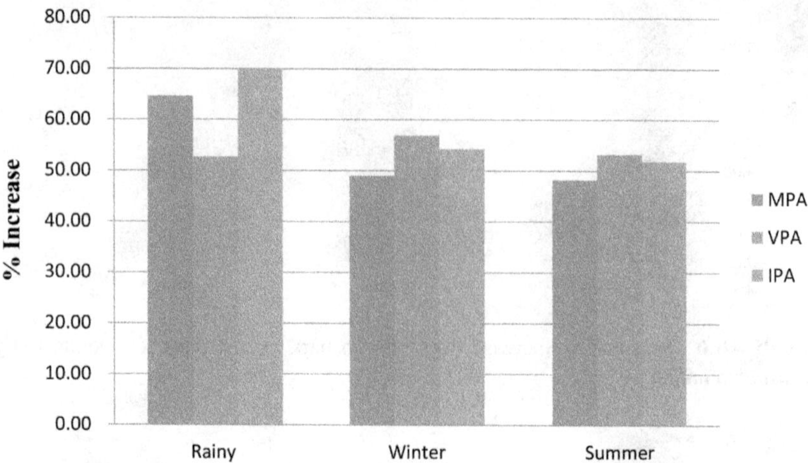

FIGURE 18.8 Seasonal % increase in sulphate accumulation over LPA in bark extract of *Azadirachta indica*.

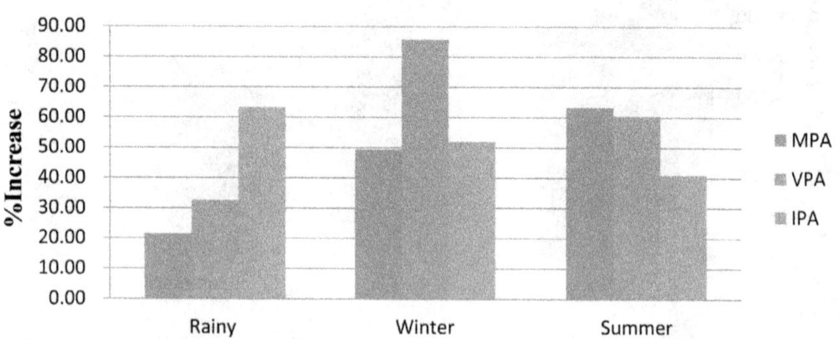

FIGURE 18.9 Seasonal % increase in nitrate accumulation over LPA in bark extract of *Acacia nilotica*.

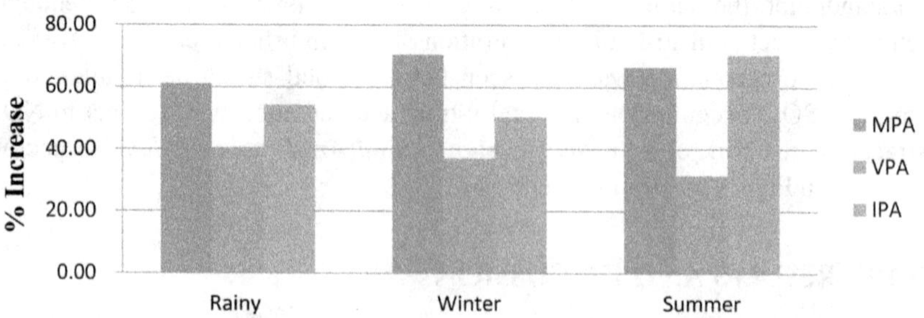

FIGURE 18.10 Seasonal % increase in nitrate accumulation over LPA in bark extract of *Azadirachta indica*.

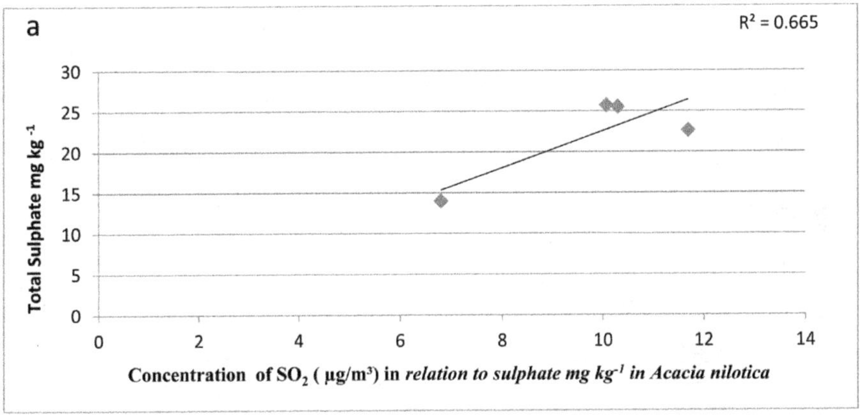

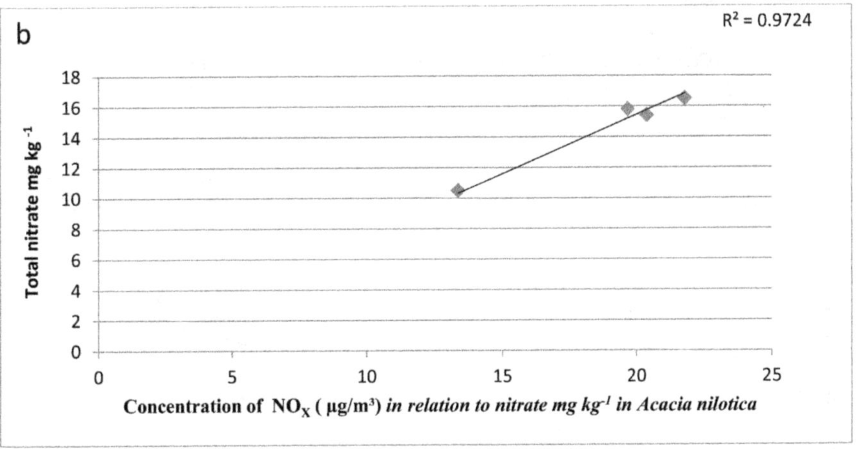

FIGURE 18.11 Relationship between concentration of SO_2 and total sulphate (a) and between NO_x and total nitrate content (b) in *Acacia nilotica* (annual average).

large amounts of PM made the air of these areas more acidic and chemically more active. However, it is yet to be determined how SO_2 and NOx interact with various bark contents. The increasing trend of sulfate accumulation may be accounted for by industrialization, urbanization, and increased traffic density. Higher nitrate accumulation may be accounted for by dry deposition. A dry climate does not allow dust to be washed off continuously, which causes a long residence time and high concentrations [34].

18.11 CHALLENGES AND POSSIBILITIES

Most of the emissions from industries, automobiles, and domestic fuels continuously pollute the city environment by adding various gaseous pollutants & particulates. Pologround, Palda, Sanwer Road, and Manglia are major industrial clusters of the

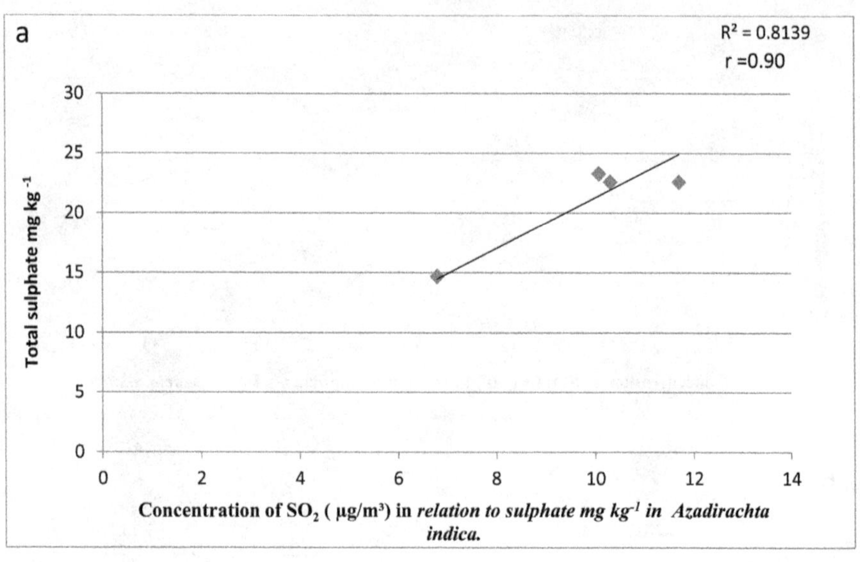

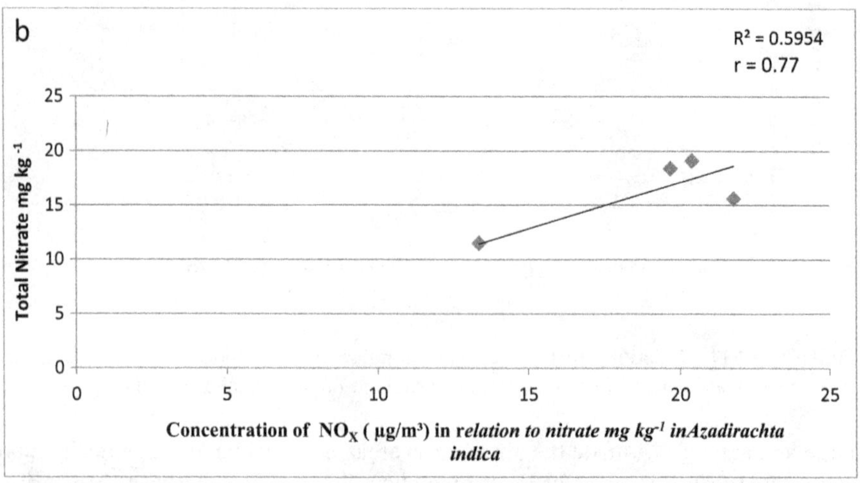

FIGURE 18.12 Relationship between concentration of SO_2 and total sulphate (a) and between NO_x and total nitrate content of *Azadirachta indica* (annual average).

city. Many surveys and study reports published from time to time in the last 15–20 years indicate that air quality in the city is not good. With recent development, the city is going to be an IT hub. Multi-national companies like Infosys and TCS have proposed developing their training centers here. This will accelerate the rate of urbanization, putting a great pressure on the environment. It's a big challenge all over India.

Air pollution can be significantly reduced by frequently monitoring those areas that are highly polluted, expanding technologies, and making people aware of the present scenario. Also, rapid urban transmission should be checked and controlled. Improved waste management with energy-efficient buildings should be designed and implemented by our governments. Electricity production from renewable power sources should be given preference.

New and innovative technology like AI should be used as a priority as an important tool to overcome the problem of air pollution. It is time to improve urban sustainability and development by using IoT and AI solutions. In connection with these techniques, a review on IoT and AI solutions in different countries has been made to address the problem of air pollution. Briefly, if we talk about AI and IoT, it is physical sensor devices that are currently associated with the web for gathering information and sharing it globally. The rapid advancement of the IoT along with AI has led to various challenges and opportunities for researchers to innovate and design memories that are low cost, small size, and have efficient power for smart handheld devices, as the basic requirement for these gadgets is memory. AI can be used as a tracking point to monitor the ambient air quality and warn the population about rising levels of PM, so that they can avoid going to places where this will affect their health adversely.

A team of researcher investigators are trying hard to combine IoT and AI along with sensor technologies to obtain reliable and appropriate environmental data to inform the population. This will definitely help to make different policy and to make our environment and earth greener [35]. The paper, written by Mihaela and co-author, describes an environment quality analysis system based on AI technique in combination with artificial neural networks and rule-based expert systems [36]. A nonprofit AI firm called WattTime, proposed by David Roberts, is going to use satellite imagery, which will help in precisely tracking the ambient air pollution from power plants coming from each part of the world. The data will then be displayed publicly [37].

In Indore, under the joint aegis of Indore School of Social Work, John Snow India Limited, and Development Corporation, low-cost air quality sensors have been designed by a researcher investigator and placed, with approval of the pollution control board, at the 20 most polluted areas at Indore city. The sensor will be able to detect the presence of small PM present in the atmosphere, which when it enters the human body causes various problems, including asthma.

Data will be collected 4 times every hour, and the mean of 24 hours will be taken to show the overall PM.. This will inform people about the level of pollution. Even pollution can be controlled. D.K. Wagela, senior scientist in the pollution department, said that development work has started to reshape the city. So, the level of air pollution is also rising. The government agency does not have sole responsibility for reducing it. The public also need to participate [38].

Despite all that has been said, for India to truly move towards the "Digital India" era, the use of innovative technologies like IoT and AI solutions, which are "low cost", for creating smart cities is the foremost priority. AI is a boon in the field of technology and can be used to monitor, reduce, and manage ambient air pollution conditions in cities where air pollution exceeds the recommended level. As well as these effective measures, the Indian government, along with researchers and investigators and different organizations, should be considering precautionary measures and foremost, planting more and more trees, which will help in controlling and reducing air pollution and will help to save a lot of lives.

18.12 CONCLUSION

The study suggests that gaseous as well as particulate pollution can be monitored and controlled by the planting of indigenous trees, particularly *Acacia nilotica* (Babul) and *Azadirachta indica* (Neem). The bark of these trees can be used as a bio-indicator of air pollution, especially for SO_2 and NO_x. The bark of these species was found to be the best accumulator of toxic pollutants. All the information gathered by this study is socially important, because the planting of more and more *Acacia nilotica* and *Azadirachta indica* trees will help to reduce gaseous as well as particulate pollution in the city. This will provide pollution-free fresh air for citizens to breathe, and it will help to conserve plant species as well.

ACKNOWLEDGMENTS

I am thankful to UGC CRO, Bhopal for financial support. I am also thankful to the Principal and Head of the Botany Department at PMB Gujarati Science College, Indore for providing the research facilities. I would like to express my special appreciation and profound gratitude to my esteemed supervisor Dr. O.P. Joshi, Retired Principal PMB Gujarati Science College, Indore, Co-supervisor Dr. Kishore Pawar, Head Department of Seed Technology, Holkar Science College, Indore, and Dr. Dilip Wagela, Senior Scientist, M.P. Pollution Control Board, Indore for their guidance, valuable suggestions, inspiration, elaborative discussions, and encouraging advice.

REFERENCES

1. Charbaszez, M. and Mroz, L. 2017. Tree bark, a valuable source of Information on air quality. *Pol. J. Environ. Stud.*, 26(2): 453–466.
2. Pawar, K., Dubey, B., Maheshwari, R. and Bafna, A. 2010. Biochemical aspects of air pollution induced injury symptoms of some common ornamental road side plants. *Int. J. Biol. Med. Res.*, 1(4): 291–294.
3. Joshi, O.P., Pawar, K. and Wagela, D. K. 1993. Air quality monitoring of Indore city with special reference to SO_2 and tree barks pH. *J. Environ. Biol.*, 14(2): 157–162.
4. Abbasy, M.B. and Quesada, E.V. 2017. Predictable influence of IoT (Internet of Things) in the higher education. *Int. J. Inf. Educ. Technol.*, 7(12): 914–920.
5. Kumar, A., Salau, A.O., Gupta, S. and Paliwal, K. 2019. Recent trends in IoT and its requisition with IoT built engineering: A review. In *Advances in Signal Processing and Communication* (pp. 15–25). Springer, Singapore.
6. Rana, A.K., Krishna, R., Dhwan, S., Sharma, S. and Gupta, R. 2019, October. Review on artificial intelligence with internet of things-problems, challenges and opportunities. In 2019 2nd International Conference on Power Energy, Environment and Intelligent Control (PEEIC) (pp. 383–387). IEEE.
7. Rana, A.K. and Sharma, S. 2021. Contiki Cooja Security Solution (CCSS) with IPv6 Routing Protocol for Low-Power and Lossy Networks (RPL) in Internet of Things

Applications. In *Mobile Radio Communications and 5G Networks* (pp. 251–259). Springer, Singapore.

8. Rana, A.K. and Sharma, S. 2019. *Enhanced Energy-Efficient Heterogeneous Routing Protocols in WSNs for IoT Application. Int J Eng Adv Technol (IJEAT), 9*(1).
9. Rana, A.K. and Sharma, S. 2021. Industry 4.0 manufacturing based on IoT, cloud computing, and big data: Manufacturing purpose scenario. In *Advances in Communication and Computational Technology* (pp. 1109–1119). Springer, Singapore.
10. Wang, Q., Zhu, X., Ni, Y., Gu, L. and Zhu, H. 2020. Blockchain for the IoT and industrial IoT: A review. *Internet Things*, 10: 100081.
11. Rana, A.K., Salau, A., Gupta, S. and Arora, S. 2018. *A Survey of Machine Learning Methods for IoT and their Future Applications. Amity Journal of Computational Sciences* 2, no. 2 (2018): 1–5.
12. Kumar, K., Gupta, E.S. and Rana, E.A.K. n.d. *Wireless Sensor Networks: A review on "Challenges and Opportunities for the Future world-LTE".*
13. Kumar, A. and Sharma, S. Demur and routing protocols with application in underwater wireless sensor networks for smart city. In *Energy-Efficient Underwater Wireless Communications and Networking* (pp. 262–278). IGI Global.
14. Pawar, K. 1982. Pollution studies in Nagda area due to Birla industrial complex discharges. Ph.D. Thesis, Vikram University, Ujjain.
15. Swamy, H. 2006. Study of air pollution effects on reproductive behaviour of some tree species at Indore city. Ph.D. Thesis, Devi Ahilya University, Indore.
16. Wagela, D.K. 1998. Assessment of air pollution effects on plants and soils in and around Indore due to automobile emulsions. Ph.D. Thesis, Devi Ahilya University, Indore.
17. Grodzinska, K. 1987. Monitoring of air pollutants by mosses and tree bark. In Steubing L. & Jager H.-J.(eds) *Monitoring of Air Pollutants by Plants* (pp. 33–42). Dr W.Junk Publishers, The Hague.
18. Marmor, L. and Randlane, T. 2007. Effect of road traffic on bark pH and epiphytic lichens in Tallinn. *Folia cryptog. Estonica Fasc.*, 43: 23–37.
19. Steindor, K., Palowski, B., Góras, P. and Nadgórska Socha, A. 2011. Assessment of bark reaction of select tree species as an Indicator of acid gaseous pollution. *Polish J. of environment Stud.*, 20(3): 619–622.
20. Berlizov, A.N., Blum, O.B., Filby, R.H., Malyuk, I.K. and Tryshyn, V.V. 2007. Testing applicability of black poplar (Populus nigra L.) bark to heavy metal air pollution monitoring in urban and industrial regions. *Sci Total Environ*, 372: 693–706.
21. Guderian, R. and Schonbeck, H. 1971. Recent results for recognition and monitoring of air pollutants with aid of plants. Prod. In 2nd International Clean Air Congress (pp. 266–273). Academic Press, London.
22. Härtel, O. and Grill, D. 1972. Die Leitfahigkeit von Fichtenborken-Extrakten als empfindlicher indicator fur luftverunrrinigungen. *Eur. J. Pathol.*, 2: 205–215.
23. Huhn, G., Schulz, H., Stärk, H.-J., Tölle, R. and Schüürmann, G. 1995. Evaluation of regional heavy metal deposition by multivariate analysis of element contents in pine tree barks. *Water Air Soil Pollut.*, 84: 367–383.
24. Wolterbeek, H.Th., Kuik, P., Verburg, T.G., Warnelink, G.W.W. and Dobben, H. 1996. Relations between sulphate, ammonia, nitrate, acidity and trace element concentrations in tree bark in the Netherlands. *Environ. Monit. Assess.*, 40: 185–201.
25. Oprea, M. and Iliadis, L. 2011. An artificial intelligence-based environment quality analysis system. In Iliadis L. and Jayne C. (eds) Engineering Applications of Neural Networks. EANN 2011, AIAI 2011. IFIP Advances in Information and Communication Technology, vol 363. Springer, Berlin, Heidelberg. https://doi.org/10.1007/978-3-642-23957-1_55
26. Madhav Mehta, Prabhu Prasad Mohanty and R. Managlagowri. 2020. Air pollution monitoring using IOT. *Int. J. Adv. Sci. Technol.*, 29(6): 2690–2696.

27. Air Pollution Status in 51 Cities of Country 2009. *Report: Central Environment and Forest Department and C.P.C.B.*, New Delhi.
28. Environmental Performance Index 2018. *The Hindu.com.*
29. Kumar, A. and Sharma, S., 2021. Internet of Things (IoT) with energy sector-challenges and development. In *Electrical and Electronic Devices, Circuits and Materials* (pp. 183–196). CRC Press, Boca Raton, FL.
30. Who, Joint, and FAO Expert Consultation. 2003. "Diet, nutrition and the prevention of chronic diseases." *World Health Organ Tech Rep Ser,* vol. 916, no. i-viii, pp. 1–149.
31. State of the Global Air 2017. *Institute for Health Matrix and Evaluation.* University of Washington.
32. Patterson, G.B. Jr. 1958. *Sulfur in Colorimetric Determination of Nonmetals* (pp. 216–308). International Science Public Inc, New York.
33. Cataldo, D.A., Harcon, M., Schrader, L. E. and Youngs, V. L. 1975. Rapid colorimetric determination of nitrate in plant tissues by nitration of salicylic acid. *Commun. Soil Sci. Plant Anal.*, 6(1): 71–80.
34. Walkenhorst, A., Hagemeyer, J., Breckle, S.W. and Markert, B. 1993. Passive monitoring of airborne pollutants, particularly trace metals, with tree bark. In Markert B (ed) *Plants as Biomonitors: Indicators for Heavy Metals in the Terrestrial Environment* (pp. 523–540). VCH Verlags-Gesellschaft, Weinheim, Germany.
35. Internet source. (https://www.who.int/airpollution/ambient/about/en/)
36. Internet source. (https://www.pdxeng.ch/2019/03/28/artificial-intelligence-for-cleaner-air-in-smart-cities/)
37. Internet source. (https://link.springer.com/chapter/10.1007/978-3-642-23957-1_55)
38. Kumar, A. and Sharma, S. 2021 IFTTT Rely Based a Semantic Web Approach to Simplifying Trigger-Action Programming for End-User Application with IoT Applications. In *Semantic IoT: Theory and Applications: Interoperability, Provenance and Beyond* vol 941, pp. 385–397. Springer, Cham,. https://doi.org/10.1007/978-3-030-64619-6_17

19 IoT-Based Smart Stick for the Blind

A Review

Varsha Vimal Sood, Kartik Bansal, and Nitish Agarwal

CONTENTS

19.1 INTRODUCTION

For a human being, vision is the most important of all the senses, as one can perceive more than 80 percent of information by seeing [1]. According to the World Health Organization (WHO), there are at least 2.2 billion people suffering from near or distant vision impairment [2]. The traditional navigation aids like white canes, guides, and walking dogs provide limited help, are ineffective in dynamic scenarios, are expensive, and require a lot of skill and training. The white cane/stick is the most frequently used navigational aid. But, it becomes challenging for users in unfamiliar environments with obstacles or potholes or if they have little knowledge regarding landmarks. There have been a number of research efforts to make the lives of the blind easier by providing an efficient and low-cost technological aid [3]. With the advancement in modern technology, there are many available devices to support the mobility of visually impaired persons, and hence, the potential for effective navigation for such people is increased. The navigational devices would help the visually disabled to live an independent life with minimal external help. The limits

DOI: 10.1201/9781003221333-19

of human capabilities become the driver for upcoming technology. The Internet of Things (IoT) [4] has already started revolutionizing the planet by automating processes. IoT is an umbrella technology incorporating various supporting technologies such as device-to-device (D2D) communications, wireless sensor networking (WSN), cognitive radios (CRs), multiple-input multiple-output (MIMO), etc., which enable physical devices to communicate with each other and the external environment as well via the Internet [5]. It has found applications in almost every field of life, ranging from healthcare, energy, agriculture, transportation, automation, education, and industry to smart cities, smart grids, smart vehicles, etc. [6–12].

The devices or machines connected in the network are made smart and intelligent in the sense that they communicate amongst each other and are able to perform erstwhile human tasks without human intervention. Sensors are the inherent basis for the IoT-based sensing, interaction, and analysis [13]. IoT is an appropriate technology to address the problems of visually impaired people by introducing a smart stick for assistance.

This chapter reviews selected articles and sources that have proposed an IoT-enabled smart stick for the visually impaired. These sticks are able to detect obstacles in the path, potholes and stairs, and surface water, and to send the location to emergency contacts. These sticks are smart enough to detect and locate obstacles and hence to inform the user, as well as those who care for her, via the Internet in a timely way by making use of various sensors. An infrared sensor detects side roads or obstacles on the left and right [14]. An ultrasonic sensor also detects obstacles; moreover, it gives many other benefits [15]. A Global System Positioning (GPS) tracker and a Global System for Mobile communications (GSM) module may be used to get live location information of the blind person and to communicate, respectively [16, 17].

The primary objective of this concept is to provide a safe and reliable technology-based assistant to the visually impaired, which would help them live an independent and dignified life. However, it is observed that there exists a vicious relationship between poverty and blindness. As per the latest reports, approximately nine-tenths of the total blind population live in the developing nations. Also, people who are economically deprived are more susceptible to visual impairments due to lack of awareness and/or inability to afford the cost of treatment [18, 19]. Hence, there is a need to develop a smart stick for the blind, which should be a portable, lightweight, and economical device with low power consumption and a quick response time.

19.2 SYSTEM MODEL

The IoT-enabled smart stick senses the obstacles within a given range using a variety of sensors depending upon the topology of the environment. The smart stick alerts the user via sound and/or vibrations of the stick through a microcontroller. Different types of sounds or strength of vibrations may provide additional information regarding the distance of the obstacle. The main parts of the smart stick include the microcontroller, sensors, GPS and GSM modules, and/or various other enhancements. A basic block diagram of the smart stick is shown in Figure 19.1.

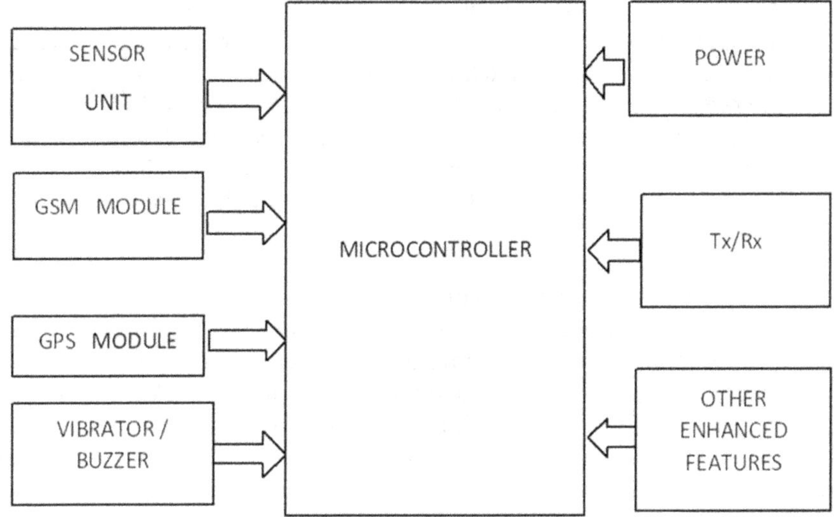

FIGURE 19.1 Block diagram of an IoT-based smart stick for the blind.

On the basis of existing literature on the smart stick for the blind, the functioning of the stick is studied in the following subsections.

19.2.1 Environment Sensing and Obstacle Detection

To provide the visually impaired user an experience of maximum perception, a number of sensors, such as ultrasonic, infrared, water sensors, etc. are employed. In the majority of works discussed here, obstacle detection and recognition are the most important objective so as to ensure a safe navigation by the user. However, there are several papers which give a general description of the surroundings as additional information so as to give the best possible substitute of the vision-based experience.

19.2.1.1 Some Commonly Used Sensors

1) *Ultrasonic sensor:* The ultrasonic sensor sends pulses as the transmitter (Tx) wave, which is reflected back from the obstacle (if there is one), detected at the receiver, and acknowledged by the echo pin. The distance of the obstacle from the smart stick can be calculated on the basis of the speed of the pulses and the time taken by the signal to return as an echo, i.e., distance = (speed of sound × total time)/2. Ultrasonic sensors are the most widely used devices in the smart stick due to their inherent simplicity and light weight. These sensors are used for measuring distance, position, and motion [20].
2) *Infrared (IR) sensor:* The IR sensor consists of a photodiode and a light-emitting diode (LED). It uses infrared rays to detect the position of obstacles on the sideways by emitting radiation. If the radiation is received back, it indicates the presence of an obstacle [21].

3) *Water sensor:* The water sensor consists of an array of exposed traces, indicating the presence of water by reading LOW. It is widely used to sense rainfall, liquid leakage, flood, etc. The level, volume, presence, or even absence of water can also be detected using water sensors [22].
4) *Gas sensor:* Gas sensors or gas detectors are electronic devices used to detect and/or identify the presence of harmful, toxic, or explosive gasses. These are used to detect gas leaks and smoke as well as their concentrations [23]

There are several works on the smart stick using one sensor or a combination of sensors. The main challenge of these devices is adhering to the power constraint. In Connier et al. [24], the authors have designed a power-saving electronic travel kit, whereby sensors can be turned off when not in use. In Varalakshmi and Kumarakrishnan [25], UV, infrared, and temperature sensors are used to detect obstacles like staircases, manholes, pitfalls, etc. The device also calculates the shortest distance to the destination for the user's convenience. Audio alerts play a very significant role in avoiding accidents. The device proposed in Yadav et al. [26] provides information on direction by voice signals using Radio Frequency Identification (RFID) technology. Additionally, it also gives voice outputs informing of the detection as well as the destination of a bus. Another feature of this RFID-based prototype is the provision of a push button so as to locate the stick in case it is misplaced.

In Yusro et al. [27], the authors have outlined a SEES (Smart Environment Explorer Stick) to enable smooth navigation for visually impaired persons. Safe and easy movement indoors as well as outdoors is made possible by the active multi-sensor context-awareness concept. The multitude of sensors incorporated in this project are UV sensors, accelerometer sensor, wheel encoder sensor, camera sensor, and compass sensor. These multiple sensors along with the GPS receiver are used to remotely monitor the visually impaired user. Additionally, a SEE-phone is included to assist the user regarding orientation. The smart stick also detects the colors of traffic lights and informs the user of this via speech.

Thus, it is observed how important is the role of the sensors in perceiving the environment. But all the information acquired by these sensors would be meaningless if not processed by a central microcontroller or processor. The microcontroller/processor serves as the brain and backbone to the IoT enabled device [28]. There are several microcontrollers available and one can choose as per the design requirements and economy.

19.2.1.2 Some of the Most Commonly Used Microcontroller Boards

1) *Arduino microcontrollers:* Arduino is one of the most widely used open-source platform for electronics hardware and software prototyping. The important features are its flexibility and ease of use. As of now, there are a variety of Arduino boards using a range of microprocessors and controllers. These boards are equipped with digital and analog input/output pins, Universal Serial Bus (USB), linear regulator, and 16 MHz crystal oscillator, and programmed using the C and C++ programming languages. The most

widely used Arduino board (Arduino Uno) is based on the ATmega328P microcontroller. In order to come up with an economically affordable stick, Barathi Kanna et al. [29] have proposed a smartphone application prototype equipped with an ESP8266 for the working-class visually impaired. The functioning of the development board is ensured using a power source and coin motors.

2) *Raspberry Pi:* Raspberry Pi is a Linux-based economical computer, which provides a set of general-purpose input/output (GPIO) pins for the physical connection of the chip with a number of things to realize the IoT. Krithiga and Prasanna [9], have proposed a smartphone app–based economical autonomous guiding system. The app is used to send information to the user to enable smooth navigation. Alerts are generated indicating forthcoming vehicles or obstacles. The sensors collect environmental information and send them to the Raspberry Pi, which sends instructions and alerts to the user.
3) *Peripheral Interface Controller (PIC):* A PIC is a small microcontroller programmed to execute a large number of operations [30]. In Lau et al. [31], two microcontrollers, Raspberry Pi acting as the master whereas the PIC 18F4525 microcontroller behaves as the slave controller, are used to develop a smart stick. This stick is further equipped with the water level sensor, ultrasonic sensor, buzzer, vibration motor, and GPS module. An additional feature of this work is the development of an emergency Android-based application (app). The app would enable carers to track and communicate with the user. In case of emergency, the user may send panic alerts to carers by shaking his/her cell phone or by pushing the power button four times in 5 s.

19.2.2 Communication Messages and Alerts

The visually impaired user may be alerted by either voice signals, vibrations, or both. In Sharma et al. [32], indoor/outdoor navigation through cluttered, noisy, and unstructured environments is ensured. The stick estimates the distance and the location of obstacles and sends audio alerts as well as vibrations. These can be sensed by the hand and ear of the user, respectively. The differing frequencies and tracks of the generated vibration and audio alerts provide an idea of the distance of the obstacle. In Sathya et al. [33], the system consists of a combination of a walking stick and a head phone. The peripheral environment is sensed using a rain sensor, an ultrasonic sensor, and a USB camera. The acquired information is sent to the Raspberry PI, which uses the RF module to send a voice-based signal to the user. In Wang et al. [34], an STM32F103 microcontroller is used to detect obstacles and communicate with the family of the user. It generates an alarm in the case of the stick falling and initiates a voice prompt. The microcontroller is connected to the SIM868 module, ultrasonic module, gyroscope accelerometer module, voice module, etc. The device has enhanced features in the form of a mobile app, which can update the family of the user regarding his location, security, and other information. Thus, the design not only ensures a safe and secure experience for the visually impaired person but also assures his carers of the same by the provision of real-time communication of information via the mobile app. Rohit et al. [35] propose a wearable real-time voice output–based scene perception system. The prototype is based on the information collected by a pi-camera after detecting, classifying, and estimating the position of the objects in an outdoor environment. The system works on a feedback mechanism, and the information gathered is conveyed to the visually impaired user by voice signals, thus giving the user the best possible perception of the environment. The proposed camera-enabled smart cane in Lavanya Narayani et al. [36] makes use of two ultrasonic sensors to measure the distance between the smart stick and stationary and falling objects. As soon as the falling object is within the range of 2 meters from the user, an alert is generated by the actuator. This system integrates a camera module ESP-32, which helps in object detection as well as acquiring images of the moving object or people. A passive infrared (PIR) sensor activates the camera module to capture the image of the moving object/person and save processor time. In the case of any emergency, the family and carers of the user are also apprised of the real-time situation.

19.2.3 Tracking

The independent movement of the visually impaired is not only a challenge for themselves but a matter of concern for their carers as well. Therefore, tracking of the user by caregivers and tracking of the smart stick by the user himself if it falls becomes a very important factor.

1) GPS (Global Positioning System): GPS is a collaboration of space-based satellite navigation systems. It is used to gain information on location and time. The information can be received regardless of weather conditions provided that there is an unobstructed line of sight among at least four GPS satellites. The smart stick may also be equipped with a GPS module, which is used to get the location of the blind user.
2) GSM (Global System for Mobiles): GSM is a multiple access (time-division multiple access [TDMA])-based digital mobile telephony system. It operates on two frequency bands, 900 and 1800 MHz, used for sending digitized and compressed data, respectively. A GSM module is used to send messages using wireless mobile communication.

Shaha et al. [37] propose a power-saving capability–oriented SWSVIP (Smart Walking Stick for the Visually Impaired People). It provides different alerts for obstacles located at different distances using ultrasonic sensors, helping users to perceive a clearer picture of their surrounding area. The GPS module attached to the smart stick provides real-time monitoring of the visually impaired user. Gaurav et al. [38] describe how the social network of things (SIoT) can play a crucial role by providing social help to people with visual disabilities. This work discusses a traffic light crossing (TLC) algorithm, which would assist visually challenged users to navigate through the traffic lights by tracking the latitudinal and longitudinal directions. By tracking a unique ID, which can be generated using the location of the user, his family members or relatives may be kept informed of the movement in real time. The work by Jasmine et al. [39] is an ideal example of image and text processing. For detection of an obstacle, an ultrasonic sensor is used, whereas a camera is used to identify objects with words written on them. These images are processed to text and finally, to speech. The device is thus able to read and convert the text to speech for the user. The GPS module provides the precise location of the visually impaired user in terms of meridian and latitude, which is shared with the carers of the user through e-mail. Divya et al. [40] describe an Arduino-based blind walking stick. The sensors incorporated in the device are the ultrasonic and infrared sensors, which detect and locate obstacles. The GSM and GPS modules are used to provide alert messages and location information. In the case of emergency, the user can message his caregivers, whereas the latter may trace the former by the GPS module. There are cases when the stick may fall from the hand of the user, and it becomes a challenge to locate it. To take care of such situations, the client is equipped with a wearable RF transmitter, whereas the stick contains an RF receiver; these work together to locate a fallen stick.

19.2.4 Other Enhanced Features

The perception of the environment may be enriched by incorporating some other enhanced features. The smart stick may be supplemented with some wearable devices such as head gear, goggles, etc. A mobile application [31, 34] may also be used in conjunction with the smart stick, which would add many other features to the navigation experience.

Mala et al. [41] propose a set of a blind stick and a wearable headset, helping users to find their way using GPS embedded in the stick. Bluetooth technology provides wireless connectivity between the stick and the headset. A smart vision stick [42] comprising an indoor/outdoor guiding system, an emergency system, and a smart detection system enables the user to interact with it via a set of voice recognition commands. These commands allow him to receive the directions and distances of any destination asked. A mobile application supported by real-time Google is used for outdoor navigation, whereas an accelerometer is made use of in indoor movements. GPS is also part of the system to take care of emergency situations.

The works discussed so far usually pertain to simple surroundings. Abhang et al. [43] use ultrasonic sensors to detect obstacles and infrared sensors to help detect elevated surfaces such as staircases. ISD1820 is used to relay speech warnings if an obstacle is encountered. When the panic button is triggered, emergency messages containing the GPS coordinates of the user are relayed to predefined emergency contacts. The design proposed in Kunta et al. [44] consists of several sensors to detect anomalies like obstacles, staircases, and wet surfaces. This anomalous terrain information is shared with the user. The whereabouts of the user are sent to concerned family members and friends via messages. This is ensured by a software application, which is an additional feature of the device helping the acquaintances of the visually impaired user to manage the stick's configuration. The device in Ray and Ray [45] makes use of a combination of pressure sensors and distance sensor to detect potholes or speed limiters in the road. The device detects the obstruction up to a certain distance in all three dimensions in space. The distance is denoted by voice messages. The current location of the user could be monitored by his caregivers via an Android app. Chang et al. [46] propose an assistive system, which is a combination of an intelligent walking stick, wearable smart glasses, a mobile device app, and a cloud-based information management platform. This package is designed to achieve the goals of aerial obstacle avoidance and fall detection. To avoid any aerial obstacle collision accidents, the visually impaired user is alerted through the vibrations produced by the proposed intelligent walking stick. In the case of a fall, an urgent notification is immediately sent to their family members.

Some other advanced technologies such as image processing, neural network, and artificial intelligence, may strengthen the capabilities of the smart stick [39, 47–50]. The work by Omoregbee et al. [47] involves the use of a convolution neural network, which was trained on the CIFAR (formerly the Canadian Institute for Advanced Research) 10 dataset. This dataset was chosen because of its composition of images of objects likely to be found as obstacles on a blind person's path. The stick uses an ultrasonic sensor for the mobility system, while the vibrating motor and audio input will provide input from the ultrasonic and the vibrating module. The vibrating motor vibrates with various strengths according to the distance of the obstacle. By using AI [48] and image processing [49], the smart stick may be made more effective and reliable. It may be able to detect, identify, and inform the user regarding familiar or unfamiliar faces, objects, and surroundings in general. In Al-Muqbali et al. [50], the project prototype investigates algorithms using Open CV and Python, which are used for detecting the objects. The visually impaired user is notified via vibrations and/or sound alerts. This research will be a positive addition to the healthcare sector by supporting blind people with the use of smart technology.

19.3 ISSUES AND CHALLENGES

1) *Portability*: Hardware and software portability is one of the most desired features of such innovations. Being a hand-held product, it should be light enough to carry and use. Also, since it is an application of IoT, the ease of shifting from one platform to another is important, keeping in mind the heterogeneity of hardware and software formats.
2) *Adaptability*: The smart stick should be designed such that it can be workable in dynamic environments as well as catering to changing software platforms while in execution.
3) *Privacy and security:* The Internet, while a boon in many ways, is highly susceptible to invasion of privacy and security [51]. Proper care must be taken to ensure the privacy and security of the client, as the device under study is supposed to elevate the quality of his life. Any kind of breach of security may prove a huge risk to his life.
4) *Improved battery life*: The battery life of the smart stick is a very crucial aspect. The device should ensure longer battery life for safe and reliable working of the smart stick.

19.4 CONCLUSION AND FUTURE WORK

The aim of this chapter is to study the various works aiming to provide a smart navigational walking stick for visually impaired persons. The aid should be smart enough to guide the user during his navigation, alert him to any obstacle well in advance, provide tracking information to his carers, give the maximum perception experience to the user, and hence, reduce his dependency on others. As future enhancements, the stick can be integrated with other wearable gadgets such as goggles and headgear specially designed for the visually impaired, using night vision flash cameras for detection in extremely low light, making the smart stick intelligent by the use of AI, and integrating the latter with suitable actuators to automatically take care of a situation without the user's intervention. Increasing the battery life of the device using energy harvesting techniques may be a future improvement.

Overall, it will be a moderate-budget navigational aid for the visually impaired that will prove to work wonders, as it will make them completely independent. The studies reviewed in this chapter give a good idea of a number of techniques that may be employed to design an economical, effective, efficient, versatile, and reliable IoT-based smart stick for the visually impaired. It will not only reduce collisions and minor accidents and increase travel speed and safety in their daily life but would also instill a sense of confidence and dignity in the long run.

REFERENCES

1. https://www.brainline.org/article/vision-our-dominant-sense
2. https://www.who.int/news-room/fact-sheets/detail/blindness-and-visual-impairment
3. F. Hutmacher, "Why Is There So Much More Research on Vision Than on Any Other Sensory Modality?" *Frontiers in Psychology*, vol. 10, no. 4, p. 2246, Oct. 2019, https://doi.org/10.3389/fpsyg.2019.02246
4. K. Ashton, "That 'Internet of Things' Thing in the Real World, Things Matter More Than Ideas," *RFID*, vol. 1, pp. 451–460, 2009.
5. K. Shafique, B. A. Khawaja, F. Sabir, S. Qazi and M. Mustaqim, "Internet of Things (IoT) for Next-Generation Smart Systems: A Review of Current Challenges, Future Trends and Prospects for Emerging 5G-IoT Scenarios," *IEEE Access*, vol. 8, pp. 23022–23040, 2020, https://doi.org/10.1109/ACCESS.2020.2970118
6. V. K. Akram and M. Challenger, "A Smart Home Agriculture System Based on Internet of Things," 2021 10th Mediterranean Conference on Embedded Computing (MECO), 2021, pp. 1–4, https://doi.org/10.1109/MECO52532.2021.9460276
7. S. Ahmadzadeh, G. Parr and W. Zhao, "A Review on Communication Aspects of Demand Response Management for Future 5G IoT- Based Smart Grids," *IEEE Access*, vol. 9, pp. 77555–77571, 2021, https://doi.org/10.1109/ACCESS.2021.3082430
8. X. Xu, L. Yao, M. Bilal, S. Wan, F. Dai and K.-K. R. Choo, "Service Migration Across Edge Devices in 6G-enabled Internet of Vehicles Networks," *IEEE Internet of Things Journal*, https://doi.org/10.1109/JIOT.2021.3089204, vol.23, pp. 11–21
9. R. Krithiga and S. C. Prasanna, "Autonomous Electronic Guiding Stick Using IoT for the Visually Challenged," in S. Patnaik et al. (eds.), *Advances in Machine Learning and Computational Intelligence, Algorithms for Intelligent Systems*, Springer, 2021, https://doi.org/10.1007/978-981-15-5243-4 59

10. H. Yu and Z. Zhou, "Optimization of IoT-Based Artificial Intelligence Assisted Telemedicine Health Analysis System," *IEEE Access*, vol. 9, pp. 85034–85048, 2021, https://doi.org/10.1109/ACCESS.2021.3088262
11. K. Paul, D. Mallick and S. Roy, "Performance Improvement of MEMS Electromagnetic Vibration Energy Harvester using optimized patterns of micromagnet array," *IEEE Magnetics Letters*, https://doi.org/10.1109/LMAG.2021.3088403, vol.230, pp. 111–121
12. F. Piccialli, N. Bessis and E. Cambria, "Industrial Internet of Things (IIoT): Where We Are and What's Next," *IEEE Transactions on Industrial Informatics*, https://doi.org/10.1109/TII.2021.3086771, vol.230, pp. 40–55
13. V. K. N. Lau, S. Cai and M. Yu, "Decentralized State-Driven Multiple Access and Information Fusion of Mission-Critical IoT Sensors for 5G Wireless Networks," *IEEE Journal on Selected Areas in Communications*, vol. 38, no. 5, pp. 869–884, May 2020, https://doi.org/10.1109/JSAC.2020.2980914.
14. C. Chiang and K. Liu, "A CMOS Wearable Infrared Light Intensity Digital Converter for Monitoring Unplanned Self-Extubation of Patients," *IEEE Sensors Journal*, vol. 19, no. 15, pp. 6430–6436, 1 Aug, 2019, https://doi.org/10.1109/JSEN.2019.2910184.
15. G. E. Santagati, N. Dave and T. Melodia, "Design and Performance Evaluation of an Implantable Ultrasonic Networking Platform for the Internet of Medical Things," *IEEE/ACM Transactions on Networking*, vol. 28, no. 1, pp. 29–42, Feb. 2020, https://doi.org/10.1109/TNET.2019.2949805.
16. A. Baksi, M. Bhattacharjee, S. Ghosh, S. K. Bishnu and A. Chakraborty, "Internet of Things (IOT) Based Ambulance Tracking System Using GPS and GSM Modules," in 2020 4th International Conference on Electronics, Materials Engineering & Nano-Technology (IEMENTech), 2020, pp. 1–4, https://doi.org/10.1109/IEMENTech51367.2020.9270120
17. R. S. Krishnan, K. L. Narayanan, S. M. Murali, A. Sangeetha, C. R. Sankar Ram and Y. H. Robinson, "IoT based Blind People Monitoring System for Visually Impaired Care Homes," in 2021 5th International Conference on Trends in Electronics and Informatics (ICOEI), 2021, pp. 505–509, https://doi.org/10.1109/ICOEI51242.2021.9452924
18. https://www.uniteforsight.org/community-eye-health-course/module13
19. https://theprint.in/opinion/you-cant-end-poverty-without-tackling-blindness/689295/
20. N. Loganathan, K. Lakshmi, N. Chandrasekaran, S. R. Cibisakaravarthi, R. H. Priyanga and K. H. Varthini, "Smart Stick for Blind People," in 2020 6th International Conference on Advanced Computing and Communication Systems (ICACCS), 2020, pp. 65–67, https://doi.org/10.1109/ICACCS48705.2020.9074374
21. A. Nada, M. A. Fakhr and A. F. Seddik, "Assistive Infrared Sensor Based Smart Stick for Blind People," 2015 Science and Information Conference (SAI), 2015, pp. 1149–1154, https://doi.org/10.1109/SAI.2015.7237289
22. R. F. Olanrewaju, M. L. A. M. Radzi and M. Rehab, "iWalk: Intelligent Walking Stick for Visually Impaired Subjects," in 2017 IEEE 4th International Conference on Smart Instrumentation, Measurement and Application (ICSIMA), 2017, pp. 1–4, https://doi.org/10.1109/ICSIMA.2017.8312000
23. Z. Saquib, V. Murari and S. N. Bhargav, "BlinDar: An Invisible Eye for the Blind People Making Life Easy for the Blind with Internet of Things (IoT)," in 2017 2nd IEEE International Conference on Recent Trends in Electronics, Information & Communication Technology (RTEICT), 2017, pp. 71–75, https://doi.org/10.1109/RTEICT.2017.8256560
24. J. Connier et al., "The 2SEES Smart Stick: Concept and Experiments," in 2018 11th International Conference on Human System Interaction (HSI), 2018, pp. 226–232, https://doi.org/10.1109/HSI.2018.8431361
25. I. Varalakshmi and S. Kumarakrishnan, "Navigation System for the Visually Challenged Using Internet of Things," in 2019 IEEE International Conference on

System, Computation, Automation and Networking (ICSCAN), 2019, pp. 1–4, https://doi.org/10.1109/ICSCAN.2019.8878758

26. A. B. Yadav, L. Bindal, V. U. Namhakumar, K. Namitha and H. Harsha, "Design and Development of Smart Assistive Device for Visually Impaired People," in 2016 IEEE International Conference on Recent Trends in Electronics, Information & Communication Technology (RTEICT), 2016, pp. 1506–1509, https://doi.org/10.1109/RTEICT.2016.7808083
27. M. Yusro, K. M. Hou, E. Pissaloux, H. L. Shi, K. Ramli and D. Sudiana, "SEES: Concept and Design of a Smart Environment Explorer Stick," in 2013 6th International Conference on Human System Interactions (HSI), 2013, pp. 70–77, https://doi.org/10.1109/HSI.2013.6577804
28. M. A. Ikbal, F. Rahman and M. H. Kabir, "Microcontroller Based Smart Walking Stick for Visually Impaired People," in 2018 4th International Conference on Electrical Engineering and Information & Communication Technology (iCEEiCT), 2018, pp. 255–259, https://doi.org/10.1109/CEEICT.2018.8628048
29. S. Barathi Kanna, T. R. Ganesh Kumar, C. Niranjan, S. Prashanth, J. Rolant Gini and M. E. Harikumar, "Low Cost Smart Navigation System for the Blind," in 2021 7th International Conference on Advanced Computing and Communication Systems (ICACCS), 2021, pp. 466–471, https://doi.org/10.1109/ICACCS51430.2021.9442056
30. J. Yun and J. Woo, "A Comparative Analysis of Deep Learning and Machine Learning on Detecting Movement Directions Using PIR Sensors," in *IEEE Internet of Things Journal*, vol. 7, no. 4, pp. 2855–2868, April 2020, doi: 10.1109/JIOT.2019.2963326.
31. N. Sahoo, H.-W. Lin and Y.-H. Chang, "Design and Implementation of a Walking Stick Aid for Visually Challenged People," *Sensors*, vol.19, no. 1, 2019, https://doi.org/10.3390/s19010130
32. S. Sharma, M. Gupta, A. Kumar, M. Tripathi and M. S. Gaur, "Multiple Distance Sensors Based Smart Stick for Visually Impaired People," in 2017 IEEE 7th Annual Computing and Communication Workshop and Conference (CCWC), 2017, pp. 1–5, https://doi.org/10.1109/CCWC.2017.7868407
33. D. Sathya, S. Nithyaroopa, P. Betty, G. Santhoshni, S. Sabharinath and M. J. Ahanaa, "Smart Walking Stick for Blind Person", *International Journal of Pure and Applied Mathematics*, vol. 118, no. 20, 2018, pp.4531–4536.
34. B. Wang, W. Xiang, K. Ma, Y. Q. Mu and Z. Wu, "Design and Implementation of Intelligent Walking Stick Based on OneNET Internet of Things Development Platform," in 2019 28th Wireless and Optical Communications Conference (WOCC), 2019, pp. 1–5, https://doi.org/10.1109/WOCC.2019.8770547
35. P. Rohit, M. S. Vinay Prasad, S. J. Ranganatha Gowda, D. R. Krishna Raju and I. Quadri, "Image Recognition Based Smart Aid for Visually Challenged People," in 2019 International Conference on Communication and Electronics Systems (ICCES), 2019, pp. 1058–1063, https://doi.org/10.1109/ICCES45898.2019.9002091
36. T. Lavanya Narayani, M. Sivapalanirajan, B. Keerthika, M. Ananthi and M. Arunarani, "Design of Smart Cane with Integrated Camera Module for Visually Impaired People," in 2021 International Conference on Artificial Intelligence and Smart Systems (ICAIS), 2021, pp. 999–1004, https://doi.org/10.1109/ICAIS50930.2021.9395840
37. A. Shaha, S. Rewari and S. Gunasekharan, "SWSVIP-Smart Walking Stick for the Visually Impaired People using Low Latency Communication," in 2018 International Conference on Smart City and Emerging Technology (ICSCET), 2018, pp. 1–5, https://doi.org/10.1109/ICSCET.2018.8537385
38. N. K. Gaurav, R. Johari, S. Seth, S. Chaudhary, R. Bhatia, S. Bansal and K Gupta, "TLC Algorithm in IoT Network for Visually Challenged Persons," in Y. C. Hu, S. Tiwari, M. Trivedi and K. Mishra (eds.), *Ambient Communications and Computer*

Systems, Advances in Intelligent Systems and Computing, vol. 1097, Springer, 2020. https://doi.org/10.1007/978-981-15-1518-7_26

39. G. S. Jasmine, D. M. Marry, S. S. Lakshmi, R. Rishiwanth, K. Sreehariprasath and J. Surendhar, "Camera Based Text and Product Lable Reading for Blind People," in 2021 7th International Conference on Advanced Computing and Communication Systems (ICACCS), 2021, pp. 1122–1126, https://doi.org/10.1109/ICACCS51430.2021.9441860
40. S. Divya, S. Raj, M. Praveen Shai, A. Jawahar Akash and V. Nisha, "Smart Assistance Navigational System for Visually Impaired Individuals," in 2019 IEEE International Conference on Intelligent Techniques in Control, Optimization and Signal Processing (INCOS), 2019, pp. 1–5, https://doi.org/10.1109/INCOS45849.2019.8951333
41. N. S. Mala, S. S. Thushara and S. Subbiah, "Navigation Gadget for Visually Impaired Based on IoT," in 2017 2nd International Conference on Computing and Communications Technologies (ICCCT), 2017, pp. 334–338, https://doi.org/10.1109/ICCCT2.2017.7972298
42. M. M. Bastaki, A. A. Sobuh, N. F. Suhaiban and E. R. Almajali, "Design and Implementation of a Vision Stick with Outdoor/Indoor Guiding Systems and Smart Detection and Emergency Features," in 2020 Advances in Science and Engineering Technology International Conferences (ASET), 2020, pp. 1–4, https://doi.org/10.1109/ASET48392.2020.9118187
43. P. Abhang, S. Rege, S. Kaushik, S. Akella and M. Parmar, "A Smart Voice-Enabled Blind Stick with An Emergency Trigger," in 2020 5th International Conference on Computing, Communication and Security (ICCCS), 2020, pp. 1–6, https://doi.org/10.1109/ICCCS49678.2020.9277202
44. V. Kunta, C. Tuniki and U. Sairam, "Multi-Functional Blind Stick for Visually Impaired People," in 2020 5th International Conference on Communication and Electronics Systems (ICCES), 2020, pp. 895–899, https://doi.org/10.1109/ICCES48766.2020.9137870.
45. A. Ray and H. Ray, "Smart Portable Assisted Device for Visually Impaired People," in 2019 International Conference on Intelligent Sustainable Systems (ICISS), 2019, pp. 182–186, https://doi.org/10.1109/ISS1.2019.8907954.
46. W.-J. Chang, L.-B. Chen, M.-C. Chen, J.-P. Su, C.-Y. Sie and C.-H. Yang, "Design and Implementation of an Intelligent Assistive System for Visually Impaired People for Aerial Obstacle Avoidance and Fall Detection," *IEEE Sensors Journal*, vol. 20, no. 17, pp. 10199–10210, 1 Sept, 2020, https://doi.org/10.1109/JSEN.2020.2990609
47. H. O. Omoregbee, M. U. Olanipekun, A. Kalesanwo and O. A. Muraina, "Design And Construction of a Smart Ultrasonic Walking Stick for the Visually Impaired," in 2021 Southern African Universities Power Engineering Conference/Robotics and Mechatronics/Pattern Recognition Association of South Africa (SAUPEC/RobMech/PRASA), 2021, pp. 1–7, https://doi.org/10.1109/SAUPEC/RobMech/PRASA52254.2021.9377240
48. Z. Chang, S. Liu, X. Xiong, Z. Cai and G. Tu, "A Survey of Recent Advances in Edge-Computing-Powered Artificial Intelligence of Things," *IEEE Internet of Things Journal*, https://doi.org/10.1109/JIOT.2021.3088875, vol.203, pp. 301–312/
49. H. K. Bharadwaj et al., "A Review on the Role of Machine Learning in Enabling IoT Based Healthcare Applications," *IEEE Access*, vol. 9, pp. 38859–38890, 2021, https://doi.org/10.1109/ACCESS.2021.3059858.
50. F. Al-Muqbali, N. Al-Tourshi, K. Al-Kiyumi and F. Hajmohideen, "Smart Technologies for Visually Impaired: Assisting and Conquering Infirmity of Blind People Using AI Technologies," in 2020 12th Annual Undergraduate Research Conference on Applied Computing (URC), 2020, pp. 1–4, https://doi.org/10.1109/URC49805.2020.9099184

Index

CPSIA information can be obtained
at www.ICGtesting.com
Printed in the USA
BVHW091744190422
634676BV00002B/21

9 781032 11737